PHYSICS AND TECHNOLOGY OF SUSTAINABLE ENERGY

Physics and Technology of Sustainable Energy

E. L. Wolf

New York University, Tandon School of Engineering

OXFORD

UNIVERSITY PRESS

OXFORD

UNIVERSITY PRESS

Great Clarendon Street, Oxford, OX2 6DP,
United Kingdom

Oxford University Press is a department of the University of Oxford.
It furthers the University's objective of excellence in research, scholarship,
and education by publishing worldwide. Oxford is a registered trade mark of
Oxford University Press in the UK and in certain other countries

The moral rights of the author have been asserted

First Edition published in 2018

Impression: 1

Published in the United States of America by Oxford University Press
198 Madison Avenue, New York, NY 10016, United States of America

British Library Cataloguing in Publication Data

Data available

Library of Congress Control Number: 2017962353

ISBN 978–0–19–876980–4

DOI: 10.1093/oso/9780198769804.001.0001

Printed and bound by
CPI Group (UK) Ltd, Croydon, CR0 4YY

Preface

This is a textbook for physics, chemistry and engineering students interested in the future of energy as impacted by depletion of fossil fuels, and in the effects of fossil fuel burning on climate. *Sustainable energy* is an innovative area in many respects and one of growing employment opportunity. The book is intended for those, including professionals, with the typical college physics course of three semesters with corresponding mathematics. The additional physics that is needed to understand the basic operation of Tokamak fusion reactors and also semiconductor devices such as solar cells, is introduced early in the text, starting with a simplified treatment of energy generation in the Sun. This is well known to require the tunneling of protons to form deuterons, as treated by Gamow and by Bethe, bringing quantum physics to the forefront. Further chapters extend quantum physics to atoms, molecules and solids, including metals and semiconductors, and to junction devices such as solar cells. The physics that underlies each of the leading forms of non-fossil-fuel energy is covered, with also a summary of the current status of the technology of each type of energy. The sustainable energy of interest in this book will be available on the long term, past the century or so remaining availability of carbon energy and is also that energy that will not tip the climate of the Earth into a warmer, ice-free, high-sea-level condition of grave economic and human consequence. In this connection, the action of the atmosphere is analyzed both for its transmission of sunlight to the surface for use in solar power, and in secondary forms such as wind, waves and hydroelectricity; but also, with greenhouse gas impurities, such as carbon dioxide, for its role in trapping energy and raising the Earth temperature. Prospects for sustainable energy and moderate climate are briefly assessed. The book contains Exercises for each chapter. I am grateful to Prof. L. A. Willson for a translation of early work (1894) on the effects of carbonic acid on atmospheric temperature by the early Swedish scientist A. G. Hogbom, preceding the seminal work by Svante Arrhenius. The author is grateful to the Applied Physics office at the Tandon School of Engineering for assistance and support, and especially thanks Prof. Lorcan Folan and Ms. Deshane Lyew in this regard. Students Cornell Anthony, Aditya Kaushal, Priyanka Mohandas and Nitish Dabas have helped a great deal in preparing this book for publication. Thanks are due to Editors Sonke Adlung and Ania Wronski of Oxford University Press for encouraging and guiding this project, and for quick assistance. The author is, as always, grateful to his wife Carol for long-term support and assistance.

E. L. Wolf
Brooklyn, New York

Contents

Some Useful Constants

Avogadro's Number	N_A	6.022×10^{23} particles/mole
Boltzmann's Constant	k_B	1.381×10^{-23} J/K
Ideal Gas Constant	$R = N_A k_B$	8.315 J/(mole K)
Fundamental Charge	e	1.602×10^{-19} C
Mass of electron	m_e	9.109×10^{-31} kg ($= 511 \text{keV}/c^2$)
Mass of proton	m_p	1.672×10^{-27} kg ($= 938.3 \text{MeV}/c^2$)
Planck's Constant	h	6.63×10^{-34} Js ($= 4.136 \times 10^{-15}$ eVs)
Planck's Constant	$\hbar = h/2\pi$	1.055×10^{-34} Js ($= 6.58 \times 10^{-16}$ eVs)
Bohr Magneton	$\mu_B = e\hbar/2m_e$	9.274×10^{-24} J/T ($= 5.79 \times 10^{-5}$ eV/T)
Coulomb Constant	$k_C = 1/(4\pi\varepsilon_o)$	8.988×10^9 Nm^2/C^2
Permeability of space	μ_o	$4\pi \times 10^{-7}$ N/A^2
Speed of light	c	2.998×10^8 m/s ($= 0.2998$ mm/ps)
Photon energy	$hc/\lambda = h\nu$	1240 (eVnm)/nm
Hydrogen atom binding energy	$k_C e^2/2a_o$	13.6eV
Bohr radius	a_o	0.0529 nm
Permittivity of space	ε_o	8.854×10^{-12} F/m
Electron Volt	eV	1.602×10^{-19} J ($= 23.06$ kCal/mole)
Stefan–Boltzmann constant	$\sigma_{SB} = 2\pi^5 k_B^4/(15 h^3 c^2)$	5.67×10^{-8} W/m^{2}K^4
Gravitational constant	G	6.67×10^{-11} J m^2/kg^2
Radius of Earth	R_E	6371 km
Distance Earth to Sun	R_{SE}	92.96×10^6 mi $= 1434 \times 10^6$ km (Aphelion 1524×10^6 km)

Glossary of Abbreviations

ABB Alstom Brown Boveri, major European steam turbine manufacturer

AGTP Absolute global temperature potential

AM-0 Solar spectrum observed above Earth's atmosphere

AM-1.5G Average solar spectrum at ground, assuming typical light path

AMU (u) atomic mass unit, (1/12) of mass of ^{12}C u = 1.661×10^{-27} kg

ARC anti-reflection coating

Bar one atmosphere pressure 101 kPa

Bbl barrel unit of volume 0.159 m^3 = 42 US gallons

BSF Back surface field, a stratagem in solar cells to reduce recombination at the back surface of the cell

BTU British thermal unit 1054 J

CAES compressed air energy storage

CAGR compound annual growth rate

CBD chemical bath deposition

CCS carbon capture and storage (sequestration)

CHP combined heat and power

CIGS copper indium-gallium selenide, a type of solar cell

CNG compressed natural gas, used in some buses

CSP concentrated solar power, solar thermal electric

CSS closed space sublimation

CVD chemical vapor deposition

D Debye, unit of electric dipole moment, 3.3×10^{-30} C-m

D_{es} Earth–Sun distance 1.496×10^8 Km

DOS density of states, per unit energy per unit volume, for electrons

E_F Fermi energy highest filled level in an electron system

E_g energy gap of a semiconductor, typically in eV

EOR enhanced oil recovery

EV electric vehicle

eV electron volt, 1.6×10^{-19} J

FACTS Flexible alternating current transmission system

fcc face centered cubic

fm femtometer 10^{-15} m, size scale of the atomic nucleus

GHG greenhouse gas, traps heat from Earth's black-body radiation

GW gigawatt 10^9 Watts

ICE internal combustion engine

HAWT horizontal axis wind turbine

HCCI homogeneous charge compression ignition, allows gasoline ICE to operate at higher compression ratio

Hcp hexagonal close packed

HEV hybrid electric vehicle

HOMO   highest occupied molecular orbital
HP   horsepower, unit of power 745.7 Watts
HRSG   heat recovery system generator
HTS   high-temperature superconductor
HVDC   high voltage direct current (transmission line)
ICE   internal combustion engine
IGBT   insulated gate bipolar transistor
IPCE   incident photon conversion efficiency
IPCC   Intergovernmental Panel on Climate Change
IR   infrared
ISS   International Space Station
ITER   international thermo-nuclear experimental reactor, large Tokamak in Cadamarche France
ITO   indium tin oxide, conductive glass
kWh   kilowatt hour = 3.6×10^6 J
LNG   liquid natural gas
LUMO   lowest unoccupied molecular orbital
MBE   molecular beam epitaxy deposits atomically perfect layers
M_e   mass of the Earth 5.97×10^{24} kg
meV   milli-electron volts
MeV   million electron volts
MPPT   maximum power point tracker
MVU   multiple valve unit, element of AC-DC conversion system
MW   million Watts Megawatt
NiMH   nickel-metal hydride, a type of battery used in the Prius hybrid auto
nm   nanometer 10^{-9} m
OMCVD   organometallic chemical vapor deposition
OMLPE   organometallic liquid phase epitaxy
OTEC   ocean thermal energy conversion
PEM   proton exchange membrane, essential unit of fuel cell
PETM   paleocene-eocene thermal maximum, global warming event 52 million years ago
Pg   Petagram, 10^{15} grams
PHEV   plug-in hybrid electric vehicle
PLiON   form of Li-ion battery
PMMA   Poly(methylmethacrylate), used as a photo-resist in patterning
PN   junction junction of P (positively doped), N (negatively doped) semiconductors.
PPV   poly(phenylene vinylene)
PV   photo-voltaic cell or module. Solar cell.
PW   PetaWatt = 10^{15} Watts
QUAD   quadrillion BTU = 10^{15} BTU = 1.054×10^{18} J
QD   quantum dot, a three-dimensionally small object, "artificial atom"
R_E   radius of the Earth = 6173 km
RRR   residual resistance ratio, measure of metal purity
R_s   radius of Sun = 0.696×10^6 km
SCR   silicon controlled rectifier, also called thyristor, used in grid-scale AC-DC conversion
SMES   superconducting magnetic energy storage
T&D   Transmission and distribution
TPa   TeraPascal, 10^{12} N/m^2 pressure, possible value of Young's modulus
TPES   total primary global energy supply, approximately 17 TW

TCO transparent conductive oxide
TVA Tennessee Valley Authority, US
TW terawatt, 10^{12} Watts
u (AMU) Atomic Mass unit, 1/12 of mass of carbon ^{12}C =1.661 $\times$ 10^{-27} kg
UHVDC ultra-high voltage direct current (transmission line)
UPS un-interruptible power supply
UTES Underground thermal energy storage
UV ultraviolet
VAWT vertical axis wind turbine
VRB vanadium redox battery, a flow battery
WEC wave energy conversion device
WECS wind energy control system
WKB Wentzel–Kramers–Brillouin approximation, applies to Schrodinger equation
XLPE cross-linked polyethylene, insulator used in high voltage cables

1

Introduction

1.1 Long-term climate-neutral energy

The definition of sustainable energy, the topic of this textbook, is straightforward. This is the energy that will be available on (after) a timescale set by the earliest benchmarks of our civilization, let us say the timescale of the earliest pyramids or the Chinese Wall, visible from space. These time spans are on the order of thousands of years. By all estimates, the main source of energy that survives this sustainable energy criterion of longevity is the dominating energy from the Sun, including its second-order consequences; wind, ocean waves and hydropower. Long-term energy is also available from the Earth as geothermal heat and from the Earth-Moon system in the motion of tides. Beyond this, we have energy that may be obtained from nuclear fission of Uranium and Thorium isotopes in the ground, and possibly nuclear fusion of the deuterium that is widely available but dilute in the oceans. The latter item is dependent on further development of a technology, terrestrial fusion, that is widely recognized but that may be not be sufficiently developed to be a viable source of long-term energy.

According to the British Petroleum (BP) Statistical Review of World Energy[1] June 2015, p. 42, the global energy consumption in 2014 was 12,892 million tons (12.892 giga tons) of "oil equivalent". These units reflect that British Petroleum (BP), is an oil company, but also that the largest portion of that consumption in 2014 was indeed of oil.

The basic conversion is 1 ton of oil = 42 GJ = 42×10^9 J, so that we can calculate the average rate of total energy consumption globally in 2014 as 17.2 terawatts (TW) = 17.2×10^{12} J/s. The graph presented by BP indicates that global consumption breaks down as 30 percent (or 5,160 GW = 5160×10^9 Watts) of coal, 2.6 percent (or 447.2 gigawatts (GW) of renewable energy (wind and solar, apart from hydroelectricity), 4.6 percent or 791.2 GW as nuclear power, 6.7 percent or 1,152.4 GW as hydroelectricity, 23.3 percent or 40,076 GW as natural gas and 32.9 percent or 5659 GW as oil. The great preponderance of the 2014 consumption, 86.2 percent, was thus of fossil fuels: oil, natural gas and coal.

---

[1] http://www.bp.com/content/dam/bp/pdf/energy-economics/statistical-review-2015/bp-statistical-review-of-world-energy-2015-full-report.pdf/ (accessed 6/3/2016).

Physics and Technology of Sustainable Energy. E. L. Wolf, Oxford University Press (2018).
© E. L. Wolf. DOI: 10.1093/oso/9780198769804.001.0001

It is widely believed, and a basic premise in this book, that this situation will substantially change, on a timescale measured in decades. For example, Bloomberg New Energy Finance[2] predicts that by 2040 zero-emission energy sources will make up 60 percent of installed electric capacity.

In Section 1.4 we quote respectable expert opinion that of total oil reserves in the world, 40 percent were used by 2009 and that 70 percent will be used by 2030. Simple extrapolations in Section 1.4 suggest that easily extractable oil as we know it today will be substantially gone sometime between 2042 and 2060. If only from a point of view of supply, it seems clear that on a timescale of decades we can expect a major upheaval in the world's use of energy. The extraction rate of oil is presently near 90 million barrels of oil per day, and rising. The recent low price has accelerated the consumption. According to BP[1], 1 million barrels of oil is equivalent to 5.41×10^{12} British Thermal Unit (BTU) or 5.708×10^{15} J, with the conversion 1 BTU = 1.055 kJ. The extraction rate thus can be converted to an energy extraction rate of $90 \times 5.708 \times 10^{15}$ J per day or 5.94 TW, not far from the consumption figure 5.66 TW for oil quoted previously from the BP chart.

U.S. energy consumption by energy source, 2014

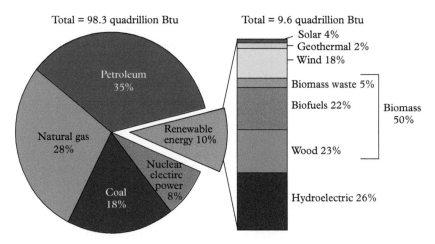

Fig. 1.1 *Approximate breakdown of power use in the US in 2014, according to the US Energy Information Administration. The percent of nuclear power in this graph is overstated, it is closer to 2.8 percent. The percentage of renewable energy is seen to be about 10 percent, if nuclear power is not counted, and of this about 48 percent is provided by hydropower, wind power and solar power. Relative to the total power usage these add up to about 4.8 percent*

Note: Some of components may not equal 100% as a result independent rounding.

Source: U.S. Energy Information Administration, *Monthly Energy Review*, Table 1.3 and 10.1 (March 2015), preliminary data.

² Bloomberg New Energy Finance, "New Energy Outlook 2016", Section 1 of *Executive Summary*.

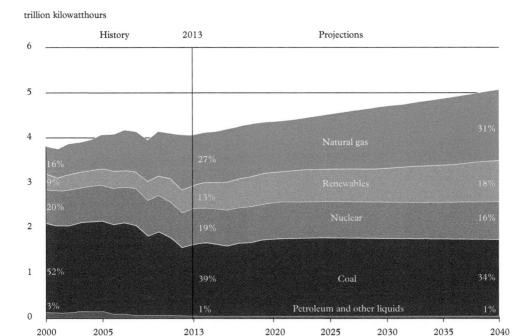

trillion kilowatthours

Fig. 1.2 *History and projection for electricity generation in US by fuel type, according to the US Energy Information Administration. In this graph it is seen that nuclear electric power is currently about 19 percent of total electric power, or about 90 GW. In 2013 the total electric power is shown here to be equivalent to 460 GW, so nuclear electric power is about 2.8 percent of the total US power, that is 3.29 TW. The graph shows that oil, apart from natural gas, is only about 1 percent of the source of electricity in the US (http://www.eia.gov/forecasts/aeo/pdf/0383%282015%29.pdf accessed 28 March 2017, p. 24*

To get some perspective on energy use in the United States, the total power used in 2014 is stated by the US Energy Information Administration[3] as 98.3 "Quads", where a "Quad" is one Quadrillion BTU = 10^{15} BTU, and one BTU is 1.055 kJ. The average total power in the US for 2014 was thus 3.29 TW, since a year is 3.15×10^7 s. Total consumer electric power use in the US for 2013 was 4.1 trillion kWh, thus an average power 460 GW, see Fig. 1.1 and Fig. 1.2. This indicates that electricity was 14.0 percent of power use in the US. So, US total power, recently, is in the vicinity of 19 percent of global power use. That the US population are heavy users of energy can be seen, since, at 320 million, the US population is only 4.3 percent of the world population of 7.35 billion. Nuclear electric power specifically for 2015 is given as 0.797 trillion kWh, or an average 91 GW. This is then 0.091/3.29 = 2.76 percent of power used in the US. According to Fig. 1.3 this is about 19 percent of total electric power in the US, see Fig. 1.3.

---

[3] http://www.eia.gov/energy_in_brief/article/major_energy_sources_and_users.cfm (accessed 03/28/2017).

billion kilowatthours

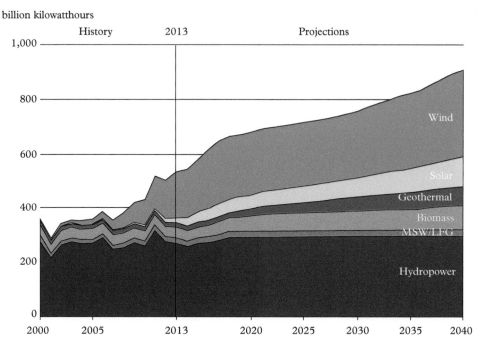

Fig. 1.3 *History and projection for components of renewable electric power, excluding nuclear electric power, in the US, according to the US Energy Information Administration. In this Figure, the acronym MSW/LFG stands for municipal solid waste/landfill gas. Biomass is considered renewable on the argument that growth of plants takes CO_2 out of the atmosphere in amounts roughly comparable to the release of CO_2 when the material is burned. In 2013 the total of these renewable electric powers adds to about 67 GW. Electric power does not strongly derive from oil, unlike the transportation energy sector. If transportation were, by edict or by consensus, to substantially drop in its future use of oil, then the increase in these categories, especially wind and solar, might be more rapid to produce more electricity for automobiles, for example, (http://www.eia.gov/forecasts/aeo/pdf/0383%282015%29.pdf accessed 28 March 2017, p. 25*

1.2   Sun-Earth system, Sun as source of Earth's energy

The mass of the Sun is M = 1.99 × 10^{30} kg, its radius R_s = 0.696 × 10^6 km, at distance D_{es} about 93 million miles (1.496 × 10^8 km) from Earth. In detail, the Earth's nearly circular orbit presently has eccentricity 0.017, and this value has varied over geologic time between 0.0034 and 0.058. It is believed by experts that such small changes have influenced the timing of the many ice ages that the Earth has undergone. It may not be widely known that about 20,000 years ago, what is now New York City was buried under a mile of ice, the Laurentide Ice Sheet, and the ocean was about 140 meters lower than at present. Motion of the glaciers carved out the Finger Lakes in New York State, and also left scars that can today be seen on rock formations in New York City's Central Park. The Earth's climate has been unstable over long timescales and is evidently sensitive to small changes in the Earth's orbit, that can slightly alter the solar influx and the gravitational forces on the Earth. The Sun's composition by mass at its core is approximately 33.97 percent hydrogen and 64.05 percent helium, plus about 2 percent of light elements up to carbon. The Sun's surface temperature is 5778 K to 5973 K, while the Sun's core temperature is estimated as 15.7×10^6 K (Clayton, 1983; Williams, 2004).[4]

The solar spectrum shown in Fig. 1.4 was measured by an automated spectrometer carried in a satellite well above the Earth's atmosphere. The sharp dips in this spectrum are atomic and ionic absorption lines, features that require elementary quantum mechanics for their explanation. The atoms and ions in question are in the Sun's atmosphere.

We are interested in the properties of the Sun, that is not only the source of all renewable energy, excluding the geothermal and tidal energies, and including biofuels that are grown renewably by photosynthesis, but which also serves as a model for fusion reactions that may be implemented on Earth. The power output of the Sun can be calculated from this measured power density shown in Fig. 1.4. If the radiation

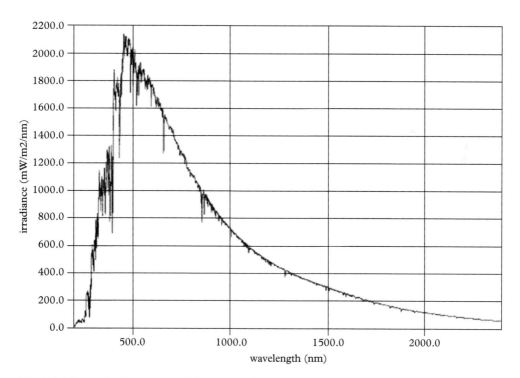

Fig. 1.4 *Measured solar spectrum of electromagnetic power density just above the Earth's atmosphere. The area under this curve is close to 1366 W/m². The overall shape is that expected for "black body" radiation for a source near 6000 K, while the sharp dips (absorptions) come from atoms and ions in the atmosphere of the Sun itself. From this measurement, knowing the radius of the Sun, we can learn that the total electromagnetic power density near the Sun's surface is 6.31×10^7 W/m² (Thuillier et al, 2003) Fig. 4*

---

[4] Williams, D. (2004). "Sun Fact Sheet" NASA http://nssdc.gsfc.nasa.gov/planetary/factsheet/sunfact.html.

power density just above the Earth is measured as 1366 W/m², then the power output of the Sun can be obtained as,

$$P_s = 1366 \, \text{W/m}^2 \times 4\pi D_{es}^2 = 3.82 \times 10^{26} \, \text{W}, \qquad (1.1)$$

taking the average Earth–Sun distance $D_{es} = 1.49 \times 10^8$ km. Similarly, we can find the total incoming power intercepted by the area of the Earth, πR_E^2, with $R_E = 6371$ km, as 174.2 PW = 1.742×10^{17} W. This is about 10,000 times the total current power consumption on Earth, around 17.2 TW in 2014, according to BP.

The 1366 W/m² can also be expressed as a time-average power over the area of the Earth, $4\pi R_E^2$, thus 341.5 W/m², averaging night and day. A more widely quoted figure is $1366 \times (1-\alpha)/4 = 240$ W/m² where the albedo or overall reflection coefficient α of the Earth and atmosphere is taken as 0.3. Confirmation of the 240 W/m² and albedo is offered by weather infrared satellite measurements, that provide nearly the same number as the total average global outgoing longwave radiation (OLR), as is expected in global radiation balance. These average figures then can be used confidently in estimates of the overall energy balance that controls the Earth's temperature.

When thinking about the available power for a solar cell directly under the Sun on a clear day, 1000 W/m² is a benchmark number. The nitrogen and oxygen that make up most of the atmosphere do not absorb in the visible range of wavelengths, but Rayleigh scattering does occur from all atmospheric molecules. This does not absorb light but re-directs some light from the direct beam and also makes the sky shine blue, because of the λ^{-4} dependence of the scattering from atmospheric molecules. Otherwise the sky would be dark, as on the Moon.

In more detail, all atmospheric molecules are polarized by the electric field of the light, and the induced electric dipoles re-radiate (scatter) a small fraction of the radiation, at the same frequency $f = c/\lambda$. The cross-section for Rayleigh (elastic) scattering is given as

$$\sigma_R = \left(2\pi^5 d^6 / 3\lambda^4\right)\left(n^2-1\right)/\left(n^2+2\right), \qquad (1.2)$$

where d is the size of the scattering center, n its refractive index, and λ the wavelength (Siegel and Howell, 2002, p. 480).

For nitrogen N_2, the largest component of the atmosphere, the cross-section was measured as $\sigma_R = 5.1 \times 10^{-31}$ m² at $\lambda = 532$ nm (Sneep and Ubachs, 2005).

The room temperature atmospheric molecular concentration N is about 2.4×10^{25}/m³, so the fraction of light scattered out of the direct beam, per meter, $N\sigma_R$, is 1.22×10^{-5}. Roughly approximating the atmospheric concentration as constant up to 30,000 feet or $\approx 10^4$ m, an altitude familiar to air travelers, the loss of direct sunlight would be ~ 0.1, to leave 1230 W/m² in direct sunlight. When the Sun is at a more typical angle $\theta = 48°$ from the vertical, the light path is enlarged to $10^4/\cos\theta = 15,000$ m, so the intensity is reduced to 1161 W/m². Near sunset, supposing $\theta = 80°$, the path length becomes $\sim 57,600$ m, and the scattered intensity becomes ~ 0.576. In this case most of the short wavelength blue light is removed from the direct beam and the remaining ~ 580 W/m² of direct sunlight becomes red, as is familiar. The red light on the mountains at sunset on a clear day is called alpenglow. Rayleigh scattering accounts for a portion, but not all, of the observed reduction from $1366\,\text{W/m}^2$ to $800\text{–}1000\,\text{W/m}^2$ in direct sunlight on Earth. A sketch of the ground level direct sunlight spectrum is given in the inset to Fig. 3.1(a), indicating additional molecular absorptions to be discussed in Chapter 3.

1.3   Quantum physics of the Sun, extended to energy technology

The origin of almost all sustainable energy available on Earth is the Sun. In a physics book about sustainable energy it is logical to start by understanding how the Sun generates this sustainable energy. It turns out that the necessary quantum physics, which is not too complicated, can be extended to aspects of solid matter as needed to understand semiconductor devices like solar cells, as well as to Earth-based fusion reactors such as the Tokamak.

In the simplest terms, the total power $P = 3.82 \times 10^{26}\,\text{W}$ (Eq. 1.1) of the Sun comes from nuclear fusion of protons, to create ^4He, in the core of the Sun. Let's find the total power density at the Sun's surface. This is $63.1\,\text{MW/m}^2 = 3.82 \times 10^{26}/4\pi R_s^2$, taking the Sun's radius $R_s = 0.696 \times 10^6$ km. These are such large energy losses by the Sun, do we need fear the Sun will soon be depleted? Fortunately, we can be reassured that the lifetime of the Sun is still going to be long by estimating its loss of mass from the radiated energy. Using the energy–mass equivalence of Einstein,

$$\Delta M c^2 = \Delta E, \tag{1.3}$$

on a yearly basis, we find $\Delta E = 3.82 \times 10^{26}\,\text{W} \times 3.15 \times 10^7$ sec/year $= 1.20 \times 10^{34}$ J/y. This is equivalent to $\Delta M = (1.20 \times 10^{34} \text{ J/y}) / c^2 = 1.337 \times 10^{17}$ kg/y. Although ΔM is large, it is tiny by comparison to the vast mass of the Sun, $M = 1.99 \times 10^{30}$ kg. Thus, we find that the fractional loss per year, $\Delta M/M$, for the Sun is 1.337×10^{17} kg/y $\div 1.99 \times 10^{30}$ kg $= 6.72 \times 10^{-14}$/y. This is tiny indeed, so the radiation is not seriously depleting the Sun's mass. On a scale of 5.4 billion years, the accepted age of the Earth, the fractional loss of mass of the Sun, during the whole lifetime of Earth, taking the simplest approach, has been only 0.036 percent.

Where does all this energy come from? It originates in the "strong force" between nucleons, that is large but operates only at short range, a few femtometers (10^{-15} m). Chemical reactions deal with the covalent bonding force. In contrast, nuclear reactions originate in the strong force, about a million times larger. The energy release from "burning" (fusing) hydrogen to make helium, is similar in principle to burning molecular hydrogen to make water, but the energy scale is a million times larger.

In more detail, the composition of the Sun is stated as 73.5 percent H and 24.9 percent He by volume, so the obvious candidate fusion reaction is the conversion of H into He.

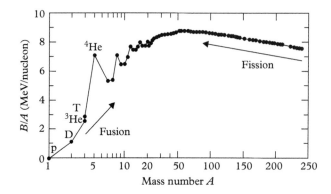

Fig. 1.5 *The Sun's radiating power comes largely from nuclear fusion of protons p into ⁴He at 15 million K. The Mass (nucleon) number A = Z + N. p, D and T, are equivalent, respectively, to ¹H, ²H, and ³H (Atzeni et al 2004) Fig. 1.1*

The basic proton–proton fusion cycle leading to helium in the core of the Sun (out to about 0.25 of its radius R_s) has several steps that can be summarized as:

$$4p \rightarrow ^4He + 2e^+ + 2\upsilon_e. \tag{1.4}$$

This says that four protons lead finally to an alpha particle (two protons and two neutrons, the nucleus of the helium atom), two positive electrons and two neutrino particles.

This is a fusion reaction of some of the elementary particles of nature, which include, as well as protons and neutrons, also positive electrons (positrons) and neutrinos υ_e. Positrons and neutrinos may be unfamiliar, but a danger is to become intimidated by unnecessary details, rather than, in an interdisciplinary field, to learn and make use of essential aspects. The important aspect here is that energy is released when particles combine to form products the sum of whose masses are less than the masses of the constituents. Further, as we will discuss again, this reaction can classically proceed only when the source particles have high kinetic energy, to overcome Coulomb repulsion when the charged particles coalesce. However, the essential process of "quantum mechanical tunneling", an aspect of the wave nature of matter, allows the reaction to proceed more easily, when the interparticle energies are in the kilo-electron Volt range, available at temperatures above 15 million degrees. From elementary physics, we recall that the average kinetic energy per degree of freedom in equilibrium at temperature T is:

$$E_{av} = \tfrac{1}{2}k_B T, \tag{1.5}$$

where Boltzmann's constant $k_B = 1.38 \times 10^{-23}$ J/K. The energy units for atomic processes are conveniently expressed as electron Volts, such that 1 eV $= 1.6 \times 10^{-19}$ J. Chemical reactions release energy on the order of 1 eV per atom, while nuclear reactions release energies on the order of 1 MeV per atom, see Fig. 1.5. A broad distribution of particle speeds v is allowed in the normalized Maxwell–Boltzmann speed distribution,

$$D(v) = (m / 2\pi k_B T)^{3/2} 4\pi v^2 \exp(-mv^2 / 2k_B T). \qquad (1.6)$$

While one may have learned of this in connection with the speeds of oxygen molecules in air, it usefully applies to the motions of protons at 15 million K in the core of the Sun.

The most probable speed is $(2k_B T/m)^{1/2}$ that corresponds to a kinetic energy $E_k = 1/2\ mv^2 = k_B T$. In connection with the probability of tunneling through the Coulomb barrier, that rises rapidly with rising interparticle energy (particle speed), one sees that the high-speed tail of the Maxwell–Boltzmann speed distribution is important. The overlap of the speed distribution, falling with energy, and the tunneling probability, rising with energy (typically as $\exp[-(E_G/E)^{1/2}]$ as we will learn later) leads to what is known as the "Gamow peak" for fusion reactions in the Sun. (In this expression E_G is the Gamow energy, to be defined later, and not to be confused with the energy gap of a semiconductor.) (The Sun's neutrino output, suggested in Eq. 1.4, has been measured on Earth, and is now regarded as in satisfactory agreement with the p–p reaction rate in the core of the Sun, see Schlattl, 2001.)

The energy release of this reaction in Eq. 1.4 can be calculated from the change in the $m_i c^2$ terms. Using atomic mass units u, we go from 4×1.0078 to $4.0026 + 2\ (1/1836) = 9.51 \times 10^{-3}$ u, and using 935.1 MeV as uc^2, we find 8.89 MeV per ^{4}He, neglecting the neutrino energy. The atomic mass unit u is nearly the proton mass, but defined in fact as 1/12 the mass of the carbon 12 (^{12}C) nucleus.

This confirms the large scale of the fusion-energy release, here nearly 9 MeV on a single atom basis. The nuclear force that binds the protons and neutrons into nuclei is indeed about a million times stronger than the typical covalent bond energies in molecules and solids. This large size is of course a driving factor toward using fusion reactors on Earth.

Returning to the Sun, it is believed that the proton–proton (p–p) cycle accounts for about 98 percent of the Sun's energy output (Bethe and Critchfield, 1938; Haxton, 1995), all occurring in the Sun's core. The energy diffuses slowly out to the outer surface with attendant reductions in pressure and temperature, the latter from 15.7 million degrees to about 5800 K.

The first reaction in the proton–proton cycle at the Sun's core (Bethe and Critchfield, 1938) is:

$$p + p \rightarrow D + e^+ + \upsilon_e, \qquad (1.7)$$

where D is the deuteron, the bound state of the neutron and proton, and has mass 2.0136 u. (The mass unit, u, is defined as 1/12 of the mass of the ^{12}C nucleus. One u is about 1.67×10^{-27} kg.) Here the energy release is 1.44 MeV, that includes 0.27 MeV to the neutrino.

This first p–p reaction occurs very frequently in the Sun, as the first step in the basic energy release process. But this reaction is impossible from the point of view of classical physics. It should not occur, from the following reasoning. Accepting the estimated temperature at the center of the Sun as 1.5×10^7 K, the thermal energy in the center

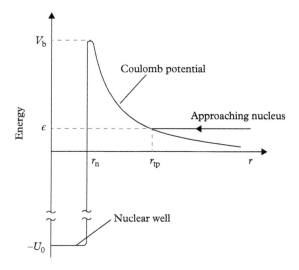

Fig. 1.6 *Sketch of fusion by tunneling through Coulomb barrier. Even at 15 Million K, the interparticle energy ε (Kilovolts) is far below the Coulomb barrier V_B (Megavolts) (Atzeni et al, 2004) Fig. 1.2*

of mass motion of two protons would be $\frac{1}{2} k_B T = \frac{1}{2} 1.38 \times 10^{-23} \times 1.5 \times 10^7$ J $= 1.035 \times 10^{-16}$ J $= 646.9$ eV.

(There will more realistically be a distribution of kinetic energies, and energies higher than 10 keV will frequently be available to colliding protons at 15 million degrees Kelvin.) But any such estimated energy is far short of the potential energy $k_C e^2/r$ that is required classically to put two protons in contact. (Here $k_C = 9 \times 10^9$ and $e = 1.6 \times 10^{-19}$C.) The radius of the proton has been measured and we will take it as 1.2×10^{-15} m. In this case the Coulomb energy $k_C e^2/r$ in eV is $9 \times 10^9 \times 1.6 \times 10^{-19}/(2 \times 1.2 \times 10^{-15}) = 0.6$ MeV. This energy is vastly higher than the kinetic energy. Classically this reaction will not occur because the two protons will never come into contact.

This fundamental discrepancy was resolved (Gamow, 1928) in the early years of the quantum mechanics, and in particular by George Gamow, an American physicist.

The resolution is that the reaction proceeds by a process of "quantum mechanical tunneling", see Fig. 1.6, and the kinetic energies near the "solar Gamow peak" the range 15–27 keV provide most of the reactions. We will return to this topic later. The process is now completely understood, and we will explore it in some detail because it is also central to experimental approaches to generating fusion energy on Earth.

A later and important reaction in the p–p cycle, that we will return to, is fusion of two deuterons. The result can be a Triton T plus a proton, or ^{3}He plus a neutron, or an α (^{4}He) plus a gamma ray (photon). (A Triton is one proton plus two neutrons, and forms Tritium atoms similar to hydrogen and deuterium atoms. Tritium, as opposed to deuterium, does not occur in nature.) This reaction is a candidate for sustainable fusion on Earth, since deuterons are quite plentiful in seawater.

1.4   Depletion: carbon fuel inventories and expected lifetimes

According to the expert Daniel Yergin (The Quest, Penguin, NY, 2011, p. 23), in the time since 1859, when large scale extraction of oil started in Pennsylvania, about a trillion (10^{12}) barrels of oil were extracted up to 2009, and, in 2009, the proven global oil reserves were 1.5 trillion barrels. Thus, one could say that 40 percent of the available oil was used in 150 years. Production has been increasing with demand, and Yergin predicted that it will increase from about 93 million barrels per day in 2010 to about 110 million barrels per day by 2030. From this, in 2030, using the same reserve number, in the period from 2009 to 2030 the further extraction will be around 0.77 trillion barrels, leaving about 30 percent of the proven reserves still to be tapped after 2030. Roughly extrapolating past 2030, assuming 100 million barrels per day extraction, this oil described by Yergin will be gone in a further 12.3 years, thus around 2042. (The BP Statistical Review of World Energy June 2015 gives 1.7001×10^{12} barrels as proven reserves at end of 2014, a slightly larger number than that given for 2009 by Yergin. Taking this number and assuming extraction of 100×10^6 barrels per day, the linear extrapolation to zero reserves occurs in 2060.) Yergin says that the observed typical pattern of oil field production is to reach a plateau followed by a gradual decline. The likely further possibility is of extracting less and less desirable oils, such as the bitumen oil available in Alberta, Canada, and oils from more and more difficult locations, such as under deep sea water, at higher costs. In that sense the oil will never run out, but will be increasingly expensive. According to Richter (2010), natural gas reserves may suffice until about the year 2100. The situation for coal is not very different, and Richter suggests that coal supply at current rates of use will run out in 2080. By comparison with the time-span of our civilization, say 5000 years since the first pyramids in Egypt, these times are short.

1.5   Effects of atmospheric carbon on climate

The need for sustainable power for civilization is clear in the long term because the fossil fuels coal, gas and oil are limited in supply. These materials were deposited over millions of years and are now being extracted so rapidly that they will last at most several hundred years. On this basis there is no question that alternative power sources will gradually be necessary to continue the world's power use on the scale of 17 TW. A second reason for a more immediate turn to non-carbon fuels is the influence on climate of the chemical residues and by-products of human activity being added to the atmosphere, primarily CO_2, but including other three-atom molecules. The essential facts were understood before 1900 from the work of John Tyndall and Svante Arrhenius, an Irish scientist and a Swede, respectively. Tyndall (1861) measured the absorption of heat radiation from a glowing hot source by many constituents of air. He found that heat radiation was essentially not absorbed by the main constituents of the atmosphere, the diatomic gases nitrogen and oxygen, nor by simple atoms such as argon and helium. The lowest energy absorptions of these gases are at much higher frequencies than the frequencies available in thermal radiation, and they therefore allow thermal radiation to pass without attenuation. On the other hand, Tyndall found that the three- and four-atom molecules were typically absorbers of heat, and he measured and catalogued the absorptions of many of these molecules. These structures have internal vibrations whose energies lie in the energy range of thermal radiation, and that can absorb energy from the thermal radiation, as will be discussed in Chapter 7. The absorption can be important even at very low concentrations. These molecules are now called "greenhouse gases", because they absorb outgoing infrared (heat) radiation from the 300 K Earth, and re-radiate some of it back to the Earth, thereby slightly raising the temperature of the Earth. Careful analysis of this effect by the Swedish chemist Arrhenius in 1896 quantified, in particular, the effect of doubling the atmospheric concentration of CO_2, in the vicinity of 280 ppm (expressed as a mole fraction, or ratio of numbers of molecules) around 1800, before the main industrial revolution. Arrhenius (1896) p. 237, calculated that it would raise the Earth's temperature by around 5°C.

His results have been verified and extended by many more recent investigations. In Chapter 7, we explain in detail how this temperature rise occurs.

1.5.1   Rise in carbon emission, CO_2 level and temperature since industrial revolution

A record of the historic atmospheric concentration of CO_2 has been constructed from measurements of gas trapped in ice cores from Antarctica. Measurements of the actual CO_2 in the air were started in Hawaii on the mountain Mauna Loa in 1957, when the measured level was about 320 ppm. At present the level approaches 407 ppm and is rapidly rising still in 2016 (Gillis, 2017). A summary of these data is shown in Figs. 1.7 and 1.8.

Fig. 1.7a shows the change in the Earth's average temperature over the period 1860 to 2010, that amounts to about 1°C. The two lower curves in the figure show modeling reported in the IPCC 2013 report of predicted temperature if only natural causes are included. In an early form of the temperature data (Mann et al (1999), see Fig. 3 therein) the small but clear upward correction to the natural cooling trend starting near 1900 was noted and politely suggested to be anomalous. The idea of anomalous warming has been attacked in the press by proponents of the fossil fuel industry, who wish to deflect any possible blame for climate change. A particularly virulent attack appeared in the National Review, that published a 270-word blog post criticizing the scientist Dr. Michael Mann as "the man behind the fraudulent 'hockey-stick' graph" (of temperature vs time), finally stimulating a lawsuit by Dr. Mann, as recently commented on (Carvin and Dick, 2017).

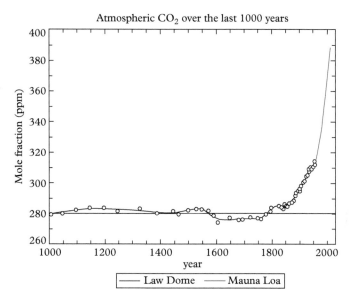

Fig. 1.7 *Atmospheric carbon dioxide concentration as molar fraction over the past 1000 years. The diamond symbols represent mean concentration data from ice cores taken at the Law Dome located near the coast of Antarctica, based on trapped gas in snow and ice samples. The curve to the right is smoothed data directly measured at the Mauna Loa Observatory in Hawaii. It is quite logical to associate the rise in the concentration starting around 1850 with fossil fuel burning by humans (P. Tans,* Oceanography *22 (#4) 2009) pp. 26–35, Fig. 1 (To convert mole fraction to mass fraction multiply by 44.0/28.97 = 1.52)*

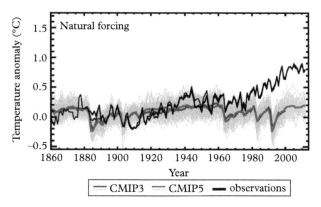

Fig. 1.7a *Time series of global- and annual-averaged surface temperature change from 1860 to 2010. The black line on the top is the observed temperature, while the lower two red and blue curves and error ranges are the result of modeling that includes only natural, not human-driven climate forcings (IPCC 2013) p. 895, Fig. 1*

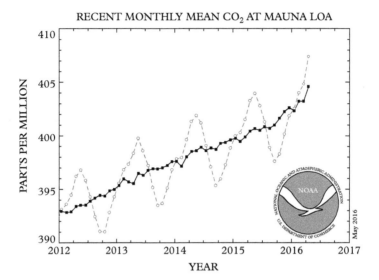

Fig. 1.8 *Carbon dioxide concentration as molar fraction at Mauna Loa Observatory, Hawaii. The square symbols represent mean concentration after removal of the annual summer to winter variation. The observatory is at an elevation of 3400 m where the concentration may differ from globally averaged concentration at the Earth's surface. (To convert to mass fraction, multiply by 44.0/28.97 = 1.52. Thus, the April 2016 value, 407.4 ppm, corresponds to 618.7 ppm mass). (US Dept. Commerce National Oceanic and Atmospheric Administration (NOAA)), http://www.esrl.noaa.gov/gmd/ccgg/trends/*

Presently the curious editorial position of one of the largest newspapers in the US is to deny the source of these upward temperature trends, whose industrial origin was understood by Arrhenius in 1895–1904[5,6] as arising from the burning of carbon.

There should be little doubt that the burning of oil, coal and natural gas, that according to BP[1] in 2014 amounted to "12,892 million tonnes of oil equivalent" released the equivalent tonnage of CO_2 into the atmosphere, and that this flux was largely absent before the industrial revolution, say 1850. More accurately, we found earlier for 2014 that 86 percent of that total global energy, thus 11,087 million tonnes (11.09 Gt) was of fossil fuels, oil, coal and gas, and all of these lead to CO_2 directly emitted into the atmosphere from tailpipes, chimneys and smokestacks. A rough estimate of the corresponding CO_2 is obtained by multiplying by 3.67 (since the mass of the fuels is principally carbon and the mass of oxygen O_2 is 32 compared to Carbon at 12, the mass of CO_2 is larger by a ratio 44/12 = 3.67) to get 43.3 Gt of CO_2. A more careful assessment of the recent fossil fuel CO_2 emission in the year 2011 is 9.5 Pg C (Peters et al, *Nature Climate Change* 3, 4 (2013)), with an estimated annual growth rate 3.1 percent per year. Since 1 Pg = 10^{15} g,

[5] http://earthobservatory.nasa.gov/Features/Arrhenius/.

[6] Svante Arrhenius, a Nobelist in Chemistry, wrote in 1904 "the slight percentage of carbonic acid in the atmosphere may, by the advances of industry be changed to a noticeable degree in the course of a few centuries". He had in 1895 published a paper showing that "the temperature of the Arctic regions would rise about 8 degrees or 9 degrees Celcius if the carbonic acid increased 2.5 to 3 times its present value".

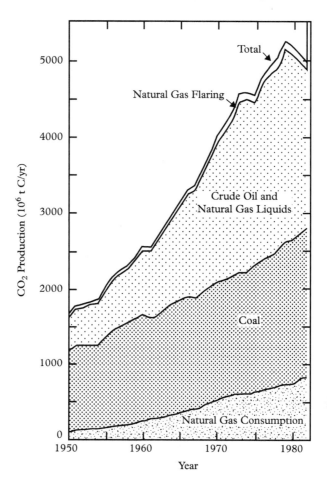

Fig. 1.9 *Annual global CO_2 production (expressed as million tons of Carbon) from fossil fuel burning, 1950–82, that totals around 5 Gt (carbon) per year by 1980. (Marland, G. and Rotty R., Tellus 36B, 232 (1984)) Fig. 5. At present the annual global CO_2 production is about 10.4 Gt (carbon) per year, based on year 2014*

and a metric ton is 1000 kg, 1Pg = 1 Gt. Thus the 2011 CO_2 emission rate is equivalent to 9.5 Gt C/yr. This corresponds to 34.9 Gt CO_2/yr (2011) or about 38 Gt in 2014, with the estimated growth, not too far from our crude estimate of 43.3 Gt CO_2.

We can find how this compares to the total CO_2 in the Earth's atmosphere at present, using the measured 407 ppm CO_2 concentration (618.7 ppm by mass). The atmospheric pressure is 101.33 kPa, and this force appears on the full area of the Earth, $4\pi R_E^2$, where R_E = 6371 km. Taking the average acceleration of gravity as 9.8 m/s² the total mass of the atmosphere is thus 5.256×10^{18} kg, (it is 5.148×10^{18} kg according to the National Center for Atmospheric Research). From this we see that the mass of CO_2 is presently about 3.25×10^{15} kg or 3250 Gt, corresponding to 885.6 Gt C. From this we can see that the correspondence between Gt C and ppm CO_2 is 885.6/407 = 2.18, although the quoted figure (CDIAC)[7] is 2.13, based on a slightly different total atmospheric mass. Using this latter figure, we can see that as the ppm has risen from 280 to 407 the atmosphere has gained 270.5 Gt C.

If we take 38 Gt as the CO_2 emission for 2014, these numbers suggest that the annual emission of CO_2 from fossil fuel burning amounts to 1.12 percent of the existing CO_2 in the Earth's atmosphere. How rapidly this will increase the concentration depends on the rate of turnover in the atmosphere of the CO_2, that is more difficult to estimate. Carbon dioxide is removed from the atmosphere by growing plants (that presumably gives the winter to summer change in concentration seen in Fig. 1.8 at the Mauna Loa location) and is also substantially absorbed by the oceans, creating carbonic acid, making the oceans more acidic. The present rate of increase of the atmospheric CO_2 concentration appears to be roughly 2.4 ppm/yr, (0.6 percent/yr) while the annual variation summer to winter at that location is about 6 ppm (1.5 percent), see Fig. 1.9. If we compare the observed CO_2 growth at 0.6 percent/year with the 1.12 percent estimated from the fossil fuel burning production as a percentage of the total CO_2 estimated to be present in the atmosphere, one possible explanation would be that about half of the annual production is absorbed in the oceans, that are known to be becoming more acidic. Clear indications that only about half the emitted carbon is retained in the atmosphere is available from CDIAC, that gives a figure 391 Gt C emitted since 1751 by fossil fuel burning and cement production, and 600 Gt C as total emission since 1850, including contributions from deforestation and land use changes (Pierrehumbert, 2016).

The effect on temperature of the added carbon is a complicated issue, as suggested by Fig. 1.10.

Referring to Fig. 1.10, Matthews et al (2009) define "carbon sensitivity" as the increase in atmospheric CO_2 concentration that results from emissions, as impeded by natural carbon sinks, including the oceans. "Climate sensitivity" is shown as a general characterization of the temperature response to atmospheric CO_2 changes. Feedbacks between climate change and the strength of carbon sinks are shown as the upper dotted arrow (climate-carbon feedbacks). The Carbon-climate response (CCR) aggregates the climate and carbon sensitivities (including climate-carbon feedbacks) into a single metric representing the net temperature change per unit carbon emitted.

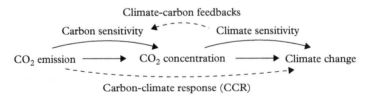

Fig. 1.10 *Schematic representation of the progression from CO_2 emissions to climate change (Fig. 1, Matthews et al, 2009)*

---

[7] CDIAC Carbon Dioxide Information Analysis Center, Oak Ridge National Laboratory http://cdiac.ornl.gov/ (accessed 3/28/2017).

In addition to the work of Mathews et al, a further literature (Allen et al (2009); Meinshausen et al (2009)) is available that argues that a reasonable estimate of the effect is a linear change in temperature as the total carbon emission increases, with an estimate of 2°C being expected for total emission 1000 Gt C. In agreement with this view, according to Pierrehumbert 2016 the amount added to reach 2°C, beyond the 600 Gt already emitted, is thus 400 Gt and will be reached in 2038 at the present rate. Interesting comments on this are available at Trillionthtonne.org.[8] It is suggested that after a pulse of carbon is emitted, there is falloff on the scale of decades, but after 1000 years about a quarter still remains, with about 60 percent having gone into the ocean and 15 percent into the land.

With a good portion having an atmospheric-lifetime of 1000 years or more, the best that can be hoped for is to hold the CO_2 near the present level by stopping further emission. Further discussion of this situation is available at Pierrehumbert 2016. (See also discussion after Fig.1.11.) The effect that such increases, common in the Earth's geologic history, have had on sea level and ice accumulation can be deduced from ice core samples as we will soon discuss. The positive aspect of the situation is that such changes in sea level occur very slowly, on the scale of 5000 years.

The legal aspect of carbon dioxide emissions in the US was recently reviewed (Davidson, 2017). The US Supreme Court in 2007, led by Justice John Paul Stevens, ruled that the US Environmental Protection Agency had an obligation to assess the risks posed by greenhouse gases, labeled as pollutants. The EPA, in turn, recognized that those emissions had contributed to a public-safety crisis of national and global scope. The ruling, known as the "endangerment finding" by the EPA was issued in 2009, in the G. W. Bush era. In spite of the Supreme Court ruling of 2007, policies based on that decision and the "endangerment finding" have recently been reversed by the latest administration, with court challenges no doubt in view to resolve the discrepancy.

1.5.2   Potential for warming and ocean rise, the glacial record

More extensive and detailed information obtained from ice cores from Antarctica are shown in Fig. 1.11. These data show temperature, CO_2 and CH_4 concentrations found in gas bubbles trapped in ice from varying depths below the surface at the Vostok location in Antarctica. The temperatures are inferred from measured isotopic ratios of oxygen 16 and oxygen 18. The data show strong resemblance of temperature and concentration changes over an approximate 100,000-year period. There have been large swings in temperature and CO_2 concentration. The minima in temperature correspond to ice ages (glaciations), the most recent ice age having existed about 20,000 before years the present. It has been stated that the periodicity is similar to periodicity believed present in the eccentricity of the Earth's orbit around the Sun. The change in eccentricity makes small changes in the amount of light received and in the gravitational force on the Earth and the oceans by the Sun.

---

[8] http://www.trillionthtonne.org/ (accessed 4/4/2017).

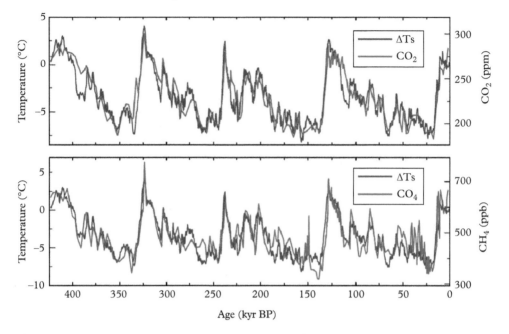

Fig. 1.11 *Antarctic (Vostok) ice core records of temperature, CO_2 (upper) and CH_4 (lower, in parts per billion) for 420,000 year-span to the present, including four glaciations. Note that the present CO_2 concentration 407 ppm is larger than at any time in this record. Historically the Earth's temperature has correlated with the CO_2 concentration. The shortest time for a full swing from glaciation to warm period on this chart appears to be about 5000 years. The warming transitions seem to occur more rapidly than the cooling transitions. The data used in this figure have come from sources including Vimeux, F., Cuffey, K. M and Jouzel, J., Earth and Planetary Science Letters 203, 829 (2002), Petit et al, Nature 399, 429 (1999) and Hansen and Sato, Proc. Nat'l Acad. Sci. 101, 16109 (2004). National Aeronautics and Space Administration, Goddard Institute for Space Studies http://www.giss.nasa.gov/research/briefs/hansen_11/*

1.5.3   Correlation of CO_2 level and temperature

The correlation of CO_2 variation with temperature variation in the data is clear, and suggests that the presently sharply rising CO_2 concentration will be accompanied by a corresponding sharp rise in temperature. The authors cited in Fig. 1.11 point out that the average global temperature does not vary as widely as the Antarctic temperatures measured from the ice cores. They estimate that the global swing in temperature is about half that observed in Antarctica. So far, the Earth's temperature on a mean basis has risen about 1°C since the start of the industrial revolution, see Fig. 1.7a. The larger temperature variations seen in the ice record are surely correlated with large changes in sea level. In the most recent ice age that peaked about 20,000 years ago, data shown in Fig. 1.12 reveal that the sea level was 120 m below the present sea level, the direct consequence of massive amounts of water being tied up in ice supported on land.

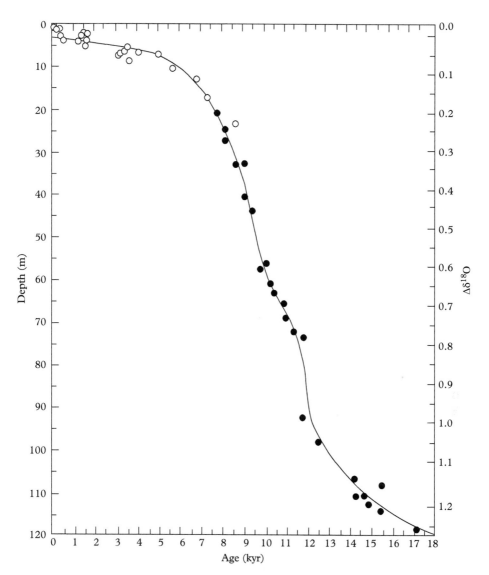

Fig. 1.12 *Measurements of the depth below the present sea level for the last 17,000 years. The data (filled circles) are obtained from core samples of the coral* Acropora Palmata *near the island Barbados in the Caribbean Ocean, while the open circles are data from A,* Palmata *cores from other locations in the Caribbean. Two steep sections in the graph indicate rapid melting events (melt-water pulses), termed MWP-1A, 13,070 years ago, and MWP-1B, 10,445 years ago. The former is characterized by Hansen et al 2007 as a period of sea-level rise "3–5 m/century for several centuries" (R. G. Fairbanks,* Nature ***342** 637: 1989) Fig. 2*

The sea level graph in Fig. 1.12 was obtained from drilled core sections of coral reef in the Caribbean Ocean, analyzing for a particular type of coral and dating the carbon in the samples obtained. The coral reefs near Barbados contain several separate well-known species of coral, one being *Acropora Palmata*. This coral is known to grow only to a depth of 5 m from the current sea level. Samples of *A. Palmata* taken from drilling cores were then dated by the amount of isotope ^{14}C.

The Laurentide Ice Sheet above North America 20,000 years ago had a thickness as great as 2 miles in some locations. At present, very extensive ice sheets remain in the Arctic and Antarctic. Recent satellite measurements of the extent of the remaining ice reveal that the equivalent sea-level rise, if the present ice were to entirely melt, would be about 80 m. Such complete melting is not believed to have occurred in the record of the past 420,000 years shown previously, but complete melting is believed to have occurred earlier in the Earth's history. According to Pierrehumbert (2010) p. 47, "evidence for warm ice-free polar conditions in the Eocene and late Cretaceous is unambiguous".

These eras were about 50 million and 100 million years ago, respectively. The Pierrehumbert text discusses "hothouse Earth" conditions in some detail. If these conditions were to occur again, perhaps triggered by the sharply rising CO_2 shown in the data of Figs. 1.7 and 1.8, it would take thousands of years for the change to completely occur. Note that the most rapid warming transitions evident in Fig. 1.11 occur over intervals about 5,000 years. In the most recent interglacial period, about 125 ky ago, it is estimated that the sea level was about 6 m higher than presently.

The sea level has risen 120 m over the past 17,000 years, as large portions of the ice of the most recent glaciation have melted. Hansen et al (2007) also mention a more recent warm era, the mid-Pliocene, about 3.5 million years ago, when the temperature was 2–3°C warmer than present, and the sea level 25 ± 10 meters higher than present. In the most recent interglacial (warm) period it is estimated that the sea level was about 6 m higher than at present, but not more than 10 m higher than at present.[9]

The sea level has changed over time, according to how much water is held on land as ice. Recent satellite observations of glaciers have been interpreted to give an estimate of 80 m for possible further sea level rise from the present if all the remaining ice, mainly in the Arctic and Antarctic, were to melt. The recent measurements of ice extent from satellite images correspond to sea level rises 64.8 m from East Antarctic ice sheet, 8.06 m from West Antarctic ice sheet, 0.46 m from the Antarctic Peninsula, 6.55 m from Greenland ice and 0.45 m from all other ice caps, ice fields and valley glaciers, giving a total of 80.32 m of sea level rise.[10]

An 80-m rise would flood most of New York City and other coastal cities, and much more, e. g., most of the state of Florida. More immediate but smaller sea rises, up to a few meters, could come from breaking-up of portions of the West Antarctic ice sheet, capable in total of 8.06 m sea rise, according to the report of Poore et al.[11]

---

[9] International Panel on Climate Change IPCC 2013, Section 5.6.2 on p. 425.

[10] Slightly different numbers are given in IPCC 2013, Table 4.1 on p. 321. These add up to 66.1 meters in sea level rise, including 58.3 m from the Antarctic ice sheet and 7.36 m from the Greenland ice sheet.

[11] Poore, R., Williams, R., and Tracey C. "Sea level and climate": US Geological Survey Fact Sheet 002–00, 2p, at http://pubs.usgs.gov/fs/fs2-00/ (accessed 28 March 2017).

Such a total melting event, that would mainly involve the East Antarctic ice sheet, would perhaps have a time progression similar to the melting of the Laurentide ice sheet that covered North America about 20,000 years (20 ky) ago, and contributed to the sea rise indicated in Fig. 1.12. The time for that to occur is indicated in that figure, on the order of 10 ky, and occurred over the interval 5 ky to 15 ky ago. In the earlier warming transitions shown in Fig. 1.10, the large warming changes occurred over intervals on the order of 5 ky, seen in the carbon dioxide and temperature records.

1.6 Physics as guide to expectations and technology

The topics of this book, energy including its availability and its impact on the Earth climate, and sea levels, have been subjects also of media campaigns supported by interested parties aiming to mold public opinion. For example, the providers of oil and coal resist suggestion that burning of fuels can noticeably change the climate. The basic science involved, notably that the Earth's temperature is basically set by the balance of incoming radiation from the Sun matched by its outgoing infrared black body radiation, and that this balance can be upset by small concentrations of impurity molecules resulting from combustion, as was clearly shown in 1896 by Arrhenius, is largely not understood in present society whose leaders have more frequently attended law school or business school than have studied engineering or science. A basic understanding of the science can help the citizen sort out what news or opinion is likely correct from a large flow of information that may be tilted to favor a particular economic interest.

1.7 Plan of this book

Following the present Introductory Chapter 1, Chapter 2 presents an inventory of the sustainable energy resources that stem principally from the Sun, and to some extent from the Earth and Moon in their orbits. Chapter 3 gives more detail on the "wireless" electromagnetic transfer of energy from the Sun through vacuum and the Earth's atmosphere. It discusses aspects of the atmosphere that are of interest from the points of view of bringing energy in from the Sun, and also for modulating the outflow of infrared energy that sets the temperature of the Earth. Chapter 4 explains how the Sun creates energy, including a simplified mathematical model. This section also introduces quantum physics, needed to properly understand nanophysically based energy conversion devices, such as solar cells. Chapter 5 expands the ideas of quantum physics to cover atoms, molecules, metals, semiconductors, and p–n junctions, the underpinning of energy technology. Chapter 6 explains three working methods that release fusion energy in laboratory situations on Earth. The power output from a Tokamak type fusion reactor is analyzed and numerically estimated by scaling the simplified reaction model, shown in Chapter 4, to predict the Sun's output, to the Tokamak realm of parameters. Chapter 7 returns to a more detailed treatment of the atmosphere, and in particular to the physics and the optical absorption of the principal impurity molecules that provide the greenhouse effect. The topics then turn, in Chapters 8–10, to the physics and technology of wind, hydro and tidal energy, to the physics and technology of solar thermal energy conversion, and photovoltaic conversion, including a survey of solar cell types. Chapter 10 deals in more detail with types of solar cells, including prospects for developing new cells with higher efficiency and lower cost. Chapter 11 deals with storage and transmission of energy by the electric grid, important as a consequence of the time-varying and geographically separated nature of renewable energy sources. Chapter 11 also deals with the greenhouse gas contributions of various parts of the global economy, with some emphasis on automobiles, and what effects are expected on temperature. Finally, Chapter 12 deals with the future of renewable energy, as a part of a global long-term energy future maintaining a tolerable climate and sea level.

2

Sustainable Energy Beyond Carbon

2.1 Introduction

We are interested in energy sources that will last on a time scale of thousands of years, and further, that will not interfere with other important aspects of life on Earth, such as clean air and water in abundant supply.

Plants grew by photosynthesis starting in the carboniferous era, about 300 million years ago, and the decay of some of these, instead of oxidizing back into the atmosphere, occurred underground in oxygen-free zones. These anaerobic decays released hydrogen but did not release the carbon as CO_2, but retained the carbon leading to the deposits of oil, gas and coal. These deposits are now being depleted on a hundred year timescale, and will not be replaced. Once these accumulated deposits are depleted, no quick replenishment is possible. The energy for future civilization will have to come from alternative sources, in the absence or abandonment of the fossil fuel deposits. The words sustainable and renewable apply to this vision of the future.

There is clear evidence that the amount of available oil is limited, and is distributed only to depths of a few miles. We already discussed the expert opinion of this in Section 1.4, without discussing the origin of the fossil fuels and their confinement to the upper few miles of the Earth. The geology of oil very clearly indicates limited supplies. It is agreed that the continental US conventional oil supplies have mostly been depleted, the term "Hubbert's Peak" applies correctly to this situation. A readable and authoritative account (Deffeyes, 2001) explains that liquid oil was formed over geologic time in favored locations and only in a "window" of depths between 7,500 and 15,000 feet, roughly 1.5 to 3 miles. (At depths greater than three miles the temperature is too high to form liquid oil from biological residues, and natural gas forms.) The limited depth and the extreme time needed to form oil from decaying organic matter (it only occurred in particular anaerobic, oxygen-free locations, otherwise the carbon is released as gaseous carbon dioxide), support the conclusion reached in Section 1.4, that the world's accessible oil is going to run out, certainly on a time scale of 100 years. Recent methods of fracking have made accessible additional oil tied up in shale rock, near the Earth's surface. This resource is also being extracted rapidly.

Physics and Technology of Sustainable Energy. E. L. Wolf, Oxford University Press (2018).
© E. L. Wolf. DOI: 10.1093/oso/9780198769804.001.0001

Further, scientists increasingly agree that accelerated oxidation of the coal and oil that remain, as implied by the present energy use trajectory of advanced and emerging economies, is fouling the atmosphere. Increased combustion contributes to changes in the composition of the rather slim atmosphere of the Earth in a way that will alter the energy balance and raise the temperature on the Earth's surface. Dramatic loss of glaciers is widely noted, in Switzerland, in the Andes Mountains, and in the polar icecaps, that relate to sea-level rises.

New sources of energy to replace depleting oil and gas, are needed. The new energy sources will stimulate changes in related technology. An increasing premium will probably be placed on new sources and methods of use that limit emission of gases that tend to trap heat in the Earth's atmosphere. New emphasis is surely to be placed on efficiency in areas of energy generation and use. Conservation and efficiency are admired goals that are being reaffirmed.

Since the present economy is 84 percent based on fossil fuels, oil, gas and coal, an interim period is inevitable before a fully sustainable regime can be established. With this in mind, we will consider ameliorations and long-term compromises in the use of fossil fuel, for example, the possibility to capture and store underground CO_2 released in burning coal and the continued necessity for oil or gas as a starting point for important industrial chemicals. Further, there seems no sustainable alternative technology to the turbines used in jet aircraft. However, it is well established that bio-fuel can be used successfully in jet aircraft engines.

2.2 Energy from the Sun

The bulk of sustainable energy is that coming directly and indirectly from the Sun. The heating effects lead to motion in the atmosphere that in turn cause wave motions that are available to be tapped. Hydroelectric power is the result of the Sun's continuing role in evaporating water from the ocean that falls inland and flows downhill and can be tapped. The heat from the center of the Earth and the energy of tidal motions, coming basically from the orbital motion of the Moon, are sustainable energy sources from the Earth itself. A summary of the global sustainable energy sources is offered in Table 2.1. By far the largest entry in the Table is the solar input onto land, 30,500 Terawatts, based on an average ground level sunlight of 205 W/m^2.

2.3 Fusion reactions on the Sun

Energy from the Sun is entirely produced by nuclear fusion, and it is important to understand how this works. There are two main aspects of a complex process that are basic. First is the strong nuclear attractive force leading to reactions of the elementary proton particles, and to formation of more stable nuclear products, such as ^{4}He (see Fig. 1.4). Since this combination of the 4 nucleons (2 protons and 2 neutrons) is more stable, a large kinetic energy Q is released. The strong interaction binding in ^{4}He amounts to around 9 MeV per particle. The strong nuclear force has a short range, on the order of a femtometer fm = 10^{-15} m, sometimes referred to simply as f, honoring physicist Enrico Fermi. This force acts equally binding protons and neutrons independent of the electric charge. The individual particles have "charge radii" on the order of femtometers. The charge radius is the size within which the electric charge is contained, about 1.2 fm for a proton, directly measured by electron scattering at high energy. (These experiments actually show substructures of the proton and neutron, each built up from 3 smaller particles, called quarks, within the overall radius 1.2 fm. These details are not needed for our discussion.) Nuclei in general have a radius, measured by high energy scattering experiments, that fit an empirical rule,

$$R = R_o A^{1/3}, \qquad (2.1)$$

where R_o is the basic nuclear size parameter, and A = Z + N, is the number of nucleons, protons plus neutrons. The value of R_o is about 1 fm, values 1.07 fm, 1.2 fm, and 1.44 fm have been used for slightly different circumstances. Formally the radius of the proton, A = 1, is just R_o, and that of the deuteron, A = 2, is 1.26 R_o.

In the most important single reaction in the Sun, described as Eq. 1.7, the joining of two protons to make a deuteron, new particles, a positron and an electron neutrino v_e, are produced. Initially an unstable ^{2}He or pp^{+2} species is briefly formed, and in an extremely small fraction T of its decays generates the first important fusion product D = ^{2}H, releasing energy, via

$$p + p \rightarrow D + e^+ + v_e. \qquad (1.7)$$

The neutrino is neutral, has nearly zero mass, and is hard to detect because it rarely interacts with matter. In spite of this difficulty, the neutrino flux from the Sun at the Earth has been measured and is now in agreement with solar models. The neutrino flux is conceptually an easy way to measure the rate of ^{4}He formation in the Sun, because precisely 2 neutrinos are formed for each helium. The final pp cycle turns 4 protons into ^{4}He + 2e$^+$ + 2v_e, with additional steps,

$$P + {}^2H \rightarrow {}^3He + \gamma \qquad (2.2)$$
$${}^3He + {}^3He \rightarrow {}^4He + p + p. \qquad (2.3)$$

The net energy release on the Sun per cycle of these reactions is 26.2 MeV, excluding the energy that goes away with the neutrinos.

2.3.1 Fusion reactor technology on Earth

Fortunately, the complication with neutrinos is not present in reactions that are used on earth in Tokamak reactors, so these reactions are easier to start than the basic p–p reaction needed on the Sun. The experts know that the complication in the p–p reaction slows down the burning of hydrogen to make helium in stars like the Sun, and has the effect of greatly increasing the life of such stars. For our purpose we primarily need to understand in some detail the mechanism by which the charged and strongly repelling constituent particles (protons) get close enough to react. This will be also our introduction to Schrodinger's equation, built upon the wave aspect of matter.

2.4 Solar influx: direct heating, thermal solar power, photovoltaics

We have seen that the full power density of the Sun at the top the atmosphere is 1366 W/m^2. What falls on the ground is modified, mostly by the rotation of the Earth, scattering by the atmosphere and climate effects. In the end the Sun's energy density varies considerably with differing cloud cover characteristic of different parts of the world. A summary of the large geographical variation is shown in Fig. 2.1 (a). The squares marked on this map represent about 0.16 percent of the Earth area and are judged to be sources of all the world's power need, about 20 TW estimated for mid-century (see Fig. 2.1), assuming the areas are covered with 10 percent efficient solar cells. This is total power consumed, not just electric power!

The units in Fig. 2.1 (b) are effective hours of sunlight per year on a flat plate collector, including weather effects. The peak value 2100 hours per year of sunlight works out an average Watts/m^2 value as 2100/(365 × 24) 1000 W/m^2 = 240 Watts/m^2. Note that in the Midwest portion of the US where the effective hours per year are shown as around 1600, this corresponds to 1600/365 = 4.4 hours per day, at around 1000 Watts/m^2. This time span, 4.4 hours, is roughly the duration of the peak electric demand, often about twice the night-time demand.

There are three primary methods of making use of the available solar power at a given location. Solar photovoltaic cells, mentioned previously, are joined in providing electricity by solar thermal plants, that use mirrors to focus sunlight on high-temperature collectors that typically heat water to steam for conversion to electricity with the conventional steam turbine. An example of such an installation is the Ivanpah Solar Electric Generating System, near Ivanpah, CA, in the Mojave Desert. This facility, with a gross capacity of 392 MW and using 173,500 heliostats, or focusing mirrors, started operation in 2014. The heated steam goes directly to turbine-generator sets, and the capacity is rated at 1.079 TW-h per year. This form of solar electricity lends itself more naturally than photovoltaics to overnight energy storage, by the stratagem of piping a portion of the hot water or other fluid, such as molten salt, to a large insulated reservoir, from which power can be extracted at night by use of a heat exchanger, although this feature is not present at the Ivanpah facility.

The third main form of solar technology is the direct solar hot water and heat system. If the objective is to provide hot water and/or hot water heating for a residence, it is vastly inefficient to generate electricity as an intermediate step, and such an intermediate

(a)

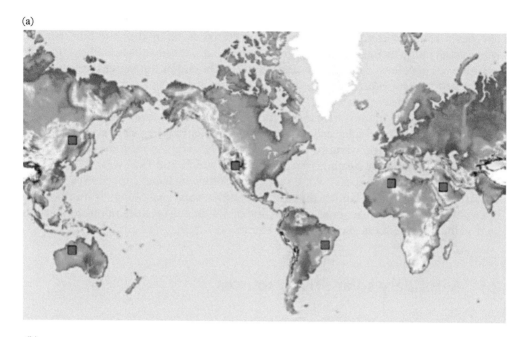

(b)

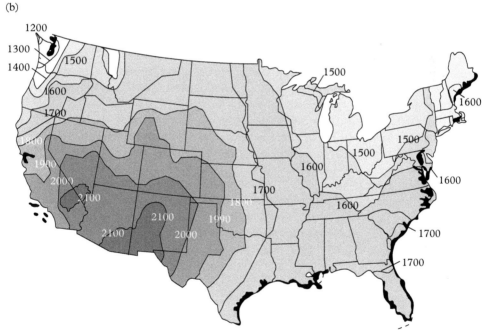

Fig. 2.1 *(a) Map of requirement of land to meet the world's total power demand in mid-century solely by solar cells. (Armaroli et al 2006). (b) Map of solar intensity across the United States. The units are described in the text (Richter, B., 2010)*

step negates the straightforward storage aspect of the direct solar system. Such a system has a collector composed of insulated water pipes exposed to the Sun, running with a small pump through a large separate storage bath, that is thus heated by exchange from the set of collector pipes. The desired hot water and/or hot water for radiators is also heated by heat exchange coils immersed in the large insulated storage hot water tank. This direct heating has unit (100 percent) efficiency compared to 20 percent at best for commercial photovoltaic installations, and has much lower capital cost for the facility than an array of solar cells. Israel, for example, requires such systems in construction of houses. According to the International Energy Agency, the total installed capacity for direct solar heating was 375 GW, of which 262 GW was in China. For comparison, it is reported that worldwide photovoltaic installed capacity reached 178 GW by the end of 2014. So, by the numbers, direct solar technology is number one, not number three. It has no emissions, is not capital intensive, does not need a grid to be distributed, and lends itself to energy storage.

2.5   Secondary solar-driven sources

Wind power, hydropower, river-flow energy and ocean waves are indirect results of heating by the Sun.

2.5.1   Wind power, a form of flow energy

A map of wind speed in the US is shown in Fig. 2.2. The largest average values, in the Midwest, are in the range 8 to 9 m/s. According to Hermann 2006, the average wind-speed at altitude 50 m is 6.6 m/s and the average power density of the wind is 336 W/m² along the wind direction.

The uneven distribution of the resource makes clear the need for a wide grid network, or for conversion to a fuel such as hydrogen that could be piped or shipped in containers.

The power that can be derived from wind or water flow is proportional to v^3. To understand this result, consider an area $A = \pi R^2$ oriented perpendicular to a flow at speed v of fluid of density ρ. In one second a length $L = v$ containing mass $M = Av\rho$ will pass through the area. This represents a flow of kinetic energy $dK/dt = dM/dt\ v^2/2$, so that power $P = \eta\ dK/dt = \eta\ dM/dt\ v^2/2$ can be obtained if the efficiency of the turbine is η. Thus,

$$P(R) = \eta\ \pi R^2 \rho v^3/2. \tag{2.4}$$

A turbine such as that shown in Fig. 2.3 with radius R = 63.5 m, and assuming v = 8 m/s, taking ρ = 1.2 kg/m³ for air at 20°C yields:

$$P = \eta\ \pi\ 63.5^2 1.28^3/2 = \eta 3.89\ \text{MW}.$$

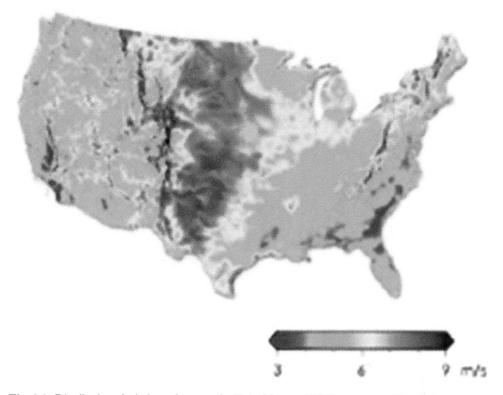

3   6   9 m/s

Fig. 2.2 *Distribution of wind speeds across the United States. (US Department of Energy)*

A good efficiency in practice is about 0.4, giving 1.56 MW/turbine at the assumed 8 m/s, that is a favorable value, shown in the dark areas of the wind map Fig. 2.2, typically in the US in a band running from Texas to Minnesota.

It is quite easy to show that the maximum efficiency is 16/27, about 0.59 (Betz's Law), by realizing that the speed v' behind the turbine is reduced, and the average speed is v_{av} = ½ (v+v'). Thus, the corrected formula is:

$$P(R) = \pi R^2 \rho \, v_{av} \left(v^2 - v'^2\right)/2 = \pi R^2 \rho \, v^3 \left(1 + r - r^2 - r^3\right)/4, \text{where } r = v'/v. \quad (2.5)$$

To find the maximum P against varying the ratio r, we set dP/dr = 0, to find that the best r is 1/3, corresponding to efficiency η = 16/27. This result was first obtained by Lanchester (1915), followed by Betz (1920). It is probably called Betz's Law because of Betz's book in 1926 that carefully extended the analysis to allow the design of wind turbines of high efficiency. We will discuss this extension in Chapter 8.

This formula provides a maximum power 16/27 = 0.593 of the unperturbed power $P_0(R) = \pi R^2 \rho v^3/2$. This corresponds to v'= v/3, so one can see why the wind turbines

Fig. 2.3 *Enercon model E-126 7.5 MW wind turbine. The hub height is 135 m. The specifications say that the machine can be set to cut off at a chosen windspeed in the range 28 to 34 m/s. From the text one would extrapolate to a power from one device, at 28 m/s, of 66.9 MW. The specifications say the blades are epoxy resin with integrated lightning protection, http://www.enercon.de/en-en/66.htm*

are best not longitudinally arranged, because the exit air velocity is quite reduced. The conservation of mass flow means also that the radius of the air column behind the turbine is increased by $3^{1/2} = 1.73$.

Consider an array of such turbines, spaced by 10 R, a very conservative spacing. Then the power per unit ground area delivered by the array of the designated turbines at 8 m/s is $1.56\ MW/(635\ m)^2 = 3.86\ W/m^2$. A rough comparison with solar cells is that an average solar power on Earth is $205\ W/m^2$ with an expected efficiency around 0.15, thus $30.75\ W/m^2$. The possibility exists of having both solar cells and wind turbines in the same area, plausible if the area is not cultivated. Questions of the installation costs are deferred, but the starting estimate of $1/(peak installed watt) generally is useful.

We can ask how large a windfarm is needed to generate 500 GW, approximately the electricity used in the US? If we take $3.86\ W/m^2$, the answer is: Area = $500 \times 10^9/3.86 = 12.95 \times 10^{10}\ m^2$, or 360 km or 223 miles, on a side. This is comparable to the area of the state of Iowa, equivalent to 237 miles on a side! The positive aspect is that the turbines do not necessarily preclude the normal use of the land, for example to grow wheat or corn. But there is no escape from the reality that wind energy, and also solar energy, are diffuse sources. An example of a large wind farm is the Jiuquan Wind Power Base at the edge of the Gobi Desert in China, where it is reported that there are more than 7000 turbines arranged in rows (Hernandez, 2017).

Or, returning to the US electricity market, we may ask how many wind turbines at 1.56 MW per turbine? That number is $N = 500 \times 10^9/1.56 \times 10^6 = 320{,}513$ turbines. At a spacing of 635 m = 0.394 miles per turbine or 2.53 turbines per mile, we could imagine turbines along 126,684 miles of highway. The total mileage in the Interstate Highway System is 46,751 miles, while the total US highways extend 162,156 miles. The cost of the turbines at \$1 per Watt is \$500 billion. The cost of the US Interstate Highway System is said to be \$425 billion in 2006 dollars. \$500 billion is approximately equal to 0.07 of the US military budget for a period of 10 years.

2.5.1.1 *Free-flow water turbines*

The same kinetic energy extraction analysis applies to water flow in a river, that benefits immediately from the factor $1000/1.2 = 833$ increase in density. A recent measurement of Mississippi water flow[1] recorded 11 mph velocity and 16 million gallons per sec. flow under a bridge near Vicksburg, MS. With conversions 1 mph = 0.447 m/s and 1 US gallon = 4.404×10^{-3} m^3 we have 4.92 m/s and $dV/dt = 7.06 \times 10^4$ m^3/s for water flow at this location. The power is then $(dM/dt)(v^2/2) = 1000$ kg/m^3 $(7.06 \times 10^4$ m^3/s$)$ $(4.92$ m/s$)^2/2 = 94.8$ MW.

The efficiency can be at most 0.59, corresponding to loss of speed by 2/3, and the resulting disruption of the river flow if the full cross section were filled with rotor blades would be prohibitive. Still it seems that tens of MW could be extracted from such a flow if it were continuous and if the installations could be sited to avoid blocking of commerce.

The size of a 1 MW river-flow or tidal-flow turbine is much smaller than a 1 MW wind turbine because of the thousand-fold increase in water density. Probably this means the water turbine would be cheaper. Water turbines, especially Francis Turbines, are well-known and highly developed for hydroelectric installations. The smaller forms of free-flow turbines for river-flow applications are not as well established commercially as are wind turbines.

2.5.2 Hydroelectricity

An energy Mgh is associated with raising a mass M at height h in Earth gravity g. In a hydroelectric plant with dam height h the ideal power output can thus be expressed, with water mass flow dM/dt, as,

$$P_{hydro} = \eta \, dM/dt \, gh \qquad (2.6)$$

---

[1] "Mississippi River Flooding: Vicksburg Company measures depth, velocity" (The Associated Press, May 21, 2011).

Water running through turbines at the base of the dam is used to generate electricity, leading, based on this simple formula, to a typical plant efficiency $\eta = 0.9$. (We will describe in Chapter 8 the most common type of utility-scale water turbine, the Francis Turbine.) It is evident from Fig. 1.2 that hydroelectric power is at present by far the largest renewable energy source, amounting to about 1.07 TW worldwide in 2008, or about 7.3 percent. These are extremely large projects typically, and the best sites, offering dam heights h as large as 100 m and 200 m, are already utilized. The situation is often for a large installation that it is near a copper mine or an aluminum smelting facility that has supported capital investment. The availability of efficient DC power transmission lines may make the benefit of these large installations more widely available.

Similar large facilities are at Niagara Falls in the US and the Akosombo Dam in Ghana, Africa. The Three Gorges dam in China at completion has capacity 22.5 GW.

Fig. 2.4 *Grand Coulee Dam is a hydroelectric gravity dam on the Columbia River in the US state of Washington. The dam supplies four power stations with an installed capacity of 6.81 GW. It is the largest electric power producing facility in the US. http://en.wikipedia.org/wiki/File:Grand_Coulee_Dam.jpg*

Fig. 2.5 *Illustration of a 500 MW pumped hydroelectric energy storage facility. The turbines can be reversed to pump water back into the reservoir. This form of energy storage in the electric grid is of larger capacity and lower cost than any known form of battery. Carters Dam in Georgia, US. http://en. wikipedia.org/wiki/File:U.S.ACE_Carters_Dam_powerhoU.S.e.jpg*

The planned Grand Inga Dam in Congo is projected as 39 GW. The Belo Monte Dam on the Xingu River, a tributary of the Amazon, has been approved[2] by Brazil.

The dam would be 3.7 mi. long and the power would be 11 GW. The Itaipu Dam between Brazil and Paraguay is rated at 14 GW. From 20 0.7 GW generators, two 600 kV HVDC lines, each about 800 km long, carry the DC power to Sao Paolo, where terminal equipment converts to 60 Hz. It provides 90 percent of electric power in Paraguay and 19 percent of power in Brazil[3].

Turbines can be made with the capacity to be reversed and to pump water back to the reservoir when demand is low. This storage capability is called "pumped hydro" and efficiency in the pumping mode can be 80 percent. Capacity on the American and Canadian sides of the Niagara River total 5.03 GW, of which 0.374 GW are pumped storage/power producing units (pumped hydro) such as shown in Fig. 2.4. The pumped storage facility Carters Dam in the state of Georgia provides a maximum power output during peak demand conditions of 500 MW. Fig. 2.5 shows the generators and power distribution from this large water reservoir created by an earthen dam.

[2] http://www.bbc.co.uk/news/world-latin-america-13614684 (accessed 03/11/17).
[3] http://en.wikipedia.org/wiki/Itaipu_Dam (accessed 03/11/17).

2.5.3 Ocean waves

The ocean waves are a vast resource of energy that derives from the Sun, clearly a sustainable resource. The conventional equation for the vertical displacement z, taking z = 0 at the quiet ocean surface above deep water is:

$$z = H/2 \sin(kx - \omega t) = H/2 \sin(2\pi x/\lambda - 2\pi t/T). \qquad (2.7)$$

The trough to peak distance is denoted H. (In detail, the motion of small parcels of water in the wave are actually elliptical or circular, to be mentioned forthwith.) The close similarity of the distribution of wave height H with local wind speed is shown in Fig. 2.6a,b. Wind is a secondary result of the Sun's heating of the Earth's surface, and these two figures demonstrate nicely that ocean waves are the result, in turn, of the winds.

The power available in wind comes from mass flow, as we have discussed previously. In the case of waves, the energy resides primarily in vertical motions seen in Eq. (2.7), oscillations of the top few meters of the ocean.

The local oscillation can be tapped by taking energy from the vertical motion of a point absorber, that might be a buoy attached to the ocean floor, or a float attached to a larger floating framework above deep sea. Locally, there is a mass M of water oscillating vertically a total excursion H, the wave height, at a frequency 1/T, with T generally on the scale of 3 to 15 seconds, allowing the local extraction of power. The simplest sinusoidal wave is shown in Fig. 2.7, after Salter (1974), showing vertical displacement

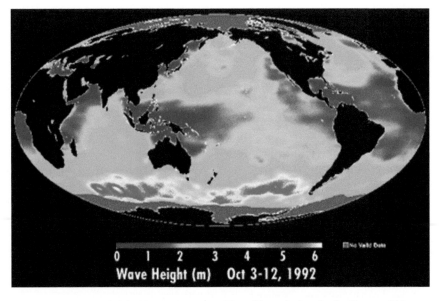

0   1   2   3   4   5   6
Wave Height (m)   Oct 3-12, 1992

Fig. 2.6a *Measured wave height H over the oceans. The wide variations in the distance H between peak and trough lie mainly between 1 m and 5 m. A rough estimate of the average might be near 3 m http:// sealevel.jpl.nasa.gov/education/classactivities/onlinetutorial/tutorial1/windwaves/*

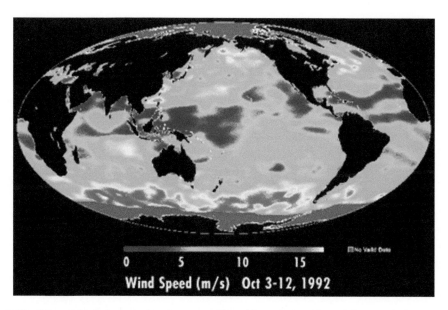

Fig. 2.6b *Measured wind speed over the oceans. The wide variations lie mainly between 5 m/s and 15 m/s. A rough estimate of the average might be near 8 m/s*

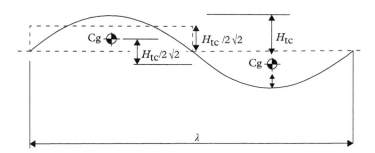

Fig. 2.7 *One wavelength of a sinusoidal wave, showing total excursion $H = H_{tc}$ and total excursion of the center of mass, $H_{tc}/2\sqrt{2}$ (Salter, 1974) Fig. 1*

z vs horizontal position x, from x = 0 to x = λ. A plot of displacement at x = 0 for time between 0 and T would be identical, showing a total height change H of the top of the water.

A more detailed picture of water motion in the wave is given by (Phillips, 1977).

Assuming a uniform depth of ocean d, the horizontal and vertical semi-axes of the elliptical motion of a small parcel of water are given, respectively, as:

$$H/2 \; \mathrm{cosh}k(z+d)/\mathrm{sinh}kd, \quad H/2 \; \mathrm{sinh}k(z+d)/\mathrm{sinh}kd. \qquad (2.8)$$

Here k is the wavenumber, in radians per meter. For deep water, depth d is much larger than wavelength $\lambda = 2\pi/k$, the motion reduces to circles of diameter:

$$H \exp kz = H \exp\left[-2\pi(\text{depth})/\lambda\right]. \tag{2.9}$$

In a typical example, with $H = 3$m and the wavelength $\lambda = 156$ m, the circular amplitude decays only slightly, about 11 percent, from the top surface to depth H. In more detail, these circles do not exactly close, there is additionally a smaller motion, a horizontal mass flow (Stokes, 1847),

$$dx/dt = \omega k\left(H/2\right)^2 \cosh 2k(z+d)/2\sinh^2 kd \tag{2.10}$$

of the top layers of water along the direction of the wind, making the motions vertical circular spirals along x, decaying exponentially down from the surface. It should be said that this Stokes motion of the water parcels is not the same as the group velocity of the wave, that is generally ½ the phase velocity, thus ½ ω/k. The drift motion along the direction of the wave linearly increases with height, the drift is negligible at large depth. This motion is responsible for the "storm surge" that accompanies a strong onshore wind as occurred in Hurricane Sandy that flooded lower Manhattan on 29 October, 2012.

The analysis of Phillips (1977) gives the mean kinetic energy T per square meter, for water of depth d, as,

$$T = \rho\omega^2(H/2)^2 \left(4k\right)^{-1} \coth kd. \tag{2.11}$$

This can be rewritten, using the dispersion relation, with k the wavenumber,

$$\omega^2 = gk, \left(\text{or } \lambda = gT^2/2\pi\right) \tag{2.12}$$

valid for "gravity waves", that neglect any influence of surface tension, as,

$$T = \rho g(H/2)^2 (\coth kd)/4. \tag{2.13}$$

For any conservative dynamical system undergoing small oscillations, the mean potential and kinetic energies are equal, namely $T = V$, so the total energy per square meter, $E = T + V$, equals to,

$$E = \tfrac{1}{2} \rho g(H/2)^2 \coth kd. \tag{2.14}$$

For most relevant situations the water is deep enough that the coth term can be set to unity.

The oscillation is similar to a child's swing, with all the energy in kinetic energy at zero displacement, and all of the energy can be captured by stopping the swing over a time T, the period of the motion. The corresponding power is:

$$P = \rho g (H/2)^2/2T. \tag{2.15}$$

If we arbitrarily choose $H = 3$ m, $T = 10$s, this power comes to 1.10 kW/m^2. According to (2.12) the corresponding wavelength is 156 m.

A "point absorber" is a device designed to capture this energy from the vertical motion of the water at a fixed location. A simple model of a point absorber is a buoy of area 1 m^2 with excluded volume H that will rise a distance H in a time T/2 and then fall a distance H in the same time. We can still treat the excluded volume of dimensions 1 m × 1 m × H m as a point if the wavelength is much larger than the size of the buoy, 1 m, so it looks like a point. We assume the buoy is matched to a load with force matching the Archimedes principle force for the fully submerged buoy,

$$F = \rho g H, \tag{2.16}$$

where the lateral dimensions of the buoy are 1 m × 1 m. In this case work,

$$W = \rho g H^2 \tag{2.17}$$

will be done by the water force on the upswing, and the same work will be done by gravity on the downswing, all in a time T. (The commercial Pelamis device to be mentioned has a similar double-acting power extraction. Pelamis is based on long pontoons connected at flexible joints, oriented perpendicular to the waves. As wave passage flexes the joints, power is extracted from both motions.) So, the power extracted by the ideal double-acting point absorber device will be:

$$P = 2\rho g H^2/T. \tag{2.18}$$

With typical numbers $H = 3$ m and period $T = 10$ s this power is 58.8 kW from one square meter of surface absorber. The deep-water gravity wave wavelength follows the rule (2.12), so the wavelength of the "typical" ocean wave chosen would be 156 m, indeed much larger than the assumed lateral dimension 1 m.

This power is expressed per square meter of a "point" absorbing device, and is different from the average power per square meter of ocean surface. Following Salter, 1974, noting that the total vertical displacement of the center of mass of the water in the wave is $H/2^{3/2}$ (see Fig. 2.7) the change in potential energy per square meter in one period will be:

$$\Delta E = \rho g H^2/16. \tag{2.19}$$

Thus, the corresponding change in total energy per sec (power) per square meter will be,

$$P = \rho g H^2/8T. \tag{2.20}$$

This average power, assuming again H = 3 m and period T = 10 s, is 1.10 kW/m², as found in Eq. 2.20. This number, taken as a global average, depending on rather arbitrarily chosen values of H and T, may be too large, as one could question that it could exceed the average solar input power 205 W/m² from which it ultimately derives. There is no question of course that locally extractable power can greatly exceed the average solar influx. Waves of height greater than 10 m are certainly observed.

A different approach to estimating wave power is commonly used in the literature, as proposed by Salter (1974). Salter characterizes the moving wave as a horizontal power per unit length along the wave crest (Salter, 1974, Eq. 2):

$$P_h = \rho g^2 H^2 T/32\pi. \tag{2.21}$$

This formula represents the horizontal flow of energy, obtained by multiplying $\Delta E = \rho g H^2/8$ by the group velocity of the deep-water wave,

$$v_g = gT/4\pi. \tag{2.22}$$

On this basis P_h is a different power conceptually than the potential power represented by the purely vertical motion in (2.13).

The horizontal power works out to P_h = 86 kW per meter length of crest, using the same parameters. If one takes the wavelength as 156 m, so 86 kW is associated with an area one wavelength by one meter, 156 m², then the corresponding power per m² is 0.55 kW/m², that is smaller by a factor 2 than the vertical power expressed in (2.13). In either case the amount of power available is large, especially so in preferred locations suggested by Fig. 2.6b, and typically along seacoasts facing oncoming waves. Wave energy conversion (WEC) technology includes several point absorber devices and also devices designed to absorb the horizontal power. In his important article of 1974 Salter describes a design for a wave energy converter in which the horizontal movement of water rotates a vane.

Unfortunately, even though wave energy conversion is in an area of conventional mechanical engineering and has been worked on for more than a century, available commercial devices are not widely installed, and firms have been driven out of business in the recent era of perhaps artificially low oil prices.

The wave approach Eq. (2.21) is almost universally used to estimate the potential power along coastlines, for example the west coast of the United States. Salter (1974) makes an estimate of 77 kW/m average wave power along the western approach to the Hebrides Islands. This is a serious estimate based on detailed observations of wave heights and periods along that coast, that Salter describes as one of the most favorable locations in the world for extracting ocean power. He remarks that "a few hundred kilometres of installation could meet the total present electrical energy requirements of the UK" (in 1974). According to Scruggs and Jacob (2009) 255 TWh/year can feasibly be extracted from wave energy around the United States coastline. This works out to an average power 29 GW. If one takes the reported figure for aggregate global coastline length as 1.25×10^6 km, and takes a power figure half that obtained for the favorable

Table 2.1 *Global natural power sources in terawatts.*

Average Global Power Consumed 2014	17.2
Solar input onto land mass[a]	30,500
Wind	840
Ocean Waves	56
Ocean Tides	3.5
Geothermal world potential	32.2
Global photosynthesis	91
River flow energy	7

[a] Solar input onto land area assuming $205 \, W/m^2$.
(Based on Table 8.1, Burton Richter, (2010). *Beyond Smoke and Mirrors: Climate Change and Energy in the 21st Century.* (Cambridge: Cambridge University Press.)

Hebrides coast, one gets a global potential power from coastline waves as 48.1 TW, that may be compared to the 56 TW that is listed in Table 2.1 for Ocean Wave Power.

In practice, all of the presently installed devices are tethered to the sea floor fairly close to shore. The failure of the industry to receive large funding stems in part from fear that any installation may be destroyed in a severe storm. Building in durability drives up the cost. Two well-known designs are the electric power buoy, that is tethered and moves up and down as waves pass, generating electricity; and the mentioned Pelamis "sea snake" device that is a series of pontoons in-line with power-extracting flexible joints between the pontoons. It appears that the devices do indeed function as intended, at least in moderate weather, but are too expensive to build and install in the present regime of oil prices. The benchmark $/Watt installation cost is approachable in wind and solar but likely not in ocean wave power.

In his early and thoughtful essay on "Wave Power", Salter (1974) asks whether ocean power could become a large power source to replace fossil fuels. His opinion is that tethered devices are not practical and that a large installation requires a self-powered floating frame of dimension ½ km to 1 km in size, on which the extraction devices are mounted. The frame is large enough that it remains stable as waves pass, allowing the bases of individual extraction devices to remain stationary against the wave motion. Such an installation may be optimally sited far from land, so Salter discusses the possibility of using the extracted ocean wave power to electrolyze sea water to produce hydrogen as an energy medium that could be delivered to port periodically. Salter proposes a very large installation: 500 m is larger than drilling platforms common in the oil industry and larger than the aircraft carrier USS Enterprise, that is 341 m long and about 40 m wide. The longest ship, an oil tanker, is 458 m long.

If we assume the hypothetical ocean power floating installation is 0.5 km × 0.5 km, then the power available can be estimated from the previous figure for average ocean wave power, 0.551 kW/m², as 137.6 MW. If we envisage the installation as an array of point absorbers spaced by 10.0 m, thus, 2500 point absorbers, then from these figures the available power, at 58.8 kW per absorber, would be 147 MW. At the competitive 1 $/Watt estimate, the allowable cost would be 147 M$. A price of 120 M$ is mentioned for the largest class of oil tanker.

The designs of wave energy conversion devices, termed WECs, to extract the wave energy, are naturally adapted to a particular situation, such as at a given depth of water beyond a shoreline, where waves are approaching land. The wave amplitude and speed increase as the open-water wave approaches land.

As waves approach land at depth d, the wave speed is:

$$v_g = \left[(g\lambda/2\pi) \tanh(2\pi d/\lambda) \right]^{1/2}. \tag{2.23}$$

Water depths in the range 40 to 100 m are typical of present installations. The potential extractable wave energy from the Pacific West Coast of the US is estimated (Scruggs and Jacob, 2009) as 255 TWh per year, and in Europe about 280 TWh per year. These numbers are equivalent to average powers of 0.029 TW and 0.032 TW, respectively. (29 GW is an appreciable fraction, about 0.06, of electric power consumption in the US.) It is not clear what the capital and operational costs of such extraction would be, but at least one commercial device, the "Pelamis" (water snake), has been subsidized by the government of Portugal and put into service.

A plausible estimate of available wave power along a coastline is in terms of power per unit length of the coastline. On the Atlantic coast of Great Britain this is estimated (Scruggs and Jacobs, 2009) as 40 kW/m of exposed coastline. This estimate depends on the height of the waves, that is a function of the windspeed and the unimpeded span of water facing the coast over which the waves can collect energy from the wind. This estimate might be compared to the estimate for the Pacific coast of the US. If that coastline is 1000 miles or 1.6 Mm, then we get, at 40 kW/m, the estimate 64 GW, fairly close in agreement.

The Pelamis (the word means "water snake") device is a linear array of four linked pontoons, each 30 m long, oriented perpendicular to the waves. The flexing motion occurring at the linking joints with wave passage is used to create electricity, and the device is classed as an "attenuator" of wave motion. Pelamis devices totaling 2.25 MW capacity have been installed in the sea near Portugal. Vertically-bobbing buoy devices anchored at modest depths are also practical. Devices may also be based on trapping water from the tops of waves, extracting energy as that water falls back into the sea.

While the potential seems appreciable for tapping wave energy in coastal regions, the much larger potential power at the open sea seems in practice unavailable, by virtue of its remoteness.

On the other hand, following Salter 1974, one might conceive of "ocean stations", large floating facilities that need not be close to land. Such stations might be used, for example, as a basis for desalinization of seawater, for extraction of deuterium from the sea, or for electrolytic hydrogen generation. Possible "ocean stations" for hydrogen production could also harvest wind and solar power. Schemes for delivery by tanker, analogous to the shipping of oil and liquid natural gas, might evolve.

An economically sound and competent city might launch its own ocean station, to capture energy for its sphere of influence, and thus reduce dependence on its surrounding grid. This scenario might extend to viable coastal cities worldwide, perhaps Dhaka or Mumbai, as well as New York City.

2.6   Long-term resources in the Earth

Some of the long-term energy that is available is stored in the Earth, or is the result of the orbital motion of the Moon around the Earth. In addition, the composition of the ocean contains enough deuterium, present from the beginning of the Earth, to constitute a long-term resource.

2.6.1   Deuterium

Fusion of light elements to release energy is the heating mechanism of the Sun. A good starting point for fusion is the deuteron, two of which can fuse to make ^{4}He with release of nearly 24 MeV of energy. The most likely products for DD fusion are actually a triton plus a proton, with 4 MeV; or ^{3}He plus a neutron, with 3.27 MeV, so that the average energy release per DD fusion is 3.7 MeV. The deuteron fusion reactions are considered important because D particles, deuterons, are present on Earth, notably in sea water. Wherever protons occur, there is about 1/6400 chance of finding instead a deuteron. Heavy water HDO therefore occurs as $1/3200 = 0.031$ percent of all water. There is enough in the ocean that this is considered a sustainable or renewable energy source. The problem is that at present there is no practical process using deuterons to actually release energy by the fusion reactions.

If we take the ocean mass as 1.37×10^{21} kg, comparing with the mass per water molecule, $18 \times 1.67 \times 10^{-27}$ kg, we find that there are $N = 4.6 \times 10^{46}$ water molecules in the ocean. This means there are 9.2×10^{46} H atoms, and therefore there are 1.42×10^{43} deuterons. The energy release if all of these deuterons were fused at the average energy release of 3.7 MeV, is therefore $1.42 \times 10^{43} \times 3.27 \times 10^6 \times 1.6 \times 10^{-19}$ J $= 7.45 \times 10^{30}$ J. If the present energy consumption is 14.7 TW, so that one year's energy consumption is 4.63×10^{20} J, the deuteron-based energy would last for 1.6×10^{10} years. So, we may say that the deuterium in the ocean, if it can be converted, is a renewable resource. In Chapter 6 of the book we will look into the possibilities for achieving this release of energy.

2.6.2   Ocean tides

Tides are caused by the motion of the Moon around the Earth, in large part. In "funnel" locations like the Bay of Fundy the flows can be large and rapid. Harvesting tidal flows can be similar to harvesting the flow energy of a river. In some cases, all of the

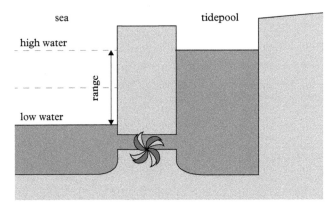

Fig. 2.8 *An artificial tide pool. As shown the pool is filled by the high tide at an earlier time, and is now able to discharge water through a turbine generating electricity (David J. C. MacKay, 2009)*

flow can be funneled into a single set of turbines, a situation more like that at Niagara Falls. This is suggested by Fig. 2.8 of the artificial tidepool shown here.

Famous optimum locations, such as the Bay of Fundy, with a tidal range of 17 m, are at least partly exploited. At present, the 20 MW tidal power plant at Bay of Fundy is the only such plant in operation. However, there is scope for more energy to be tapped in this category.

An example of a much larger potential is shown in Fig. 2.9, based on tidal flows in the British Isles. In the analogy to the tidal basin, the North Sea roughly plays that role in the example of the British Isles as the gateway between the Atlantic and the North Sea. The energy flow can be taxed on the intake and the exhaust of the cycle.

Calculations of the available power, up to 190 GW, are indicated on the diagram in Fig. 2.9.

On smaller scales, the common occurrence of sandbars parallel to a beach suggests many locations that could be utilized. In the Atlantic coast of the US, the Outer Banks of North Carolina enclose Pamlico Sound, an area about 2000 square miles, or 5.18×10^9 m^2. The tidal excursion at Cape Hatteras, on the ocean side, is 3.6 feet, while the tidal excursion on the inside, for example, at Rodanth on Pamlico Sound, is only 0.72 feet. So, it appears that the interior, Pamlico Sound, is decoupled from the tidal excursion on the Atlantic side, by the relatively small openings, through the Outer Banks, between the Sound and the open Atlantic Ocean, where the tides are over 3 feet. Nonetheless, the energy exchanged every 12 hours, $U = Mgh$, where M is the mass of the water in Pamlico Sound to a depth of $h = 0.72$ feet, we can estimate to be quite large. Namely, $U = (5.18 \times 10^9$ m$^2 \times 0.22$ m$) \times 1000 \times 9.8 \times 0.22 = 2.46 \times 10^{12}$ J. In terms of an average power $P = dU/dt$, this is 57 MW. The annual market value of the entirety of this potential power, at 0.14 \$/kWh, would be $57 \times 10^6 \times 3.15 \times 10^7 \times (3.6 \times 10^6)^{-1} \times 0.14 = \$69.8 \times 10^6$. In an age of governments needing to raise taxes, this might be an incentive to install water turbines.

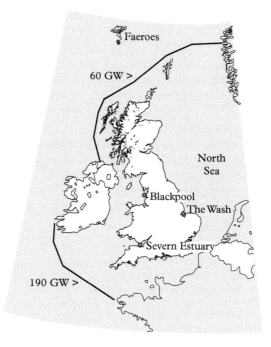

Fig. 2.9 *Map suggesting locations of optimal tidal energy flows from the Atlantic Coasts of the British Isles (David J. C. MacKay, 2009: p. 81)*

This situation is present in numerous smaller scale examples. In New York City, the TV host will speak of the danger, on a given day, at a particular beach, of "rip currents", to swimmers. "Rip currents" are tidal flows of water through such constrictions (between open sea and a tidal pool) as we have discussed. There are many locations where sandbars or "keys" are located just off the mainland.

It would seem that constructing artificial entrapments of this sort, for example, a sandbar ("key") extended by levees (dams) to trap tidal flows, augmented with water turbines and grid connections, could be a new activity for the illustrious US Army Corp of Engineers, which has installed numerous bridges, levees and other water-related engineering projects in the US.

2.6.3   Ocean thermal energy conversion (OTEC)

The Sun's radiation on the ocean leads to a small rise in temperature at the surface relative to deeper water. In tropical oceans the first 100 m or so of the ocean is reliably in the range 26 to 28°C, while the temperature at depths near 1000 m is 4°C to 5°C. This small temperature difference, perhaps 20°C, corresponding to 25°C at the surface and 5°C at a depth of 1000 m that can be reached by a long pipe, allows a heat engine to extract kinetic energy with small efficiency, limited because of the small temperature difference. A guide to the efficiency is provided by the Carnot efficiency, $\eta = 1 - T_c/T_h$ that gives 6.7 percent for 278 K and 298 K. A working OTEC plant in the republic of Nauru, drawing cold water from 700 m depth, was operated continuously for 10 days producing 31.5 kW of OTEC power connected to a grid (Mitsui et al, 1983).

A working laboratory scale example of this type of energy converter is described by Faizal and Ahmed (2013). These authors conclude that the available efficiency is about half the Carnot efficiency, and on this basis, suggest that the working efficiency of such a plant is limited to 3.5 to 4 percent, see Fig. 2.10.

This resource may be attractive nonetheless in small island communities as a practical alternative of importing diesel fuel for generators. The ocean temperature gradient is steady over time, as one practical advantage over wind or solar energy collection. A global estimate of 30 TW for this resource has been provided by Rajagapolan and Niehous (2013).

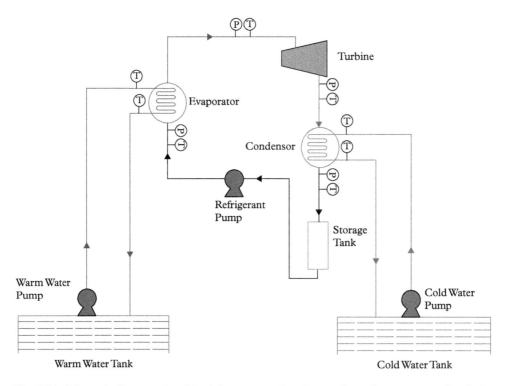

Fig. 2.10 *Schematic diagram of working laboratory version of ocean thermal energy conversion device. The turbine is run using a closed cycle of refrigerant fluid. Heat exchangers provide the high and low temperatures obtained from differing ocean depths. The working device provided a measured efficiency of about 1.5 percent. In the diagram, the symbols P and T indicate locations of pressure and temperature sensors (Faizal and Ahmed, 2013) Fig. 2*

2.6.4 Geothermal

Geothermal potential according to Table 2.1 is 32.2 TW with a higher value, 44.3 TW from a different source.[4]

The core of the Earth is molten, and heat leaks out to the surface. The energy release actually comes from two sources. One is radioactive decay of elements like uranium and thorium in the outer layers of the Earth. The second is the heat from the Earth's core that remains molten, at a much higher temperature. While the trend is of cooling of the core from its primordial high temperature, it has been mentioned that some heat input, a continual heating of the Earth's core, comes from the motion of the Moon, that continually distorts the shape of the Earth, as well as driving the tides.

From a physics point of view, the condensation energy in forming the Earth from a dispersed cloud of dust to a condensed sphere of radius R,

$$E = -3/5 \ GM^2/R, \tag{2.24}$$

is a benchmark value, easily calculated. Here G is the universal gravitation constant, $G = 6.67 \times 10^{-11}$ J m^2/kg^2, so that, with $M = 5.97 \times 10^{24}$ kg, and $R = 6.37 \times 10^6$ m, we find:

$$E = -2.24 \times 10^{32} \ J.$$

This released kinetic energy is vast, comparable to the present rate of consumption extended for 4.84×10^{11} years! It is clear that most of this energy has already been lost, mostly by radiation shortly after the condensation. If we were to attribute this full energy to heating of the Earth, we can estimate what the temperature would have been. In a simple model of a solid or liquid, the thermal energy is,

$$U = 3Nk_B T \tag{2.25}$$

If we attribute all the mass $M = 5.97 \times 10^{24}$ kg to iron atoms, atomic mass $55.85 \times 1.67 \times 10^{-27}$ Kg, then $N = 6.4 \times 10^{49}$ atoms, and $T = U/3Nk_B = 84.5 \times 10^3$ K. The radiation power from the surface of the early Earth at that temperature would be,

$$P = 4\pi \ R^2 \sigma_{SB} T^4, \tag{2.26}$$

where the Stefan–Boltzmann constant $\sigma_{SB} = 5.67 \times 10^{-8}$ W m^{-2} K^{-4}. This evaluates as,

$$P = 1.47 \times 10^{27} \ W.$$

In the simplest view, this suggests that the original heat energy could be radiated away in about 42 hours, since $42 \times 3600 \times P = U$. But the radiative cooling quickly slows as

---

[4] D. L. Turcotte and G. Schubert, "Geodynamics" (Cambridge: Cambridge University Press 2002).

the temperature falls, and the linear approach fails. At present, the inner core temperature has been estimated as 5700 K, while lava (magma) at temperatures ~ 1500 K are present at some locations as close as 10 km to the Earth's surface. The remaining heat energy in the Earth's core is of course enormous and certainly can be regarded as a renewable resource.

Practical extraction of the Earth's heat is accomplished at locations where molten lava extends close to the surface, providing regions of hot rock that are used to heat injected water to produce steam. US. capacity of this type is 3.09 GW, with the largest facility at The Geysers field in CA.[5]

Iceland is known for warm water lakes with people swimming in the winter, and has exploited its geothermal energy to a great extent. The first power plant to use heat directly from magma[6] was built in Iceland.

A map of locations in the United States where magma exists within 10 km of the surface reveals sites concentrated in western states and along the Aleutian Islands.

2.6.5   Fission fuels, expected lifetimes

According to the World Nuclear Association, "Uranium Mining Overview", February 2016, the current rate of mining of uranium is about 67,000 tons per year, Canada, Kazakhstan and Australia are leading producers. The known resource in the ground of available uranium is estimated as 5.9 Mt, so the lifetime of the resource at the current rate of extraction is 88 years. A much larger total amount of uranium is dissolved in the oceans, estimated as 4.5 Gt. This is extremely dilute, estimated as 3.3 ppb, but approaches to extracting it are described in the literature. As explained in the report of the World Nuclear Association,[7] mining of uranium in the ground is difficult because the deposits are typically dilute and also because the mining operations leave behind long-lived toxic radioactive materials once the uranium compounds are extracted. The potential sea water source would not suffer from these environmental problems but is undoubtedly more expensive.

[5] http://www.energy.ca.gov/tour/geysers/ (accessed 03/21/17).

[6] http://www.dailymail.co.uk/sciencetech/article-2552418/The-LAVA-fuelled-power-station-Iceland-harnessing-energy-directly-molten-magma-time.html (accessed 03/21/17).

[7] http://www.world-nuclear.org/information-library/nuclear-fuel-cycle/mining-of-uranium/uranium-mining-overview.aspx (accessed 03/21/17).

3

Solar Radiation through the Atmosphere

The Sun's light arrives at Earth through the vacuum of space and then the Earth's atmosphere as electromagnetic waves described by Maxwell's equations. In contemporary electrical engineering jargon, this is "wireless", like the connection of cellphones. The temperature of the Earth is largely set by the balance between power of absorbed electromagnetic waves from the Sun and the power of infrared waves emitted as thermal or "black body" radiation. The interaction of these waves with impurity molecules in the atmosphere modifies this balance and is at the heart of the phenomenon of global warming.

Maxwell and others showed the experimental laws of electricity and magnetism, associated with Coulomb, Ampere, Faraday and Gauss, to imply the propagating electromagnetic field, $\mathbf{E}(x,t)$, corresponding to radio and light waves, including the correctly predicted speed of light,

$$c = (\varepsilon_0 \mu_0)^{-1/2},$$ (3.1)

$$2.998 \times 10^8 \, \mathrm{m/s}.$$

This is so important to our topic that we give a review of Maxwell's equations.

3.1 Maxwell's Equations and Black-Body radiation

To start, note that the radially diverging Coulomb electric field of a point charge Q, $E = k_C Q/r^2$, leads to Gauss's law of electrostatics for the outward electric flux:

$$\Phi_E = \int \mathbf{E} \cdot \mathbf{dS} = Q/\varepsilon_0,$$ (3.2)

where Q is the total electric charge inside the closed surface S (and $\varepsilon_0 = 8.85 \times 10^{-12} \, \mathrm{F/m}$, usually expressed here in the Coulomb constant $k_C = 1/(4\pi\varepsilon_0) = 8.99 \times 10^9 \, \mathrm{Nm^2/C^2}$). This law can be restated as:

Physics and Technology of Sustainable Energy. E. L. Wolf, Oxford University Press (2018).
© E. L. Wolf. DOI: 10.1093/oso/9780198769804.001.0001

$$\nabla \cdot \mathbf{E} = \rho/\varepsilon_0, \tag{3.3}$$

where ρ is the charge density. $\nabla \cdot \mathbf{E} = \partial E_x / \partial x + \partial E_y / \partial y + \partial E_z / \partial z$, is called the divergence of $\mathbf{E}$: (div$\mathbf{E}$). We see that the divergence of $\mathbf{E}$ is ρ/ε_0.

Turning to the magnetic field, $\mathbf{B}$, the for the statement analogous to Gauss's law we find:

$$\nabla \cdot \mathbf{B} = 0, \tag{3.4}$$

since nature includes no free magnetic charges. The basic sources of $\mathbf{B}$ are spin magnetic (dipole) moments and electric currents I, frequently in the form of an electron in an atomic orbit to form a current loop, that leads to a magnetic moment and a dipole magnetic field. The magnetic field $\mathbf{B}$ circling a long wire carrying current I is easily calculated using Ampere's Law:

$$\int \mathbf{B} \cdot \mathbf{dL} = \mu_0 I, \tag{3.5}$$

where the integral follows a closed path L enclosing the current I. The magnetic field vector $\mathbf{B}$ curls around the wire:

$$\mathbf{curlB} = \nabla \times \mathbf{B} = \mu_0 \mathbf{J}, \tag{3.6}$$

where $\mathbf{J}$ is the current density vector, of units A/m². The **curl** of vector $\mathbf{B}$, $\nabla \times \mathbf{B}$, is again a vector. It is usually presented (taking $\mathbf{i, j, k}$ as unit vectors along x,y,z) as a determinant whose rows are $\mathbf{i, j, k}$; $\partial/\partial x, \partial/\partial y, \partial/\partial z$; and B_x, B_y, B_z. For example, if we assume $\mathbf{J} = \mathbf{k}J_0$, then $\mathbf{k}\mu_0 J_0 = \mathbf{k}\left(\partial B_y/\partial x - \partial B_x/\partial y\right)$, leading to the expected circling B vector field in the x,y plane.

Further, Faraday's Law states that an electric field $\mathbf{E}$ appears along a path enclosing a changing magnetic flux:

$$\int \mathbf{E} \cdot \mathbf{dL} = -\partial\left(\Phi_M = \int \mathbf{B} \cdot \mathbf{dS}\right)/\partial t. \tag{3.7}$$

This circling electric field is an analog of the magnetic field circling the current mentioned previously. By analogy, the corresponding differential statement of Faraday's Law is:

$$\mathbf{curlE} = \nabla \times \mathbf{E} = -\partial \mathbf{B}/\partial t. \tag{3.8}$$

J. C. Maxwell discovered (by restoring a missing current density term, $\varepsilon_0 \partial \mathbf{E}/\partial t$, into the Ampere's law expression) that propagating electromagnetic waves are implied by these experimental laws. The missing current density is called Maxwell's "displacement current", $\varepsilon_0 \partial \mathbf{E}/\partial t$. The expanded (corrected) Ampere's Law expression is:

$$\mathbf{curlB} = \nabla \times \mathbf{B} = \mu_0 \mathbf{J} + \varepsilon_0\mu_0 \partial \mathbf{E}/\partial t. \tag{3.9}$$

The new term provides a current density between the plates of a charging capacitor, making the charging current continuous around the loop, as is sensible.

3.1.1 Waves propagating in free space (vacuum)

Following Maxwell, in free space, the relevant equations (assuming no currents and no charge density) are (1) $\nabla \times \mathbf{E} = -\partial \mathbf{B}/\partial t$, and, (2) $\nabla \times \mathbf{B} = \varepsilon_0 \mu_0 \partial \mathbf{E}/\partial t$. Forming the **curl** of Eq. (1):

$\nabla \times [\nabla \times \mathbf{E} = -\partial \mathbf{B}/\partial t]$, and then using the 2nd equation, we find $\nabla \times \nabla \times \mathbf{E} + \varepsilon_0 \mu_0 \partial^2 \mathbf{E}/\partial t^2 = 0$. The mathematical identity $\nabla \times \nabla \times \mathbf{E} = \nabla(\nabla \cdot \mathbf{E}) - \nabla^2 \mathbf{E}$, plus the fact that $\nabla \cdot \mathbf{E} = 0$ for vacuum, lead to the important Maxwell wave equation for free space,

$$\nabla^2 \mathbf{E} - \varepsilon_0 \mu_0 \partial^2 \mathbf{E}/\partial t^2 = 0. \tag{3.10}$$

This equation has traveling plane wave solutions with phase velocity in vacuum $c = (\varepsilon_0 \mu_0)^{-1/2} = 2.998 \times 10^8 \, m/s$. An example of such a wave is:

$$E_y = E_{y0} \cos\left[2\pi(z - ct)/\lambda\right] = E_{y0} \cos(kz - \omega t), \tag{3.11}$$

where $\lambda = c/f = 2\pi/k$ is the wavelength and f the frequency. (The polarization direction and the propagation direction are at right angles in these waves.)

These results explain the flow of electromagnetic power (1366 watts/m² at the top of the Earth's atmosphere), from the Sun.

3.1.2 Thermal radiation in equilibrium

The spectrum of electromagnetic waves in equilibrium with a hot body, known as black body radiation, well describes the radiation that comes from the Sun. To understand this, going beyond the waves propagating in free space that are mentioned previously, it is useful to first consider radiation in equilibrium inside a cavity held at temperature T. Following Eq. (3.11), describe an electromagnetic wave propagating in the unrestricted x-direction as $\mathbf{E} = \mathbf{E}_o \sin(kx - \omega t)$, where $\mathbf{E}_o$ will be along the y or z, direction. In a 3-D case, say a cubic cavity of side L, imagined to have metallic perfectly conducting walls, $\mathbf{E}$ can have no component parallel to the wall. Suitable standing waves with nodes at the walls occur with an integer number of half wavelengths across the cavity in each direction.

3.1.2.1 Counting electromagnetic modes

This gives the propagation vector k of suitable waves as,

$$\mathbf{k} = (\pi/L)\,(l\mathbf{i} + m\mathbf{j} + n\mathbf{k}), \tag{3.12}$$

with positive integers l, m, n and unit vectors **i, j, k**. Because the electric field has two orthogonal polarizations, there are two modes for each propagation vector. The frequency is $\omega = ck$, that leads to:

$$\omega^2/(c\pi/L)^2 = R^2 = \left(l^2 + m^2 + n^2\right). \tag{3.13}$$

In a space with coordinates l, m, and n, the modes out to a frequency ω fall inside one octant of a sphere of radius R. The number of modes $N(\omega)$ out to frequency ω is:

$$N(\omega) = (1/8) \times 2 \times 4\pi R^3/3 = \pi\omega^3 L^3/3c^3\pi^3 \tag{3.14}$$

and becomes,

$$N(v) = 8\pi\, v^3\, L^3/\left(3c^3\right), \tag{3.15}$$

where $v = \omega/2\pi$. The number of modes per unit frequency is:

$$dN(v)/dv = 8\pi v^2 L^3/c^3. \tag{3.16}$$

If we assume L is very large, and divide by L^3, we can interpret this as the number of modes per unit volume at frequency v in empty space. We wish to find the energy density $u(v)$ per unit frequency and per unit volume. To do so, we assume that each mode is occupied according to the Bose–Einstein distribution function,

$$f_{BE} = \left[\exp(hv/kT) - 1\right]^{-1}, \tag{3.17}$$

namely, the number of photons in the mode.
    Thus, we find the energy density to be:

$$u(v) = \left[8\pi\, hv^3/c^3\right]\left[\exp(hv/kT) - 1\right]^{-1}. \tag{3.18}$$

This is the Planck thermal energy density in equilibrium at temperature T, describing the spectrum of black body radiation. Since the radiation moves at speed c, the corresponding power density is c/4 u, so

$$P(v) = \left[2\pi\, hv^3/c^2\right]\left[\exp(hv/kT) - 1\right]^{-1}, \tag{3.19}$$

that can also be written as:

$$P(\lambda) = \left[2\pi\, hc^2/\lambda^5\right]\left[\exp(hc/\lambda kT) - 1\right]^{-1}. \tag{3.19a}$$

The peak in this spectrum $P(\lambda)$ occurs at:

$$\lambda_{max} = c/v_{max} = \left(2.90 \times 10^6\, nm\ K\right)/T. \tag{3.20}$$

In the idealized spectral power density $P(\lambda)$ (3.19a) choosing T = 5778 K actually is a good representation of the observed Sun's power spectrum, see Fig. 1.4. In that Figure the ordinate, the spectral power density, is called "irradiance" and is expressed as mW/m²/nm, with a peak value about 2100 near $\lambda = 500$ nm. The area under the measured curve Fig. 1.4 is close to $1366 \, \text{W/m}^2$, that corresponds to a total incoming power intercepted by the area of the Earth, πR_E^2, with $R_E = 6371$ km, as $174.2 \, \text{PW} = 1.742 \times 10^{17} \, \text{W}$. This is about 10,130 times the total current power consumption on Earth, around 17.2 TW. Since we have a good estimate of the Sun's surface temperature T from the peak position in Fig. 1.4, we can use this power density to estimate the emissivity ε, using the relation $P = \varepsilon \sigma_{SB} T^4$. This gives emissivity $\varepsilon = 0.998$, that seems reasonable.

3.2 Stefan–Boltzmann law of radiation from hot matter

The total energy per unit volume, U is obtained by integrating $u(v)$ from $v = 0$ to $v = \infty$, made easier by changing variables to $x = hv/kT$, that gives, since $\int_0^\infty x^3/(e^x - 1) \, dx = \pi^4/15$,

$$U = \int u(v) dv = \left[8 \, \pi^5 k^4 / (15 h^3 c^3) \right] T^4 = (4/c) \, \sigma_{SB} T^4, \text{where } \sigma_{SB}$$
$$= 5.67 \times 10^{-8} \, \text{W}/\text{m}^2\text{K}^4. \tag{3.21}$$

Thus, we have a derivation of the Stefan–Boltzmann law (2.26) for the radiation power density (energy per unit time and per unit area) from an equilibrium surface at temperature T, that is:

$$dU/dt = Uc/4 = \sigma_{SB} T^4, \tag{3.22}$$

where the Stefan–Boltzmann constant is $\sigma_{SB} = 2 \, \pi^5 k^4 / (15 h^3 c^2) = 5.67 \times 10^{-8} \, \text{W/m}^2\text{K}^4$. The pressure P exerted on the wall of the cavity containing the radiation is:

$$P = U/3 = (4/3c) \, \sigma_{SB} T^4 \, \text{N/m}^2 \tag{3.23}$$

This equation is of the form $P = bT^4$, that resembles the equation of state of a gas. (The units in (3.23) are correct, if we recall that U is energy per unit volume.) The origin of this pressure is the momentum of the photon, $p = E/c$, and the fact that $F = dp/dt$.

An adiabatic compression or expansion is one in which the volume is changed without allowing energy to flow in or out of the system. An example is the rapid compression of the gas in the cylinder of a diesel engine, that results in ignition of the gas and injected fuel without the need for a spark plug. It can be shown, for thermal radiation, that the temperature–volume relation is $VT^3 = $ constant under adiabatic conditions.

The thermal radiation described by Eqs. (3.17–3.23) will exit from a small hole in an empty cavity containing radiation (photons) in equilibrium at temperature T. It is called "black body" radiation, because any radiation falling on the small hole to enter the cavity makes many internal reflections and has no chance of coming back out: in a limit, the small opening is perfectly absorptive of radiation. The radiation out from the small hole is a flux of photon particles whose directions will be perfectly random, as they are bouncing around experiencing equally all 4π directions inside the cavity.

So black body radiation is characterized by a uniform angular distribution into a full hemisphere of directions, a solid angle 2π. The corollary of blackness, or full absorption, is unit emissivity, corresponding to Eqs. (3.17–3.23).

As mentioned, radiation from the Sun is observed to fit the black body power density for a temperature near 5778 K and emissivity near 1.0. The concepts of black body radiation are idealized, and it is perhaps remarkable that the resulting mathematical description applies so widely to real physical surfaces not obviously in equilibrium. Not surprisingly, absorptivity/emissivity of observed surfaces can be quite variable. For example, a mirror has low values of absorption and emission, corresponding to large reflectivity. As mentioned earlier, black body radiation out into space from the Earth basically controls the Earth's temperature, in balance with the fixed radiation coming in from the Sun. (There is some heat coming out from the interior of the Earth, but it is not large enough to alter the overall temperature.) The reflectivity of portions of the Earth's atmosphere for the outgoing black body radiation is referred to as the greenhouse effect, and has been known, since the nineteenth century, to be sensitive to small changes in the molecular composition of the atmosphere. For that reason, we will cover molecular bonding and absorption of radiation in Chapter 7. The detailed properties of molecules in the Earth's atmosphere, including the absorption of such molecules in the visible and infrared regions of the spectrum, are needed to understand in detail what controls of the Earth's temperature.

3.2.1 Aspects of variable emissivity and reflectance

A recent article (Raman, et al, 2014) demonstrates an engineered layered thin film structure, analyzed within the framework of black body radiation, that has such low reflectivity in the visible and such high emissivity in the infrared that it acts as a passive *refrigerator* even when exposed to full Sunlight. This material might be compared to the well-known inexpensive aluminized Mylar sheet, usually 2 mils = 5.1 μm in thickness, marketed as a "first aid rescue blanket". The reflectivity of the thin Al layer deposited by sputtering onto the polymer Mylar is replaced in the new cooling structure with a "dielectric mirror", composed of alternating insulating layers, SiO_2 and HfO_2, both transparent in the visible, but of different refractive indices n. (These layers are also chosen to have high emissivity in the infrared, a feature not desired for the "space blanket".)

The new layered structure provides radiative cooling below ambient air temperature under direct Sunlight. As indicated in Fig. 3.1, the layered medium has extremely low absorption in the visible and high absorption/emission in the infrared, particularly in the range 8 μm–13 μm (the atmospheric "window"), where the Earth's atmosphere does not offer much absorption of outgoing radiation. Reflectivity of incident Sunlight is said to be 97 percent in the constructed thin film. Under direct Sunlight, measured as 850 W/m², the cooling power was found to be 40.2 W/m² and the temperature of the cooler falls 4.9°C below ambient. Note that the peak wavelength for blackbody radiation at 300 K occurs at 9.67 μm, as given by λ_{max} = (2.90 × 10⁶ nm K)/T. The upper (a) inset of the figure indicates the ground-level solar spectrum known as AM 1.5, air

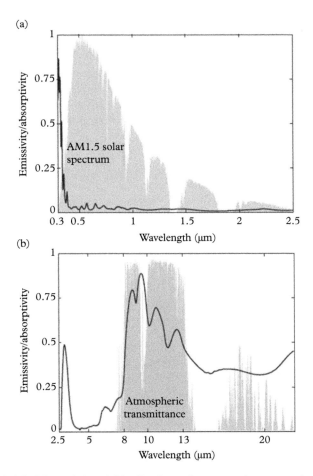

Fig. 3.1 *(a) Emissivity/absorptivity of thin film layered structure demonstrated as passive radiation cooler. (b) The sharp dip in the atmospheric transmittance shown in the lower panel at about 9 µm, interrupting the "window", can be attributed to ozone in the stratosphere. (Raman et al 2014)*

mass 1.5, an approximation to the 150 percent greater length of travel through the atmosphere if the Sun is a typical 48.2° from the vertical in the sky. The large modifications in this spectrum, compared to the spectrum seen above the Earth (Fig. 1.4), arise from molecular absorptions as we will discuss in Chapter 7.

To make a back-of-the-envelope analysis of the cooling property of the film structure, we can use Eq. (3.19) to estimate the outgoing power in the atmospheric window infrared range 8–13 µm, taking the film emissivity average at 0.7 as suggested in Fig. 3.1 panel (b). The atmospheric absorption is shown also in Fig. 3.2, where the absorption, in the window range varies approximately from 0.13 to 0.3. In that window range the average value of the irradiance or spectral power density $P(\lambda)$ (3.19a) is approximately

$$29.2 \times 10^{6} \left(W/m^{2} \right)/m = 29.2 \left(W/m^{2} \right)/\mu m$$

at 300 K. For the atmospheric window range of width 5 μm, assuming atmospheric transmission 0.731 (consistent with Fig. 3.2, upper trace near 1 μm) multiplying by the film emissivity 0.7 seen in Fig. 3.1, we find outgoing power is 5 × 29.2 × 0.7 × 0.731 = 74.7 (W/m²). If the incoming power from the Sun is 850 (W/m²), then we can solve for the needed reflectivity R, to obtain the stated cooling power 49.2 (W/m²). On these assumptions, one finds the condition, in W/m²:

$$49.2 = 74.7 - 850\left(1 - R\right), \tag{3.24}$$

that gives R = 0.97 for the composite structure in the visible, agreeing with the authors' statement.

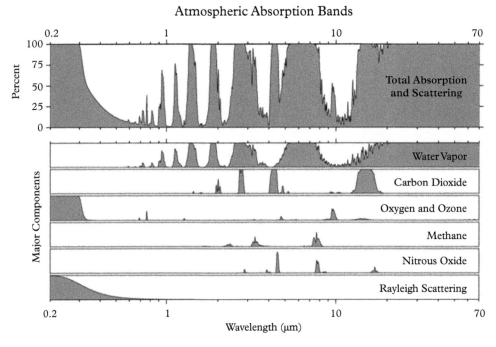

Fig. 3.2 *Absorption and scattering spectra for major impurity gases in the atmosphere from the visible to the far infrared. This suggestive Figure shows that with typical concentrations of water vapor and carbon dioxide, the Earth's atmosphere becomes opaque, completely absorbing, for wide ranges in the infrared. Note the absorption of ozone near 9.4 μm and that of CO_2 near 14.85 μm. At the extreme left of the Figure one sees beginning of absorption from the diatomic molecule oxygen in short wavelengths near 2000 A or 0.2 μm. Also shown in the short wavelength limit is the effect of Rayleigh scattering as discussed in Chapter 1, see Eq. 1.2 http://www.globalwarmingart.com/images/4/4e/Atmospheric_Absorption_Bands.png*

This high reflectivity, corresponding to extremely small absorptivity, is shown in the visible region in the upper panel of Fig. 3.1. This is achieved, in the passive cooling structure, by alternating layers of larger and smaller indices of refraction n. The underlying reflection r at a single interface between transparent media of indices $n_2 > n_1$ is reflectivity

$$r = (n_2 - n_1)^2 / (n_2 + n_1)^2 \qquad (3.25)$$

for normal incidence.

In this cooling structure the transparent dielectrics SiO_2 and HfO_2 have index values $n_1 = 1.46$ and $n_2 = 2.11$, respectively, in the visible. In the present case, from Eq. (3.25), r = 0.051 for the internal reflectivity. To get the desired high overall reflectivity, the design makes use of the "distributed Bragg reflector" or "dielectric mirror" criterion of having an odd number of alternating layers, with the outer layers of larger index, satisfying

$$\lambda/4 = n_2 t_2 = n_1 t_1, \qquad (3.26)$$

with t the layer thickness.

The desired large emissivity in the infrared of the dielectrics used in the new structure arises from material properties. Namely, internal vibrations excitable by light in the 10 μm range, lead to peaks in absorptivity, and also correspond to peaks in emissivity. The background atmospheric absorption by impurity molecules leads notably to the "window" near 10 μm, as shown in Fig. 3.2. The molecules that lead to the absorption in Fig. 3.2 include water, carbon dioxide, methane, ozone and others. The amount of water is quite variable at different locations on the Earth, depending on the temperature, but the other concentrations, apart from ozone, can be taken as nearly uniform.

To understand such effects that are important also, in the case of the molecular absorptions leading to the greenhouse effect, we will address in later chapters, such considerations as material properties.

3.3 The Atmosphere: absorption, scattering, re-radiation: energy balance, equilibrium temperature

It is quite common, but mistaken, to think of the Earth's atmosphere as limitless, so that any activity of humans can be neglected. The height of the atmosphere is in reality about 0.1 percent of the Earth's radius, and its mass is extremely small relative to that of the Earth. The atmosphere is mostly molecular nitrogen and oxygen, with molar masses 28 and 32, respectively. It can be treated as an ideal gas, that, as mentioned in connection with Eq. (1.2), has a sea-level particle concentration 2.4×10^{25} m^{-3} at the nominal temperature 300 K and pressure, 101 kPa. Taking the average mass as M = 30 AMU per molecule, the mass density at ground level is then $\rho = 2.4 \times 10^{25} \times 30 \times 1.67 \times 10^{-27}$ kg/m^3 = 1.20 kg/m^3.

3.3.1   Atmospheric basics

Considering the pressure change $\delta P = -\rho g \, \delta z$ with a change of height δz in the Earth's gravity, $g = 9.8$ m/s^2 we can find the pressure as a function of height z. Thus,

$$\partial P / \partial z = -\rho \, g \qquad (3.27)$$

and

$$P = \rho RT/M \qquad (3.28)$$

(the ideal gas law, where the gas constant $R = N_{Av} \, k_B$) lead to the differential equation

$$\partial P / P = -(gM/RT) \, \partial z. \qquad (3.29)$$

Thus, we find:

$$\ln(P/P_0) = -z/h, \qquad (3.30)$$

where,

$$h = RT/gM \qquad (3.31)$$

can be called the height of the atmosphere for molecular mass M. Thus, the pressure falls off exponentially:

$$P = P_0 \exp(-z/h). \qquad (3.32)$$

The height h evaluates as h = 8.48 km for the atmosphere as a whole, but, e.g., is reduced to 5.78 km for CO_2, that has mass 44 compared to the atmospheric mass 30. So at the height of the airliner, 30,000 feet, taken as 10,000 m, the air pressure will be 101 kPa exp(−10,000/8480) = 31 kPa.

Compared to the radius of the Earth, 6371 km, the atmosphere's height of 8.5 km is 0.13 percent, small indeed. The mass of the atmosphere is 5.1×10^{18} kg, that is an even smaller fraction, 8.5×10^{-7}, of the mass of the Earth, 5.97×10^{24} kg. The derivation we have made assumed a constant temperature, that is incorrect but not badly enough to invalidate the pressure prediction. In fact the temperature falls with height up to about 12 km. If we guess that the temperature T varies linearly with the pressure P, i.e., as T = 300 exp(−z/h), so at 10,000 m the temperature would be reduced by the same factor as the pressure, we get T = 92 K. This guess is seriously wrong, the temperature at 10,000 m is approximately 220 K. An approximate understanding of the "lapse rate" −dT/dz, the temperature variation with height z, can be obtained from the constancy of the quantity $c_p T + gz$, called the "dry static energy" (Randall, 2012, p. 67), where c_p, the specific heat at constant pressure, $c_p = 1004$ J/(K kg).

Thus, from Randall, (2012) we find

$$T(z) \approx 300 - gz / c_p, \tag{3.33}$$

that we can expect to be nearly correct up to about 10 km, a region called the troposphere, that holds about 80 percent of the atmospheric mass. (The temperature reaches a minimum value, 218 K, nearly constant between 12 km and 20 km, and rises with altitude above 20 km, in air heated by sunlight absorption by ozone. The temperature at the top of the stratosphere, $\approx$ 50 km, is around 270 K.)

Our Eq. (3.33) gives for altitude z = 10 km an estimate,

$$T(10 \text{ km}) = 300 - 9.8 \ 10,000 / 1004 = 202.4 \text{ K}.$$

This is not too far off, and the discrepancy with the stated 220 K (Randall, 2012), exposes the fact that $c_p T + gz$ is only approximately constant, and actually increases with height.

The ideal gas law continues to apply locally, so we can use our approximate results to get an air density, proportional to $1/V = P/RT$, for the dry conditions,

$$\rho(z, T) \approx 1.2 \text{ kg/m}^3 \exp(-z/h) / \left[1 - gz/300 c_p \right]. \tag{3.34}$$

This analysis applies to air with no water vapor. Allowing humidity, the chance for cloud formation makes the situation more complicated.

3.3.1.1 *Aspects of cloud formation in a humid atmosphere*

-dT/dz, the "dry adiabatic lapse rate" in meteorological terms, from (3.33) is g/c_p, and is about 10 K per kilometer. The lapse rate is reduced to about half this value near the Earth's surface in locations where the atmosphere is moist, and this has to do with the release of kinetic energy, warming, as moist air rises and condenses into vapor to form a cloud. In addition, since the molecular mass of H_2O is only 18, vs the 28 and 32 of oxygen and nitrogen, respectively, a parcel of moist air is slightly less dense and will be buoyed upward by surrounding dry air. The "specific humidity" is defined as the mass fraction of the air due to water vapor, ρ_{vapor}/ρ, that is typically parts per thousand. The vapor pressure of water is an increasing exponential function of temperature, this is purely a molecular property. In a crude approximation valid only near 288 K, the function is roughly

$$P_{Sat} \sim 1.7 \text{ kPa} \exp\left[(T-288)/15 \right]. \tag{3.35}$$

The value of the pressure of water vapor in equilibrium at the accepted average Earth surface temperature 288 K is about 1.7 kPa, compared to the 101 kPa of the atmosphere, but this 1.7 kPa increases by 7 percent per degree Kelvin at 288 K. Above a warm tropical ocean the saturation vapor pressure may approach 4 kPa, corresponding to 28°C, that would correspond to about 2.5 percent by mass of the atmosphere being water vapor. The relative humidity is defined as the water vapor pressure compared to the saturation value, 100 percent relative humidity, that implies the formation of tiny droplets, too small to fall downward, thus creating a floating cloud.

In the formation of a cloud, warmed less-dense moist air continues to rise in an environment of cooler more dense air. The temperature falls with increasing altitude z, and the bottom of the cloud occurs where the available water vapor pressure reaches the saturation vapor pressure, that will fall with increasing altitude. So the cloud bottom is higher when the humidity is lower. The cloud contains tiny water particles, but they are too small to fall and remain suspended. The long time of suspension of small particles in air is a consequence of the viscosity of air, in the regime of motion of small particles governed by Stokes' law.

Namely, the force needed to move a sphere of radius R at a velocity v through a viscous medium is given by Stokes' law,

$$F = 6\pi\eta Rv. \tag{3.36}$$

This is valid only for very small particles and small velocities, under conditions of streamline flow. The relevant property of the medium, air, is the viscosity η, defined in terms of the force $F = \eta vA/z$ necessary to move a flat surface of area A parallel to an extended stationary surface at a spacing z and relative velocity v in the medium in question. The unit of viscosity η is the Pascal-second (one Pascal is a pressure of 1 N/m^2). The viscosity of air is about 0.018×10^{-3} Pa-s, while the value for water is about 1.8×10^{-3} Pa-s. The traditional unit of viscosity, the Poise, is 0.1 Pa-s in magnitude.

The fall, under the acceleration of gravity g, of a tiny particle of mass m in this regime is described, following Stokes' law, by a limiting velocity obtained by setting F equal to mg. This gives, for small particles in air,

$$v = mg/6\pi\eta R = 2/9\ R^2\rho\ g/\eta = 1.21 \times 10^5\ R^2\rho \tag{3.37}$$

As examples, a particle of 10 μm radius and density 2000 kg/m^3 falls in air at about 24 mm/sec, while a 15 nm particle of density 500 kg/m^3 will fall in air at about 13 nm/sec. The latter example resembles a tiny soot particle known to come from jet engine exhaust and known to nucleate water vapor in the familiar contrail from the jet airplane. On this basis a rain drop of radius R = 0.2 mm will fall at 5 m/s that is about right for a large raindrop.

Clouds are powerful engines to raise energy upward from the surface of the Earth. The energy is largely in the form of the latent heat of water. The "cumulus instability" is described as the tendency of humid air to float upward under the influence of positive buoyancy generated through the release of latent heat. The rising less dense warm humid air forms a cloudy tower that can be many kilometers tall. This can be accompanied by precipitation of water, and lightning and thunder may occur when ice crystals are present. Frictional effects can release electric charge from ice but not from liquid water. The energy release of water condensation is described by the latent heat of vaporization that is 2260 J/gram, a large number. (A further heat of fusion, if the water freezes, is 337.4 J/g.) A rainfall of 1 inch per hour under a thunder cloud is common and up to 6 inches per hour have been recorded. We can suppose the latter figure in the interior of the thunder cloud. In an hour a water volume $6 \times 2.54 \times 10^4$ cc/square meter will accumulate, corresponding to release of 344 MJ of latent heat. 344 MJ/3600 sec-m^2 is a large power density, 95.4 kW/m^2 and can lead to a strong updraft inside a thunder cloud. Such an updraft is a known phenomenon, and is needed certainly to explain the formation of hailstones. Consider a hailstone on the order of 2 cm in diameter, with several internal onion-like layers, indicating several rises and falls before hitting the ground. The terminal velocity of hailstones of 2 cm diameter is about 12 m/s by measurements of Knight and Heymsfield (1983). (The Stokes Eq. (3.37) no longer applies because the air flow is turbulent.)

Raising the hailstone to a height H where it would be cold enough to freeze could only occur in a strong updraft, faster than 12 m/s, based on the measurement of Knight and Heymsfield. At the lapse rate dT/dz ≈ 10 K/km, assuming 30°C at ground, the height H to freeze the hailstone would have to be at least 3 km. If we interpret the power $P = 95.4$ kW/m^2 as raising a column of air of height H = 4 km with cross section 1 m^2, assuming an average density $\rho = 0.5$ kg/m^3 (from (3.34), taking rough account of the falling density with altitude, thus mass M = ρH per square meter) at an unknown velocity $v_{updraft}$, we can write,

$$P = Mgv_{updraft}. \tag{3.38}$$

These numbers give $v_{updraft}$ = 4.9 m/s or about 11 mph. This speed is about half the estimated terminal velocity and clearly is not enough to lift a hailstone. Indeed, according to Randall, 2012, p. 62, updraft speeds in thunderclouds can be 20 m/s or more.

In fact, the main origin of the updraft is the buoyancy of the less dense warmer air produced by release of latent heat. The heated column of air, as water condenses, rises like a hot air balloon, because its mass is smaller than the mass of the displaced air. The upward force, by Archimedes principle, is the weight of the displaced air, that is heavier than the warmed air. (Ascent rates for hot air balloons are reported as typically 6 m/s.) Consider a 1 m^3 "parcel" of air (in the column of height H) that has mass 1.2 kg at 300 K, at the mentioned 6-inch per hour rate of rainfall, releasing 95.4 kW/m^2 under the cloud of height H, taken as 4 km. The 1 m^3 parcel of would proportionally release 1m/H = 1m/4km of that power, that would be 23.85 W. According to the governing equilibrium condition c_pT + gz = constant. But we are adding energy, and at constant height z the energy change 23.85 J/s for the parcel shows up as a change in c_p T. Since c_p is 1004 J/kg K, that would give a temperature change dT/dt = 23.85/1004 = 0.0238 K/s . If we assume the column of air is rising with speed v, then the relevant heating time would be 4000/v sec, so the total rise in temperature ΔT of each parcel, as it rises H = 4000 m to the top of the column, would be 0.0239 K/s × 4000/v = (95.6/v) K. (If the speed v were 10 m/s, then the relevant time would be 4000 m/10 m/s, or 400 sec., but we do not know the right speed v, and treat v as an unknown.) Using the ideal gas law, density 1/V = P/RT, the density with heating would be fractionally reduced by 95.6/v

$\Delta T /300K = 0.319/v$. According to Archimedes' law, the upward buoyant force on the column would correspond to gravitation g acting on a fraction $(0.319/v)$ of the mass of the column. If the density of the column were constant, this would be $(0.319/v)$ g ρH. With $H = 4$ km and $\rho = 1.2$ kg/m^3, the force per square meter forcing the column upward is $(1.5 \times 10^4/v)$ N/m^2. How fast will the column rise with this force? (This is similar in principle to how fast the hot air balloon rises when its interior temperature is raised, and depends on the drag force offered by the air to the moving balloon.) The terminal updraft velocity can be approximated assuming a classical drag force, proportional to the square of the velocity, on the updraft column. This gives a terminal velocity formula, where A, the projected area is 1m^2 and ρ is the density of the fluid taken as 1.2 kg/m^3:

$$v = [2mg/\rho AC_D]^{1/2}. \tag{3.39}$$

Taking the dimensionless drag coefficient $C_D = 1.0$, and identifying mg in the formula as the driving buoyant force, we find $v = [21.5 \times 10^4/v\rho AC_D]^{1/2}$ or $v^3 = [21.5 \times 10^4/1.2]$ for the limiting speed. This gives 29.2 m/s, about 65 mph, that seems plausible. The latent heat release from condensation of water warms the column to make it less dense, and the force of gravity, by Archimedes Principle, raises the column. While this is a rough analysis, it illustrates that the latent heat of water is the origin of energetic phenomena that occur in humid atmospheres, including vortices in tornados, with air moving at more than 100 mph.

The "convective available potential energy" (CAPE), in Joules per kg, is the meteorologist's basis to understanding updraft formation. (Randall, 2012, p. 86 ff) This is the amount of energy a parcel of air would have if lifted a distance vertically through the atmosphere. A mathematical definition is,

$$CAPE = \int g\left[\left(T_{parcel} - T_{env}\right)/T_{env}\right]dz, \tag{3.40}$$

usually given in Joules/kg. The possible conversion of this potential energy to kinetic energy in the updraft would suggest a limiting velocity estimate, setting:

CAPE $= [1/2 \, mv^2]/m$, with m the mass: thus, the limiting velocity is $v_L = [2 \times CAPE]^{1/2}$. Measurements of temperature vs height z near thunderstorms have given CAPE values as large as 5500 J/kg to 8000 J/kg, that correspond to extremely high limiting velocities 105 m/s to 126 m/s.

3.3.2   Optical absorption in a uniform medium

The atmosphere is almost entirely, 98.6 percent, composed of diatomic nitrogen and oxygen molecules, that do not absorb visible light. There are only parts per million of impurity triatomic molecules, that are strong absorbers at some wavelengths, as shown in Fig. 3.2. Water, H_2O, can be a much larger proportion, and has to be treated separately, because its concentration is highly variable by location and by temperature, and is often time varying. The atmosphere is quite transparent in the visible and supports essentially the vacuum speed of light. We have mentioned in Chapter 1 the effects of

Rayleigh scattering in Eq. (1.2), that redirects a portion of the light but does not absorb it. The first effect of a uniform non-absorbing medium is to reduce the speed of light as described by an index of refraction, n. The atmosphere is not dense enough for this to be important, so the speed of light in the atmosphere is given by (3.1) reduced by an index 1.0003. There are spectral regions where the atmosphere does absorb strongly due to triatomic and larger molecules, including CO_2 at about 400 ppm by volume and ozone O_3 that is generated in the upper atmosphere (the stratosphere, between 20 km and 40 km, at concentrations 2 ppm to 8 ppm) by UV light. Absorption of light in the atmosphere is usually described as a decaying intensity I

$$I = I \exp(-\kappa M), \tag{3.41}$$

with an absorption coefficient $\kappa(\lambda)$ expressed in units m^2/kg with M denoting the absorber mass in kg/m^2 at the base of an absorbing layer. (In a semiconductor, like silicon, light absorption is conventionally described as intensity

$$I = I_0 \exp(-\alpha x), \tag{3.42}$$

in W/m^2 with $\alpha(E)$ in m^{-1} units and strongly energy dependent. In Si, the absorption is essentially zero for photon energy hc/λ less than the "bandgap" energy 1.12 eV, and rises sharply above that value.) In both cases the absorption is sharply energy-dependent and represents loss to specific excitations of the absorber, such as generating electron-hole pairs in the semiconductor, and exciting specific motions, such as vibration and bending modes in a linear triatomic molecule like CO_2. The diatomic molecules, representing 98.6 percent of the atmosphere, do not have strong absorptions, as mentioned.

In the atmosphere, CO_2 has strong absorption, $\kappa(\lambda) \sim 10^3$ m^2/kg, centered at $\lambda = 15.4$ µm in the infrared. Following Pierrehumbert 2011, p. 35, this peak, composed of many sharp absorptions, is broadened by rapid collisions with air molecules, and is so strong that, even at the low 400 ppm CO_2 concentration, a width of about 3.4 µm centered at 15.4 µm, the atmosphere becomes opaque. Similarly, ozone, O_3, absorbs strongly near 9.4 µm, and H_2O absorbs between 6.3 µm and 8 µm [$\kappa(\lambda) \sim 10^4$ m^2/kg at $\lambda = 6.3$ µm] and also between 18 µm and 50 µm [$\kappa(\lambda) \sim 10^4$ m^2/kg at $\lambda = 50$ µm]. In these wavelength ranges, where all light is absorbed by the opaque minority gases, the layer in question has to be treated as a distinct black body layer that has its own temperature and radiates up and down according to the black body radiation formula. All of these ranges are in the infrared, so opacity does not occur for incoming visible light, but only for the outgoing infrared long-wavelength light generated by the Earth's surface, near temperature 288 K on average. One feature previously mentioned (see Fig. 3.2) is that in the spectral range 7.7 µm to 12.5 µm, apart from the narrow ozone absorption at 9.4 µm, there is reduced absorption, and this is referred to as the infrared "window" allowing direct radiation from the ground to outer space. According to Trenberth et al (2009), (see their Fig. 1) a realistic value for this is about 40 W/m^2 on a global average.

This 40 W/m² is less than half the value we would estimate, 106.7 W/m², following the text prior to Eq. (3.24), from the black body radiation power Eq. (3.19a) of 29.2 W/m² μm at 10 μm, applied to a 5 μm range centered at 10 μm, with an assumed window transmissivity through the atmosphere of 0.731. That value was for a clear sky and 300 K, while the Trenberth value is a global average, keeping in mind that a satellite picture shows the Earth at least half covered in clouds. At 288 K, the accepted average temperature of the Earth, the black body radiation power at 10 μm is slightly reduced to 25.8 W/m² μm, that gives 94.3 W/m² for our estimate of Earth radiation from the surface out the 10 μm atmospheric "window", unimpeded by clouds.

The summary of Trenberth et al (2009) is shown in Fig. 3.3.

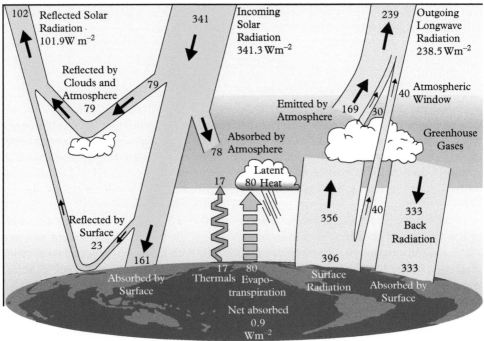

Global Energy Flows Wm^{-2}

Fig. 3.3 *Global energy flows in W/m². The Incoming Solar Radiation of 341.3 W/m² attributed to all the Earth's area $4\pi R_E^2$ corresponds to the 1366 W/m² that falls on all the projected area πR_E^2 as shown in Fig. 1.4. The numbers shown in this figure are averaged over the period between March 2000 and March 2004. The direct radiation from the Earth's surface through the "atmospheric window" is indicated as 40 W/m². The composition of longwave outgoing radiation is shown as 40 W/m² directly from the surface, 30 W/m² from tops of clouds and 169 W/m² from the atmosphere including greenhouse gases. The net of the radiation inflow from space is 0.9 W/m². The albedo, the fraction of incoming light directly reflected back by clouds and the Earth's surface is 0.299. The Sun's clear-sky power density at the Earth's surface from this figure is 1052 W/m² obtained as 341.3 W/m² minus 78 W/m² (absorbed by atmosphere), all multiplied by 4 (Trenberth et al, 2009) Fig. 1*

3.3.3 Ground (surface) as heat source at T_g

Since the atmosphere is quite transparent, the ground is primarily heated by absorbed Sunlight. According to Fig. 3.3, the average absorbed visible radiation is $161\,W/m^2$. So the atmospheric temperature, at least up to about 10 km, is largely set by the ground temperature, although the lapse rate $-dT/dz$ is locally strongly influenced in particular by water vapor and clouds. The average value carefully compiled, for example, by Trenberth et al, (2009) is $161\,W/m^2$ from directly absorbed Sunlight. The temperature of the ground reaches its average value T_g as a balance of energy flows, in addition to the absorbed $161\,W/m^2$, that include $396\,W/m^2$ of upward infrared radiation (including the $40\,W/m^2$ estimated previously into outer space through the atmospheric window) and a further $333\,W/m^2$ infrared radiation coming down from the atmosphere. Directly related to the atmosphere are estimates of surface energy losses by convection: upward "thermals", that float hawks above warm fields, and evaporation and cloud/thunderstorm formation as related to Eqs. (3.39, 3.40). These convective and evaporative losses heating the atmosphere are estimated to total to $97\,W/m^2$ surface energy loss on a global, night and day, seasonal average.

3.3.4 Reflectivity of ground

The $161\,W/m^2$ absorbed by the global surface is accompanied by $23\,W/m^2$ that is reflected back, according to Trenberth et al, so the average reflectivity is $23/184 = 0.125$. This can be called a surface albedo, and the $184\,W/m^2$ if multiplied by 4 to represent the incident Sunlight, rather than the 24-hour average, is $736\,W/m^2$. A direct measurement of $850\,W/m^2$ was mentioned in connection with Eq. (3.24). The global surface albedo represents an average over many different locations, sea and land, ice or snow, vegetation or desert. In large measure light reflection occurs as suggested by Eq. (3.25) at an interface between air of index 1.0 and insulating materials such as sand, often silica SiO_2 with index 1.46; titania TiO_2 index 2.6, or rock such as limestone $CaCO_3$. Such insulating materials have an energy gap of 5 eV or more and do not absorb light, although coloration with impurities such as iron may re-introduce absorption and color the material. Water, covering 67 percent of the Earth, has an index of refraction near $n = 1.34$ at visible wavelengths, so the reflection coefficient against air at $n = 1$, at normal incidence, according to Eq. (3.25) is 0.064, a low albedo value, making the ocean look blue. The albedo or reflection coefficient of snow or sea ice on the other hand is on the order of 0.8. While ice formally reflects like water, the much larger albedo value for snow may come from the microcrystals that are oriented at high angles to the light, although two reflections may be needed to get the light to return on the same path. So, the sunlit snow looks brightly white from any angle.

3.3.5 Blackbody radiation from $T_g \approx 288$ K, peaks at 9.7 μm

Let us consider thermal radiation *from* the Earth, introducing the question of the energy balance for the Earth itself.

The simplest assumption to make is that the Earth's temperature is set by the balance between the 1366 W/m² that comes in from the Sun and the black body radiation going outward from the whole area of the Earth. If we neglect any atmospheric effect and take the emissivity of the Earth as 1.0 then this implies

$$4\pi R_E^2 \sigma_{SB}(T_E)^4 = \pi R_E^2 1366 \text{ W/m}^2,$$ (3.43)

and the resulting temperature is 278.6 K.

The Earth's surface is 67 percent ocean, and it seems the average temperature T_E must be at least 271 K, the freezing point of seawater, so this initial estimate is in the right ballpark. To make the estimate more realistic, but still basing the analysis entirely on balance of radiation, the first effect of the Earth's atmosphere and the white clouds covering the Earth is to reflect some radiation directly back to space. This effect is called the albedo α, and an accepted value for the global albedo is $\alpha = 0.3$. Correcting our energy balance for the albedo, we find,

$$4\pi R_E^2 \sigma_{SB}(T_E)^4 = \pi R_E^2 (1-\alpha) \times 1366 \text{ W/m}^2,$$ (3.44)

and the resulting corrected temperature is now 254.8 K, much too cold. This is definitely too low, and indicates the importance of the greenhouse effect, reflecting or re-radiating some infrared radiation back to the Earth's surface.

Using R_E = 6371 km the power absorbed is 174 PW (1 – 0.3) = 121.9 PW, where a petawatt PW is 10^{15} W. To resolve the discrepancy presented by the 254.8 K value, it seems most plausible that the radiated energy does not all actually leave Earth directly to outer space, as calculated prior to Section 3.3.1, but a portion is re-radiated back by the absorbing impurity atmosphere gases mentioned in Section 3.3.1. This allows a more realistic higher Earth temperature $T_E = T_G$, the ground temperature, close to T_G = 288 K (Trenberth et al, 2009), that is an accepted value.

Thus, the "Greenhouse effect" gives a modified energy balance, in the simplest estimation, as:

$$P = 4\pi R_E^2 \sigma_{SB}\left[(T_G)^4 - (T_{GR})^4\right] = 121.9 \text{ PW}.$$ (3.45)

To match the agreed ground temperature T_G = 288 K in Eq. (3.45) we get the "Greenhouse" temperature T_{GR} (occurring in the atmospheric layer containing the absorbing molecules) as 227 K. This is the crudest model to capture the essential idea of heat trapping by the atmosphere, a more detailed picture will be developed in Section 3.4.1 below. The ground temperature T_G = 288 K is a carefully considered value, but the T_{GR} = 227 K is only suggestive, in fact several different layers in the atmosphere, at several different temperatures, contribute to the radiation back to Earth and to outer space. As already mentioned in Sect. 3.3.1, the most important greenhouse gases are CO_2 and water vapor. Ozone also plays an important role.

It is reasonable to neglect, in the energy balance, heat energy coming up from the core of the Earth. It is estimated that heat flow up from the Earth's center is $Q = 4.43 \times 10^{13}$

W = 0.0443 PW, that is relatively small. Of this, 80 percent is from continuing radioactive heating, and 20 percent from "secular cooling" of the initial heat of condensation of the Earth. 44.3 TW is a large number, important as a renewable source of energy, but so small on the scale of the solar influx that it is not important in influencing the ground temperature T_G (Turcotte and Schubert, 2002).

3.4   Layered radiative model of atmospheric temperatures

It was mentioned earlier that carbon dioxide, water and ozone are absorbing in the infrared, and that available concentrations of carbon dioxide render layers of the atmosphere opaque in the wavelength ranges of the absorption. A similar situation occurs at a much higher altitude for ozone. Motivated by the known importance of CO_2 and O_3 we turn to discuss a three-layer model, extending our earlier equation $P = 4\pi R_E^2 \sigma_{SB} \left[(T_G)^4 - (T_{GR})^4 \right] = 121.9$ PW that involved only the ground and a single unspecified higher layer; to three layers. We can identify these three: Ground, CO_2 and Ozone O_3 layers by their temperatures T_G, T_C, and T_O.

3.4.1   Three emitting layers in radiation equilibrium: T_G, T_C, T_O

This model might also be called a two-layer model, as it assumes two atmospheric layers described by separate emissivities and temperatures, with a fixed ground temperature. The CO_2 and Ozone O_3 layers are assumed well enough formed to absorb, come to a new temperature, and then radiate up and down according to the Stefan–Boltzmann law. Evidence that this is the case is shown below in Chapter 7, see Fig. 7.17. (The assumed layers may be realistic only in the infrared range centered at 10 μm where the emission from the ground is peaked, according to the rule $\lambda_m T = 2.9$ mm-K, and near the specific absorptions of the individual impurity molecules.) The assumption is that the emissivity is the same as the absorption of each layer, so each layer transmits a fraction $1 - \varepsilon$ of radiation it receives from below or above. The emissivity parameter is the only means to suggest the concentration or thickness of the layer in question in this model.

The upward radiated power from the ground is:

$$P_{G\uparrow} = \sigma_{SB} (T_G)^4 \; W/m^2,\tag{3.46}$$

taking the emissivity of the ground as 1.0. In a similar fashion, the upward power above the assumed CO_2 layer can be written:

$$P_{C\uparrow} = \sigma_{SB} (1 - \varepsilon_C)(T_G)^4 + \sigma_{SB}\varepsilon_C(T_C)^4.\tag{3.47}$$

Here we have used the fact that the absorption of a layer is equal to its emissivity, a consequence of Kirchoff's law.

Finally, upward power density above the top ozone layer is:

$$P_{O\uparrow} = (1 - \varepsilon_O)\left[\sigma_{SB} (1 - \varepsilon_C)(T_G)^4 + \sigma_{SB}\varepsilon_C(T_C)^4 \right] + \sigma_{SB}\varepsilon_O(T_O)^4.\tag{3.48}$$

Considering now the downward power densities, we have below the ozone layer:

$$P_{O\downarrow} = \sigma_{SB}\varepsilon_O(T_O)^4, \tag{3.49}$$

(no power is coming in from outer space in the long-wave range), and onto the ground:

$$P_{G\downarrow} = \sigma_{SB}(1-\varepsilon_C)(T_O)^4 + \sigma_{SB}\varepsilon_C(T_C)^4. \tag{3.50}$$

Following Randall, 2012, pp. 42–44, we can make some simplifying assumptions to estimate the unknown temperatures T_C and T_O, assuming $T_G = 288$ K. Requiring that the net flow of power, up vs down, must be constant at each layer, that is to assume a steady-state condition with no extra power coming in from direct solar absorption or latent heat release, one can write $P_{O\uparrow} - P_{G\downarrow} = P_{O\uparrow} - P_{O\downarrow} = P_{O\uparrow} - P_{O\downarrow}$. This leads to the following equations:

$$\sigma_{SB}(T_C)^4 = \{[2+(1-\varepsilon_C)\varepsilon_O / [(4-\varepsilon_C\varepsilon_O]\} \, \sigma_{SB}(T_G)^4 \tag{3.51}$$

$$\sigma_{SB}(T_O)^4 = [(2-\varepsilon_C)/(4-\varepsilon_C\varepsilon_O)] \, \sigma_{SB}(T_G)^4 \tag{3.52}$$

If the emissivity values are set at 0.5 and T_G is taken as 288 K, then Eqs. (3.51, 3.53) reduce to $T_O = (0.4)^{1/4} \, 288$ K $= 228.9$ K and $T_C = (0.6)^{1/4} \, 288$ K $= 253.4$ K. These are plausible values for the upper and middle atmosphere. When these numbers for T_C and T_O are inserted into Eq. (3.48), representing the upward power density from the upper ozone layer, the result is 233.8 W/m², that is quite close to the net outgoing long-wave radiation ("OLR") shown in Fig. 3.3, that is 238.5 W/m². The greenhouse effect that keeps the Earth comfortably warm is seen in the comparison of the 238.5 W/m² with the value $P_{G\uparrow} = \sigma_{SB} \, (T_G)^4$ W/m² from Eq. (3.46) that evaluates as 390 W/m². Another way of describing the situation is by defining a bulk emissivity of the Earth, ε_B, that is the ratio $238.5/390 = 0.612$.

Actually, an expression for this bulk emissivity can be obtained from the present 3-layer model, by inserting into Eq. (3.48) the expressions for T_C and T_O in terms of T_G. This result, following Randall, 2012, is:

$$\varepsilon_B = [(2-\varepsilon_C)(2-\varepsilon_O)+2\varepsilon_C\varepsilon_O]/(4-\varepsilon_C\varepsilon_O). \tag{3.53}$$

If the nominal assumed emissivity values 0.5 are inserted into this model equation, the resulting bulk emissivity is 0.73, rather than the observed 0.61.

An approximate assessment of the sensitivity of the Earth's temperature to changes in values in the model can be based on the Earth's net incoming power density N (that is nearly zero in present observation, see Fig. 3.3):

$$N = (P_{sun}/4)(1-\alpha) - \sigma_{SB}\varepsilon_B(T_G)^4 \tag{3.54}$$

Here P_{Sun} is the power density $1366\,\text{W/m}^2$ measured above the Earth's atmosphere, and the albedo is 0.3, leading to $239\,\text{W/m}^2$ inflow, so the net inflow is about $1\,\text{W/m}^2$. It may be reasonable to assume that a change in one of the parameters, e.g., the bulk emissivity ε_B, will be compensated eventually by a change in the temperature T_G so as to maintain a constant balance N. We can write the differential of the Eq. (3.54):

$$\delta N = 0 = -(P_{sun}/4)\,\delta\alpha - \sigma_{SB}(T_G)^4\,\delta\varepsilon_B - 4\sigma_{SB}\varepsilon_B(T_G)^3\,\delta T_G. \tag{3.55}$$

If we assume the albedo α is constant, we find,

$$\delta T_G = -T_G\,\delta\varepsilon_B/4\varepsilon_B. \tag{3.56}$$

If we ask for the change in emissivity, physically related to pollutant concentrations, to change the temperature by 4 K,

$\delta\varepsilon_B/\varepsilon_B = -0.056$, so starting with 0.61 (the actual value, see Fig. 3.3), ε_B would become 0.576.

On the model (3.53), we might ask what change in ε_C would be needed to make a 5.6 percent decrease in ε_B? From Eq. (3.53) we get 0.696 rather than 0.73, a 4.6 percent change, if we change the respective emissivity values from $(0.5, 0.5)$, to $(0.4, 0.8)$.

Another aspect of the model is that it gives for the direct outward transmission $(1-\varepsilon_C)$ $(1-\varepsilon_O)$ (see Eq. 3.48) that is 0.25 for the initial emissivity model values and 0.3 for the modified values. These values would be $97.5\,\text{W/m}^2$ and $117\,\text{W/m}^2$, directly radiated into space through the two assumed intervening layers. A different approach to this was discussed after Fig. 3.2. These values are comparable to the $94.3\,\text{W/m}^2$ that was found in the text before Fig. 3.3. We will return to aspects of this topic in Chapter 7.

4

Fusion in the Sun
A Primer in Quantum Physics

4.1 Protons in the Sun's core

As was mentioned in Chapter 1, the Sun is primarily composed of hydrogen and helium, namely 33.97 percent H and 64.05 percent ^{4}He, by mass in the core, with only 2 percent of other light elements (Bahcall et al, 2001).

These atoms are ionized in the Sun to appear as protons, alpha particles (^{4}He^{2+} nuclei) and electrons, to make the system neutral. The neutron is an unstable particle, with a lifetime of 880 s. So, the Sun has no free neutrons, and nuclei (that generally contain more neutrons than protons) have to be formed by fusion as we will discuss. The neutron is stable within a nucleus, but not stable as a free particle.

Small amounts of light elements, up to carbon or so, are present, but are not needed for a simple description of the energy release. This simplified picture is adequate to address the fusion process that produces the energy. The temperature ranges from about 5900 K at the Sun's outer surface to 15.7 million K at the core. The state of the matter is a dense ionized gas or plasma, and the density and pressure fall strongly going from the center to larger radius. The core is said to be the region out to $R_s/4$, and to produce 0.99 of the total power. At its surface the Sun faces vacuum, zero pressure and zero density, but at the core the maximum density is estimated as 1.527×10^5 kg/m^3.

The motions of the particles are prescribed primarily by the temperature and density, that vary from the surface of the Sun to the interior. The properties of the dense ionized gas at the core can be estimated.

The core mass composition of the Sun is 33.97 percent H and 64.05 percent ^{4}He, (Bahcall et al, 2001) having masses, respectively, 1.67×10^{-27} kg and 6.68×10^{-27} kg and the core mass density is 1.527×10^5 kg/m^3. Therefore,

$$N_p = 0.3397 \times 1.527 \times 10^5 \text{ kg/m}^3/1.67 \times 10^{-27} \text{ kg} = 3.106 \times 10^{31} \text{ m}^{-3}. \qquad (4.1)$$

$$N_{He} = 0.6405 \times 1.527 \times 10^5 \text{ kg/m}^3/6.68 \times 10^{-27} \text{ kg} = 1.46 \times 10^{31} \text{ m}^{-3}. \qquad (4.2)$$

Physics and Technology of Sustainable Energy. E. L. Wolf, Oxford University Press (2018).
© E. L. Wolf. DOI: 10.1093/oso/9780198769804.001.0001

Note that hydrogen thus is about 66.4% of the sun by volume. The electron concentration, then is $N_e = [3.106 + 2\,(1.46)] \times 10^{31}$ m^{-3} = 6.026×10^{31} m^{-3}. (For comparison, the free electron and positive ion densities in a metal are on the order of 5×10^{28} m^{-3}, about 1000 times smaller.)

From $N_p = 3.106 \times 10^{31}$ m^{-3} we can infer that the interproton spacing is $N_p^{-1/3} = 3.18 \times 10^{-11}$ m. Compared to the hydrogen atom radius $a_o = 0.0529$ nm $= 5.29 \times 10^{-11}$ m, this spacing is about 0.6 a_o, but on a more relevant femtometer scale it is large, 31,800 fm. From an atomic point of view, the spacing less than the Bohr radius would mean that the Mott transition (Chapter 5) has occurred, electrons are free to roam away from their protons, even at low temperature. The protons, however, can be thought of as a dilute system, because their spacing greatly exceeds their charge radius. This means that only two-particle collisions will be at all likely to occur. (We will see in a moment that the classical approach distance at the available energy is 1113 fm, too great a spacing for a nuclear reaction to occur.)

We will need to estimate the total number of protons, N_{cp} in the Sun's core, defined as $0 \le r \le R_S/4$. Since the Sun is not a solid but a dense gas, its density strongly varies with radius. It is reported (http://archive.is/http://mynasa.nasa.gov/worldbook/Sun_worldbook.html) that the density at $R_S/4$ is 20 g/cc, about 0.133 relative to the density at $r = 0$. It is also reported (Zirker, 2002) that the density $\rho(r)$ decays exponentially as $\rho(r) = \exp(-\alpha r)$. With radius in units of R_S we have $\rho(0.25)/\rho(0) = 0.133 = \exp(-0.25\,\alpha)$ that gives $\alpha = 8.06\ R_S^{-1}$. This function will apply to the proton density, N_p, with value at $r = 0$, $N_p(0) = 3.106 \times 10^{31}$ m^{-3}.

Using this information, we can write:

$$N_{cp} = N_p(0) \int_0^{0.25} 4\pi r^2 \exp(-8.06\ r)\,dr = 2.51 \times 10^{56}\ \text{m}^{-3}. \tag{4.3}$$

It is also reported (Zirker, 2002, p. 11) (Phillips, 1995, p. 47) that the *total* number of protons in the Sun is 8.9×10^{56}, suggesting that (4.3) is an overestimate.

Recall from Chapter 1 that the contact distance for two protons is about 2.4 fm, so the protons in the Sun's core are far apart from this point of view. Since the nuclear attractive force has a range of only about 2 fm, the particles have to approach each other to within a few femtometers to react. We can reasonably apply the classical speed distribution function (Eq. 1.6) to the protons at the Sun's core. Making use of the core temperature 15.7 million K, the most probable speed is $(2\,k_B T/m)^{1/2} = 0.498 \times 10^6$ m/s, and the corresponding kinetic energy is 1293 eV. Again, taking the core temperature of 15.7 Million K, the closest approach, call it r_2, requires $k_B T = k_c e^2/r_2 = 1293$ eV, that gives $r_2 = 1113$ f. Since this is much larger than $r = 2.4$ fm, twice the charge radius of a proton, this thermally available approach distance is far too large for any reaction to occur. *So, the classical particle picture is inadequate to explain the heating of the Sun. A large inter-proton Coulomb barrier exists and classically particles cannot cross such a barrier, fusion could not occur.*

4.2 Schrodinger's Equation for the motion of particles

A completely different approach, *quantum physics or nanophysics*, is needed to explain the fusion events that occur in the core of the Sun. In classical physics these protons have no chance of reacting to form deuterons as must happen, for this requires a spacing about 2.4 fm, while the closest approach possible at 1293 eV is 1113 fm, from the equality $k_c e^2/r = 1293$ eV. The situation is as shown in Fig. 1.6, where the available energy is far below the barrier height.

To repeat, the process of fusion in the Sun, with proton density and temperature approximately as outlined, has led to all of the accumulated energy on Earth in the form of deposits of coal, oil, and natural gas, as well as the continuing flow of 170 PW (Petawatts, this is 170×10^{15} W) to the Earth by direct radiation. It is also a prototype or existence proof for controlled fusion on Earth. So, this is a key process, worth some thought.

The new approach needed is based on a wave aspect for all matter particles. A hint that a wave aspect for matter particles is needed to allow the particles to cross the barrier in which they are not classically allowed to exist, comes from optics.

An analogous "evanescent wave" phenomenon occurs in optics, the evanescent wave allows light to cross or "tunnel through" an air gap between two high index of refraction glass plates, see Fig. 4.1. In the gap region the light intensity falls off exponentially with spacing. Light waves obey Maxwell's equations, that are second order differential equations.

A direct verification of de Broglie's predicted wave property of matter, see Fig. 4.2,

$$\lambda = h/p \qquad (4.4)$$

(where h is Planck's constant as in Eq. (3.17) and $p = mv$ is the momentum of the particle) was found by Davisson and Germer, (1927) who reported in a paper entitled "Diffraction of electrons by a crystal of nickel" magic reflection angles for a monochromatic electron beam shone on a Ni metal crystal. Knowing the spacing of atoms in Ni, their experiment verified the prediction of de Broglie, Eq. (4.4).

A more familiar geometry is the two-slit experiment shown in Fig. 4.2. The condition for a constructive interference peak behind the two slits, is that the path difference $d\sin\theta = n\lambda$, where $\lambda = h/p$, h is Planck's constant 6.6×10^{-34} Js and $p = mv$ is the momentum of the particle.

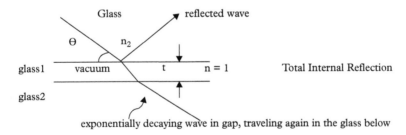

Fig. 4.1 *Sketch of "evanescent" light wave crossing air barrier between two glass plates. The intensity (electric field squared) of the evanescent wave decreases exponentially with air gap spacing t. This can be viewed as quantum mechanical tunneling of photons, light particles, with the role of the wave function being played by the electric field in the light wave*

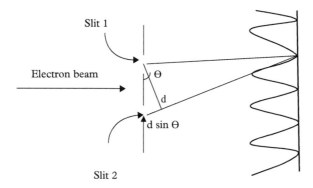

Fig. 4.2 *Sketch of electron diffraction in a two-slit geometry. The de Broglie condition $\lambda = h/p$ is found to predict the angles of maxima*

To return to the proton in the core of the Sun, we find $\lambda = h/p = 6.6 \times 10^{-34}/(1.67 \times 10^{-27} \times .498 \times 10^6) = 794$ fm. This is much smaller than (about 3 percent) of the inter-proton spacing 31,800 fm, so we can say that the motion of the solar core protons on the whole is as free classical particles. (When they slow down approaching contact, this is no longer true, and the Schrodinger treatment is essential to understand the fusion interaction of two protons.) In comparison, the electrons in a metal, even though much less dense, are completely quantum in their motion, because the calculated de Broglie wavelength there exceeds the interatomic spacing. This leads to a characteristic energy, the Fermi energy, that is larger than the classical energy $\frac{1}{2} k_B T$, with many consequences. We will need to learn about this to understand, for example, solar cells in a competent fashion.

Schrodinger found a second order differential equation to describe matter waves. Schrodinger's equation describes in practical and accurate terms the behavior of matter particles, protons in the Sun, protons in atomic nuclei, as well as electrons in atoms, metals and semiconductors. The first appearance of the wave aspect was the relation $\lambda = h/p$, the de Broglie wavelength. Evaluated for the proton in the solar core, where we have found speed v = 0.498×10^6 m/s, we find $\lambda = h/p = 6.6 \times 10^{-34}/(0.498 \times 10^6$ m/s $\times 1.67 \times 10^{-27}$ kg) = 794 fm. This is small compared to the interproton spacing, but large compared to the actual measured charge radius of the proton, which we have taken as 1.2 fm. 794 fm, the proton de Broglie wavelength, is seen also to be closer to the minimum classical spacing of two protons, 1113 fm, as found for kinetic energy corresponding to 1.5×10^7 K. A second de Broglie relation gives a frequency $\omega = E/\hbar$ (here expressed in radians per sec) to a particle of kinetic energy E.

The observation of electron diffraction in agreement with the de Broglie relations (see Fig. 4.2) means that a matter wave described mathematically by $\Psi(x,t) = \exp(ikx - i\omega t)$, must satisfy any more general equation as was produced by Schrodinger. The wave quantity $\Psi(x,t)$ predicts the location of the particle by the relation $P(x,t) = \Psi^*(x,t)$ $\Psi(x,t)$, where $\star$ indicates complex conjugate. Schrodinger's equation is then a statement of conservation of energy, where $k = 2\pi/\lambda$ and $\hbar = h/2\pi$.

$$[\hbar^2 k^2 + U - \hbar\omega]\Psi(x,t) = 0. \tag{4.5}$$

Based on this correct statement of conservation of energy, and knowing the solution, $\Psi(x,t) = \exp(ikx - i\omega t)$ in case $U = 0$, the equation has to involve $\partial^2\Psi(x,t)/\partial x^2$, to generate the $\hbar^2 k^2$. In addition, the first time-derivative $\partial\Psi(x,t)/\partial t = -i\omega\Psi(x,t)$ is needed, in order to produce the $\hbar\,\omega$ in the statement of conservation of energy.

4.2.1 Time-dependent equation

On this basis, the Schrodinger Equation in one dimension, with time-dependent potential $U(x,t)$, is:

$$-(\hbar^2/2m)\,\partial^2\Psi(x,t)/\partial x^2 + U(x,t)\,\Psi(x,t) = i\hbar\,\partial\Psi(x,t)/\partial t = \mathcal{H}\Psi. \tag{4.6}$$

The left-hand side of the equation is sometimes written $\mathcal{H}\Psi$, with $\mathcal{H}$ the operator that represents the energy terms, and is seen to represent the time derivative of the wavefunction.

4.2.2 Time-independent equation

In the common situation that the potential U is time-independent, a product wavefunction,

$$\Psi(x,t) = \psi(x)\phi(t), \tag{4.7}$$

when substituted into the time-dependent Eq. (4.7), yields:

$$\phi(t) = \exp(-iEt/\hbar). \tag{4.8}$$

Similarly, one obtains the *time-independent Schrodinger Equation*,

$$-\left(\hbar^2/2m\right)d^2\psi(x)/dx^2 + U\psi(x) = E\psi(x), \tag{4.9}$$

to be solved for $\psi(x)$ and energy E. The solution $\psi(x)$ must satisfy the equation and also boundary conditions, as well as physical requirements.

The physical requirements are that $\psi(x)$ be continuous, and have a continuous derivative except in cases where U is infinite. $\psi(x)$ is zero where the potential U is infinite.

A second requirement is that the integral of $\psi(x)\psi^\star(x)$ over the whole range of x must be finite, so that a normalization can be found.

The solutions for the equation are traveling waves when E > U, as would apply for protons free in the solar core, where we can take U = 0. On the other hand, for E < U, the solutions will be real exponential functions, of the form $A\exp(\kappa x) + B\exp(-\kappa x)$.

In such a case the positive exponential solution can be rejected as non-physical because it becomes indefinitely large at great distances. The decay constant can be seen to be:

$$\kappa = \left[2m\left(V_B - E\right)\right]^{1/2}/\hbar. \tag{4.10}$$

We return to the simplest possible treatment of proton fusion in the core of the Sun, suggested by Fig. 1.6. This model dates to Gamow in explaining the systematics of alpha particle decay of heavy nuclei. In the decay process the alpha particle is imagined as moving freely inside the nucleus, which is treated as a square well potential. The intermediate barrier corresponds to the Coulomb potential k_C $(Ze)(2e)/r$, where r is the spacing between the two charges. The potential barrier has a peak at the closest spacing, $r = r_z + r_a$. In this figure we see three regions: an inner region where bound states composed of oppositely traveling waves describe a particle bouncing around inside a fused nucleus; an intermediate range where the potential barrier exceeds the kinetic energy (the wavefunction will decay exponentially), and an outer region where traveling waves again exist. *These regions can be described accurately with Schrodinger's equation in spherical polar coordinates.* To simplify, the main features are usefully understood in a one-dimensional model, which we will also use for discussion of the state of electrons in a metal.

4.3 Bound states inside a one-dimensional potential well

In the context of nuclear fusion, the potential U inside the nucleus will be strongly negative, on an MeV scale, compared to the potential outside the nucleus. For the purpose of simple 1-d calculation, we will take the zero of energy to be that inside the nucleus, and for purpose of simple calculation, assume the potential at the outer radius L of the nucleus is infinite. In one dimension, suppose U = 0 for 0 < x < L, and U = ∞ elsewhere, where $\psi(x) = 0$. For 0 < x < L, the equation becomes:

$$d^2\psi(x)/dx^2 + \left(2mE/\hbar^2\right)\psi(x) = 0. \tag{4.11}$$

This has the same form as the classical equation for the motion of a mass on a spring, the "simple harmonic oscillator", so the solutions can be adapted from that familiar example. (For a mass on a spring, F = ma, for F = –Kx, gives the differential equation $d^2x/dt^2 + (K/m) x = 0$, with solution x = sin $[(K/m)^{1/2} t]$. The spring constant is K, in Newtons/m. The role of K/m is taken on by $2mE/\hbar^2$ in the potential well.) Thus, one writes:

$$\psi(x) = A\sin kx + B\cos kx, \text{where} \tag{4.12}$$

$$k = \left(2mE/\hbar^2\right)^{1/2} = 2\pi/\lambda. \tag{4.13}$$

The infinite potential walls at x = 0 and x = L require $\psi(0) = \psi(L) = 0$, that means that B = 0. Again, the boundary condition $\psi(L) = 0 = A\sin kL$ means that,

$$kL = n\pi, \text{ with } n = 1, 2, ...\tag{4.14}$$

This, in turn, gives,

$$E_n = \hbar^2 (n\pi/L)^2/2m = n^2 h^2/8mL^2, \; n = 1, 2, 3, ...\tag{4.15}$$

and wavefunctions, normalized to one electron per state,

$$\psi_n(x) = (2/L)^{1/2}\sin(n\,\pi x/L).\tag{4.16}$$

We see that the allowed energies increase as the square of the integer quantum number n, and that the energies increase also quadratically as L is decreased, $E \propto (1/L)^2$.

The condition for allowed values of $k = n\pi/L$ is equivalent to $L = n\lambda/2$, the same condition that applies to waves on a violin string.

These formulas are easily extended to the 3-dimensional box of side L, by taking a product of wavefunctions (4.16) for x, y, and z, and by adding energies in Eq. 4.15 according to $n^2 = n_x^2 + n_y^2 + n_z^2$. This will be discussed in Section 5.4.

This *exact solution* of this simple problem illustrates typical quantum behavior in which there are discrete allowed energies and corresponding wavefunctions. The wavefunctions do not precisely locate a particle, they only provide statements on the probability of finding a particle in a given range.

The conversion of the 1-d picture to a *spherical finite potential well V(r)* of radius a, in the case of zero angular momentum gives,

$$E_{n0} = \hbar^2 (n\pi/a)^2/2m = n^2 h^2/8ma^2, \; n = 1, 2, 3,...\tag{4.17}$$

$$\text{and } \psi_{n00}(r) = (2\pi a)^{-1/2}\left[\sin(n\,\pi r/a)\right]/r.\tag{4.18}$$

The *spherical polar coordinates* are:

$$x = r \sin\theta \cos\phi, \; y = r \sin\theta \sin\phi, \; z = r \cos\theta.\tag{4.19}$$

Here we are looking at portions of the solutions that are independent of the angles, fully *spherically symmetric* solutions.

If the value of the potential at $r = a$ is a finite value V_0 then it is found that there are no bound states for:

$$V_0 < \hbar^2/8ma^2.\tag{4.20}$$

Alternatively, we can take this as a statement that *to have a single bound state, the potential strength V_0 must be at least $\hbar^2/8ma^2$.*

This is not an obvious result to obtain, but it is simple, similar to the 1-d case for its energy formula, and easily remembered. The solutions for such a *finite* potential well, in either the 1-d or the spherical cases, are more difficult mathematically than the infinite potential well previously described. Here the allowed quantum states (wavefunctions and energies E < 0) are obtained by requiring that the wavefunctions match in amplitude and in slope at r = a. Inside we have ψ_1 (r) = A [sin(kr)]/r, (Eq. 4.18) and outside we have:

$$\psi_2(r) = D\exp(-\kappa r)/r. \tag{4.21}$$

Here k = $\hbar^{-1}$ $[2m(E + V_0)]^{1/2}$ and κ = $\hbar^{-1}$ $[-2mE]^{1/2}$. The solutions of a transcendental equation are needed, and must be found numerically or graphically: $-\cot z = [(z_0/z)^2 - 1]^{1/2}$, where z_0 = $\hbar^{-1}$ $[2mV_0]^{1/2}$ a. There is no solution if z_0 = $\hbar^{-1}$ $[2mV_0]^{1/2}$ a < $\pi/2$, which is equivalent to (4.19). For more details the reader is referred to a text (Griffiths, 2005, p. 141).

This is a treatment applicable to the trapped alpha particle inside the nucleus, the left portion of Fig. 1.6, an exact solution of Schrodinger's equation at the level used originally by Gamow to explain alpha particle decay.

4.4   Protons and neutrons in nuclei

The smallest nucleus is the deuteron. From a simple point of view a deuteron is a proton confined in a spherical finite potential well of radius a = 1.2 fm $2^{1/3}$ = 1.51 fm. The potential well V_0 is generated by the two-body attraction, in a useful simplification. So, we can describe the deuteron as an example of Eqs. (4.16–4.19). We can estimate the *minimum* barrier height as V_0 = $\hbar^2/8$ ma^2 taking radius a = 1.51 fm. This gives:

$$V_0 = \left(6.6 \times 10^{-34}/2\pi\ 1.51 \times 10^{-15}\right)^2 /\left[8 \times 1.67 \times 0^{-27} \times 1.6 \times 10^{-19}\right] = 2.26 \text{ MeV.}$$

(In fact, the *binding energy* of the deuteron is known to be 2.2245 MeV, while the simple estimate we just performed would correspond to a bound state at near zero binding energy $E_0 \approx 0$. We could go through Eqs. (4.16–4.19)) again to find V_0 (a larger number would be needed) such that the bound state energy E_0 = –2.2245 MeV. The binding energy is the result of the strong or nuclear force, whose claim on existence is indeed provided by the known nuclei, the deuteron being the smallest nucleus.

The binding energy of a nucleon when surrounded by other nucleons, say six in a cubic local environment, will be a multiple of the energy here estimated for a nucleon in contact with one other nucleon. If there are six nearest neighbors, then we would estimate 2.26 MeV × 6 = 13.6 MeV per nucleon. As we will see, this is a reasonable value for the binding energy per nucleon in nuclear matter.

The proton-proton reaction that we have discussed can be put into this framework. Two protons approaching definitely experience the Coulomb barrier. The nuclear reaction is known to produce D + e$^+$ + ν_e suggesting that an initial doubly charged nucleus ^{2}He is formed, which in some cases decays to a Deuteron, a positron and neutrino. This complicated decay process (in which a proton somehow mutates into a neutron (when bound to the proton), positron and neutrino) makes the whole reaction less likely, and in most cases the ^{2}He reverts to two separate protons, but the fact that this reaction occurs indicates that the Gamow tunneling process operates.

4.5   Gamow's tunneling model applied to fusion in the Sun's core

To consider the fusion interaction of two protons, the incoming distant proton can be treated as a spherical wave $\exp(-ikr)/r$, which is valid beyond the classical turning point r_2. In the region between the outer turning point $r_2 = r_{tp}$ and the point of contact $r_1 = r_n$ (see Fig. 1.6), is the *forbidden barrier region* where the solution to the Schrodinger equation, for kinetic energy less than potential energy, is a real decaying exponential function, quite analogous to the exponentially decaying light wave sketched in Fig. 4.1.

In classical physics the particle will never exist in this region, but nanophysics allows it in precise terms, see Fig. 4.3. If the barrier were constant at V_B, the wavefunction would be given by $\exp(-\kappa r)$, with $\kappa = [2m(V_B - E)]^{1/2}/\hbar$, in the range $r_1 = r_n < r < r_2 = r_{tp}$. In this case, the tunneling probability of the particle of energy E through the barrier of

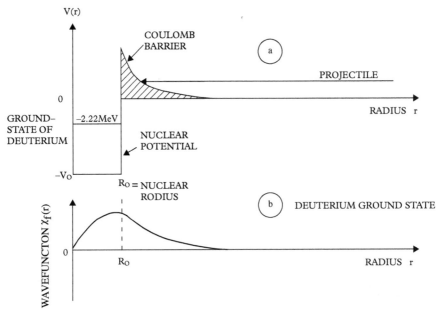

Fig. 4.3 *(a) Sketch of incoming proton upon Coulomb barrier, arrow terminates at classical turning point. The tunneling probability is T to reach inner radius R_o, and the further probability is TT to achieve the Deuteron bound state (b). Note that (b) is described by wavefunctions (4.18, 4.20), matched together at R_o. The matching in amplitude and slope in fact determines the bound state energy, shown as −2.22 MeV (Rolfs and Rodney, 1988) Fig. 6.1*

height, $V_B > E$ is $\exp(-2\kappa t)$, where $t = r_2 - r_1$ is the barrier thickness. The *transmission probability* is defined as:

$$T = \left| \psi(r_1) \right|^2 / \left| \psi(r_2) \right|^2. \tag{4.22}$$

Here $\left| \psi(r_1) \right|^2 = \psi^*(r_1)\, \psi(r_1)$. In the real case the height of the barrier follows a $1/r$ dependence as indicated in Fig. 1.6. It is difficult to solve the Schrodinger equation in case of an arbitrary barrier shape $V(r)$ and the practical approach is the simplifying WKB approximation. This useful approximation, applied to our case, gives:

$$T = \exp(-2\gamma) \tag{4.23}$$

$$\text{with } \gamma = \hbar^{-1} \int_{r1}^{r2} \left\{ 2m_r[V_B(r) - E] \right\}^{1/2} dr. \tag{4.24}$$

Here r_1 and r_2 are the turning points where $E = V$, and m_r is the reduced mass,

$$m_r = m_1 m_2 / (m_1 + m_2). \tag{4.25}$$

It is clear in physical terms that the value of the tunneling probability T is independent of which way the particle is going. An incoming wave will be large on the outside, while an outgoing wave will be large on the inside, but the penetration probability will be the same. So, the same formalism applies to our incoming wave case as to the outgoing wave case represented by the alpha particle decay previously mentioned. With $V_B(r) = k_c e^2/r$ and $m_r = m_p/2$, for two protons, with incoming energy $E = k_c e^2/r_2$ the resulting formula is:

$$\gamma = \hbar^{-1} \left(2m_r E \right)^{1/2} \int_{r1}^{r2} \left(r_2/r - 1 \right)^{1/2} dr \approx \hbar^{-1} \left(2m_r E \right)^{1/2}$$

$$\left[\pi\, r_2/2 - 2(r_1 r_2)^{1/2} \right], \text{ for } r_1 < r_2. \tag{4.26}$$

(This formula is reached by substituting $r = r_2 \sin^2 u$ in $\int_{r1}^{r2} \left(r_2/r - 1 \right)^{1/2} dr$ to get:

$[r_2(\pi/2 - \sin^{-1}(r_1/r_2)^{1/2}) - (r_1(r_2 - r_1))^{1/2}]$. For $r_1 < r_2$, using the small angle formula $\sin x \approx x$, one gets (Griffiths, 2005) Eq. (4.26).)

In this expression for $\gamma \approx \hbar^{-1} (2 m_r E)^{1/2} [\pi r_2/2 - 2(r_1 r_2)^{1/2}]$, for two protons $r_2 = k_c e^2/E$, note that the $\pi r_2/2$ term in the square bracket dominates. Thus, γ is nearly proportional to $E^{-1/2}$, and the literature has adopted the notation $2\gamma = (E_G/E)^{1/2}$ with E_G called the Gamow energy, so the fusion rate will be expressed as $\exp[-(E_G/E)^{1/2}]$.

One can easily see the factors that appear in E_G, by inspecting the formula $\gamma \approx \hbar^{-1} (2m_r E)^{1/2} [\pi r_2/2 - 2(r_1 r_2)^{1/2}]$, noting that in general $r_2 = k_c Z_1 Z_2 e^2/E$ and neglecting the term in r_1.

Let us apply this formula to the 2 protons approaching and assume $E = 1.293$ keV, so $r_2 = 1113$ fm. (It has been mentioned that the correct energy is $E = k_B T$, in this case 1.293 keV, that corresponds to the peak in the velocity distribution in the center of mass frame of the two nucleons (Section 3.5 in Rolfs and Rodney, 1988).) The reduced mass is $m_p/2$, $r_1 = 2.4$ fm. Then:

$$\gamma = \hbar^{-1}(2m_r\,E)^{1/2}\left[\pi r_2/2 - 2(r_1 r_2)^{1/2}\right]$$

$$= \left[2\pi/6.6 \times 10^{-34}\right]\left[1.67 \times 10^{-27} \times 1293 \times 1.6 \times 10^{-19}\right]^{1/2} 10^{-15}$$

$$\left[1113\pi/2 - 2\,(1113 \times 2.4)^{1/2}\right] = 9.21 \tag{4.27}$$

So, $T = \exp(-18.42) = 1.0 \times 10^{-8}$, that we may refer to as the tunneling probability or the *Gamow probability* for the p–p reaction, evaluated at $E = k_B T = 1.293$ keV.

The rate of geometric collisions per proton can be estimated from the mean free path Λ. If we imagine one proton, Λ represents how far it will go before making a collision, of the type we have described, with another proton. The basic formula for the mean free path is:

$$\Lambda = 1/(n\sigma) = 1/(n\,\pi r_2^2), \tag{4.28}$$

where $n = N_p$ is the number per unit volume of scattering centers (protons) and σ is an area if intercepted will lead to a geometric collision. We need a further probability T that a geometric collision actually leads to a fusion reaction (this is sometimes called the astrophysical S factor). This T is expected to be small based on the complicated nature of the p–p reaction , which needs to generate a positron and neutrino. Per proton, then, we can write the rate of collisions as:

$$f_{coll} = (v/\Lambda) \tag{4.29}$$

and the rate of fusion reactions per proton as:

$$f_{fus} = TT f_{coll} = TT(v/\Lambda). \tag{4.30}$$

We will calculate Λ on the assumption that the geometric collision corresponds to coming within the classical turning point r_2 of the second proton, this spacing was found to be 1113 fm, and, for simplicity, we the ignore the He ions. We earlier found that the density of protons in the core of the Sun is $N_p = 3.106 \times 10^{31}$ m^{-3} and that the most probable velocity was 0.498×10^6 m/s at 15.7 million K.

So, $\Lambda = 1/(n\,\sigma) = 1/n\,\pi r_2^2 = 8.28$ nm. (In terms of interparticle spacings this is large, about 260 spacings.) So, $f_{coll} = v/\Lambda = 0.6 \times 10^{14}$/s and $f_{fus} = TT f_{coll} = 1.0 \times 10^{-8}\,T \times 0.6 \times 10^{14}$/s $= 0.6 \times 10^6\,T$/s. We can use this to estimate the power radiated by the Sun, by multiplying by the number of participating protons N_{cp} in the core of the Sun and by the energy release per reaction, and in this way, get a value for the reaction probability T.

The number of protons is $N_{cp} = 6.87 \times 10^{56}$, taking the core as radius $r \le R_s/4$, if we take the proton density as constant at 3.106×10^{31} m^{-3}. This number is too large, and we earlier estimated, in text following Eq. (4.3), the number of protons in the core as $N_{cp} = 2.51 \times 10^{56}$, but mentioned that the number was still too large. Here we take the effective number of participating protons as 1.58×10^{56}. Then, with energy release per fusion as 26.2 MeV, we get $P = 3.96 \times 10^{50}\,T$ Watts, where T is unknown. [The p–p cycle is (Eq. (1.4)) initiated by the fusion reaction we are considering. The first reaction is quickly followed by several others with larger T (reaction probability) values, with a total release of energy in the Sun of 26.2 MeV per initial fusion.] By forcing our resulting power P = 3.96×10^{50} T Watts equal to the known power 3.82×10^{26} Watts we estimate that the fusion reaction probability T as 9.6×10^{-25} (but we will adjust it to 8×10^{-24} later). So, this reaction, we predict, is 23 to 24 orders of magnitude slower than more straightforward reactions such as D + D and D + T.

It turns out that this estimate is a reasonably accurate prediction of T for the p–p reaction! We have made a simplified analysis, on the other hand it has been stated that the reaction probability T, which is also referred to as the astrophysical S factor, is much smaller for the p–p reaction than for straight reactions not involving generation of neutrinos. To quote Atzeni and Meyer-Ter-Vehn (2004), Section 1.3.3, "the p–p reaction involves a low probability beta-decay, resulting in a value of S about 25 orders of magnitude smaller than that of the DT reaction".

In the literature this is described as a small "astrophysical cross-section", reflecting the nature of the nuclear reaction that is involved, an inverse beta decay, known to be a slow process. The first product, after the Gamow tunneling step, is an excited state, ^{2}He, of two protons, that must stabilize by emitting a positron and neutrino (inverse beta decay), else return to two separate protons. Note that the energy difference in the stabilization is positive energy 1.293 keV going to –2.22 MeV, the latter is the binding energy of the deuteron, see Fig. 4.3.

A simplified view of what may happen is suggested by the known lifetime for decay of the free neutron, 880 sec., into a proton, electron and neutrino. If one assumes that the decay of the proton (into neutron, positive electron, and neutrino) has a similar time as the known decay of a neutron into a proton, electron and neutrino, then the chance of this occurring before the excited p–p state decays can be estimated as $T = \Delta t/880$, where Δt is the lifetime of the excited p–p or ^{2}He state.

How can we estimate the lifetime Δt of the unstable excited state?

One estimate might be the oscillation period of the proton crossing the ^{2}He proto-deuteron, 4×1.51 fm/(0.5×10^6 m/s) = 1.21×10^{-20} s. In this case we get $T = 1.38 \times 10^{-23}$.

A second estimate might be from the Uncertainty Principle. There are two forms of the Uncertainty Principle, both originate in the wave aspect of particle behavior. The more familiar form is:

$\Delta p \, \Delta x \geq \hbar/2$, where p and x refer to the momentum and position of the particle. The less familiar form (Mandelstam and Tamm, 1945) is:

$\Delta t \, \Delta E \geq \hbar/2$, where t and E are time and energy. We can apply the second form to estimate the lifetime as:

$\Delta t = \hbar/(2 \, \Delta E)$, where we can take $\Delta E = 1.293$ keV + 2.22 MeV (see Fig. 4.3). In this case we find $\Delta t = 2.95 \times 10^{-22}$ s, and probability $T = 2.95 \times 10^{-22}$ s /880 s = 3.35×10^{-25}. These estimates of T are quite close to our earlier estimate $T = 9.6 \times 10^{-25}$, and the consensus from the literature as summarized by Atzeni and Meyer-Ter-Vehn, 2004.

So, from this point of view our simple analysis is reasonably accurate! The implication is that our simplified method might work quite well in cases where $T \sim 1$ as in the case of DT fusion of interest for terrestrial fusion machines. The p–p reaction is really the hardest one to understand.

To return to and to emphasize our main interest in this analysis, *in classical physics the Gamow tunneling factor T would be zero, there would be no fusion!* So, by explaining how the Sun generates energy we have shown the necessity for Schrodinger's wave treatment of matter particles, that we will extend to atoms and solids.

You can see that we came out quite well in this simplified calculation. To compare with a more standard approach, amenable to a wide variety of fusion reactions, we mention that the major deficiency in our analysis has been to overlook the *distributions* of speeds and tunneling probabilities (cross-sections) by replacing them by their most probable values. The rate of fusion is proportional to v × σ, and in a more accurate analysis the calculated property is the expectation value <vσ>, that has units m³/s. We have seen that there is a distribution of speeds v, and the cross section will vary as the speed varies. So, integrations over variables are needed. A standard framework for carrying this out gives specifically for the p–p reaction (Angulo et al, 1999):

$$<v\sigma> = 1.56 \times 10^{-43} T^{-2/3} \exp\left(-14.94/T^{1/3}\right)$$
$$\times \left[1 + 0.044T + 2.03 \times 10^{-4} T^2\right] m^3/s, \tag{4.31}$$

where temperature T is expressed in keV. Evaluating this for T = 1.293 keV we find, for the p–p reaction at 1.5 × 10⁷ K,

$$<v\sigma> = 1.56\ 10^{-43} \times 0.843 \times \exp[-14.94/1.089]$$
$$= 1.456 \times 10^{-49} [1 + .057 + .0003] = 1.54 \times 10^{-49} m^3/s$$

(Note that the exponential factor here is exp[−14.94/1.089] = exp[−13.71] = 1.107 × 10⁻⁶ as compared to tunneling probability T = 10⁻⁸ in the simplified analysis. Compared to the simplified analysis the tunneling probability is 111 times larger. This suggests that the optimal fusion events involve higher energy particles, that are fewer in number.)

We can compare this to our simplified result, by setting,

$$< v\sigma > = v\sigma. \tag{4.32}$$

Using our earlier numbers, we find, taking σ = T $T \times \pi r_2^2$ = 9.6 × 10⁻²⁵ × 1.0 × 10⁻⁸ × πr_2^2 = 3.75 × 10⁻⁵⁶ m² where r_2 = 1113 f and v = 0.498 × 10⁶ m/s, so,

$$v\sigma = 0.498 \times 10^6\, m/s\ 3.75 \times 10^{-57}\, m^2 = 1.86 \times 10^{-50}\, m^3/s. \tag{4.33}$$

So, our simplified result is 0.121 = 1/8.3 of the Angulo formula, result, setting <vσ> = v σ. If we take the view that our initial approximate result is low by a factor of 8.3, that implies that the reaction constant *T* should be increased by the same factor, 8.3, to get *T* = 8.0 × 10⁻²⁴. This value is understood as the factor by which the crucial p–p reaction is slowed down by the necessity of turning a proton into a neutron, first estimated by Bethe and Critchfield, 1938.

We can pause for a moment to summarize what we have learned, in looking at Fig. 4.4.

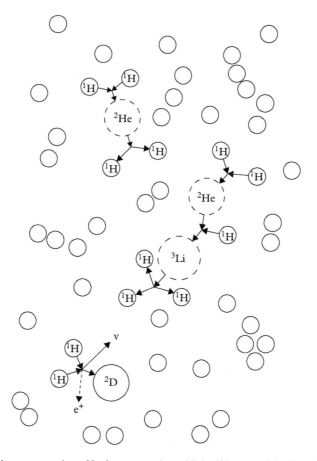

Fig. 4.4 *Schematic representation of hydrogen reacting with itself in core of the Sun. The successful process is shown in the lower left (Rolfs and Rodney, 1988, Fig. 6.4)*

In Fig. 4.4, small circles are protons (hydrogen) denoted by turning-point radius at 1293eV, r_2 = 1113 fm. Mean interproton spacing is 31,800 fm = 49.9 r_2, in Sun's core, so this picture shows more protons than a scale diagram would (at 1.5×10^7 K) by a factor of about 5^2 = 25 if we represent the proton size by its turning point radius, 1113 fm. Proton turning point collisions (e.g., lower-left pair) occur for a given proton at rate f_{coll} = v/ Λ = 0.6×10^{14}/s, where Λ = 1/(n σ) = 1/n πr_2^2 = 8.3 nm = 8.3×10^6 fm = 7438 r_2. Transient proto-deuteron events (one shown in upper left of the field, achieved by tunneling at Gamow probability T = 10^{-8}, but again releasing two protons) thus occur at rate 0.6×10^6 s^{-1} per proton. Finally, deuteron formations (see one event in lower-left portion of field, releasing one electron and one neutrino) *require further reaction probability T* = 8.0×10^{-24}. Thus, deuteron formation, per proton, occurs at rate f_{fus} = TTf_{coll} = 47.8×10^{-19}/s, or once every 0.67×10^{10} years. So it may be said that the lifetime of a proton (against forming a deuteron) in the Sun is about 10^{10} years (Clayton, 1983, see Problem 5.4, p. 369).

Yet it is precisely these rare decays that provide the 170 Petawatts heating the Earth over billions of years. This is an example of a situation where rounding off 8.0×10^{-24} to zero is a serious error!

This factor $T = 8.0 \times 10^{-24}$ will not apply to reactions more likely to be used in terrestrial fusion reactors such as the DD and DT reactions. Fusion on Earth, for this reason, should be a lot easier (by a factor of 10^{24}) than on the Sun, because we can start with deuterons mined from the ocean, and do not have to assemble them from protons as is done on the Sun. Deuterons are a necessary step on the way to making helium from hydrogen. (Again, as on the Sun, there are no free neutrons that could fuse together or with protons with no Coulomb barrier, neutrons are unstable.)

It is useful to consider this reaction in a broader context. In general terms, we consider a reaction of A and B to make C, in the present case A and B are both protons. In general, the rate R at which fusion product C is formed is:

$$R = N_A N_B (1 + \delta_{AB})^{-1} <v\sigma>_{AB} . \tag{4.34}$$

Here $\delta_{AB} = 0$ unless A = B, when it equals 1.0. The units of the rate R are $1/(m^3 s)$.

If the energy release in the reaction is Q, the power density is P = RQ, which, for $N_A = N_B = N_p$ is:

$$P = 0.5 N_p^2 <v\sigma>Q, \tag{4.35}$$

with units Watts/m^3 if Q is expressed in Joules.

We find, taking $<v\sigma> = 1.54 \times 10^{-49}$ m^3/s, P/m^3 = $1.54 \times 10^{-49} \times 0.5 (3.106 \times 10^{31})^2 \times 26.2$ MeV $\times 1.6 \times 10^{-19}$ = 313 watts/m^3 for the core of the Sun. This value is quite close to a published value 276.5 Watts/m^3 at the center of the Sun (Clayton, 1983, Table 6.6, p. 483). It is known that the power density falls off rapidly with increasing radius, and is 19.5 Watts/m^3 at 0.2 R_S. We can check this value using the total power and assuming it is generated in the core, $R \le R_S/4$. So, P/volume = $3.82 \times 10^{26}/[(4/3)\pi (R_S/4)^3]$ = 17.3 W/m^3 averaging over the whole Sun.

We will adopt 313 Watts/m^3 as a reasonable basis for scaling to a Tokamak reactor situation, in Chapter 6, using our approximate analysis summarized in Eq. (4.28) by fusion rate per proton $f_{fus} = TT f_{coll} = TT (v/\Lambda)$. The working values for the center of the Sun at 15 million K (1.293 keV) are T (Gamow factor) = 10^{-8}, T (reaction probability in p–p reaction) = 8×10^{-24}, thermal velocity of proton v = 0.498×10^6 m/s.

This value is a small power density, smaller than human metabolism, in agreement with other estimates. It shows that the large power output from the Sun derives from its immense size. A fusion reactor on Earth could be made to operate at a much higher power density, remember after all that a hydrogen bomb is a fusion reactor of a sort. For comparison, the power densities available in commercial processing tools such as *gas tungsten arc welding* (10^8 W/m^3) and *plasma torches* ($10^8 - 10^{10}$ W/m^3) are much higher (Venkatramani, 2002).

We will return in Chapter 6 to analysis of a DD fusion reactor, that we approach by adaptation of our analysis of the situation at the center of the Sun.

4.6   A survey of nuclear properties

The binding energy per nucleon BE has been plotted in Fig. 1.5. Nuclei are character-ized by a line of stability approximating $Z = N$ up to Z near 100 as shown in Fig. 4.5. The stability of large nuclei disappears beyond $Z = 100$ as the Coulomb repulsion energy increases. As a function of nucleon number, A, the binding energy BE per par-ticle rises from zero to a maximum near $A = 56$ and then gradually decreases. Nuclei beyond $A = 260$ or so are unstable.

The basic properties of the nuclei can be understood in simple terms, based upon the strong nuclear attractive force, leading to a binding energy U_o in the vicinity of 12 MeV per nucleon. Nuclei are nearly spherical, as expected with a short range attractive force, similar in principle to the intermolecular force in a drop of water. Water drops are spherical to minimize the number of molecules at the surface, where they do not have a full set of near neighbors. A liquid drop model was used early to understand aspects of nuclear fission. As in a water droplet, particles near the surface are less strongly bound, and this effect can be described as a surface energy.

It is easy to use this idea to understand the rising portion of the BE curve in Fig. 1.5. If we imagine a large spherical nucleus of radius R, with A nucleons, the binding energy is $U = A \, U_o$ and the volume $V = 4/3 \, \pi \, R^3 = (4/3) \, \pi \, R_o^3 \, A$, since $R = R_o \, A^{1/3}$. The binding energy per unit volume is then $U_V = U_o / [(4/3) \, \pi \, R_o^3]$. However, the nucleons within a distance δ of the surface, $(R - \delta) \le r \le R$, where δ is a measure of the range of the

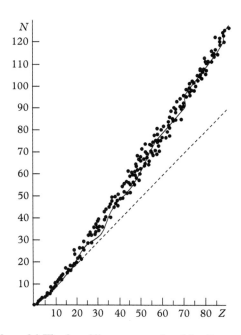

Fig. 4.5 *A survey of stable nuclei. The dotted line corresponds to $N = Z$ or to $A = 2Z$. Nuclei for Z much larger than 100 are unstable because of Coulomb repulsion*

nuclear force, will see only *half* of their binding energy, since there are no nucleons present beyond R. The "surface energy" is related to the volume $\Delta V = 4\pi R^2 \delta = 4\pi R_o^2 A^{2/3} \delta$, and the energy loss is $\frac{1}{2} U_V \Delta V$.

Then a formula for the binding energy, BE, with surface correction, is:

$$BE = A U_o - \frac{1}{2} U_V \Delta V = A U_o - \frac{1}{2} \left\{ U_o / \left[(4/3)\pi R_o^3 \right] \right\} 4\pi R_o^2 A^{2/3} \delta$$

$$= A U_o \left[1 - 3\delta / \left(2 R_o A^{1/3} \right) \right] \tag{4.36}$$

and the corrected BE per particle:

$$BE/A = U_o \left[1 - 3\delta / \left(2 R_o A^{1/3} \right) \right]. \tag{4.37}$$

With the empirical choices $3\delta/2 R_o = 1.08$ ($\delta = 0.72 R_o$) and $U_o = 12$ MeV, the working formula

$$BE/A = 12\,\text{MeV} \left[1 - 1.08 / A^{1/3} \right] \tag{4.38}$$

fits reasonably well the trend of BE values shown in Fig. 1.5 for A = 2, 3, 4 and 56.

A second basic aspect is the Coulomb repulsion between the Z positive charges. The electrostatic repulsive energy of a uniform spherical distribution of charge Q = Ze with radius R is $U_{Coul} = 3/5\, k_C\, (Ze)^2/R$.

A subtler aspect influencing the stability of a nucleus tends to promote the nearly equal number of neutrons and protons. It is believed that this is similar to the filling of states in atoms as guided by a rule that will allow an electron of spin up and an electron of spin down in the same state, but will not allow two electrons of the same spin. In the context of nuclei, the charge, neutron vs. proton, is similar to spin up or down, and the lowest energy is achieved with similar numbers of each.

The cost of the Coulomb energy is important for large Z, and is a reason why the proton number Z is less than the neutron number N. For large nuclei, looking at Fig. 4.5, the ratio Z/(Z+N) = Z/A is about 0.38.

We can incorporate these two ideas into an estimated empirical Coulomb repulsion energy per particle U_{Coul}/A, (in MeV), taking Z = 0.38 A, to get,

$$U_{Coul}/A = \left(3/5\, k_C e^2 \right) (0.38A)^2 / R_o A^{4/3} = 0.104\,\text{MeV}\, A^{2/3} \tag{4.39}$$

If we evaluate this for A = 238 (Uranium) we get 3.99 MeV per nucleon as the Coulomb repulsion.

We can estimate the value of A at maximum in the BE curve, from the derivative of the sum of two energy terms.

$$dU/dA = d/dA \left\{ -0.104\,\text{MeV}\, A^{2/3} + 12\,\text{MeV} \left[1 - 1.08 A^{-1/3} \right] \right\} \tag{4.40}$$

$$dU/dA = -(2/3)0.104\,\text{MeV}\, A^{-1/3} - 1/3(12\,\text{MeV})1.08 A^{-4/3}. \tag{4.41}$$

Multiplying by $A^{4/3}$ and setting $dU/dA = 0$ we find $A = 62$. This is not far from the observed value, usually quoted as Fe at $A = 56$.

The total BE/A function is then approximately

$$BE/A = 12\left[1-1.08\,A^{-1/3}\right]-0.104\,A^{2/3}\,(\text{in MeV}). \tag{4.42}$$

The first term is the attractive short-range force with a surface correction that roughly describes the increase in BE per nucleon as the number of nearest neighbors increases. The second term is the Coulomb repulsion that is long range and is seen to increase on a per nucleon basis as $A^{2/3}$. This function describes approximately the data curve shown in Fig. 1.5.

Using this we can estimate the energy release in a hypothetical fission reaction ^{236}U $\rightarrow$ 2 ^{118} Pd using the formula (4.40). The BE for ^{236}U is 1400 MeV, while the energy for two ^{118} Pd nuclei is 1617.9 MeV. The energy release is 217.9 MeV, close to a typical figure quoted as 200 MeV per fission. It is clear that the primary change is in the Coulomb energy. This is not a practical reaction, although ^{236}U is a starting point reached by capture of a slow neutron by ^{235}U. The fission products are typically two nuclei of different A, for example Kr and Ba, plus an average of 2.5 neutrons.

On the other hand, for the fusion reaction 2 D $\rightarrow$ ^{4}He we find from our approximate formula an energy release of 8.89 MeV. This is seriously wrong, since the quoted reaction 2 D $\rightarrow$ ^{4}He $+ \gamma$ lists $Q = 23.85$ MeV. It is clear also from the plot in Fig. 1.5 that the ^{4}He or alpha particle is exceptionally stable. Shell structures in nuclei, such as this strongly bound unit of 4 nucleons, play a role beyond the general picture of a liquid of closely interacting nucleons.

While our understanding makes clear that the size of atomic nuclei is limited to A values less than about 240, with radius around 7.5 fm, extended neutron matter is believed to exist in neutron stars. This neutron matter is apparently stabilized, in the absence of protons, by pressure.

We can find the *mass density* of nuclear matter from our working radius formula, since the mass is $A\,m_p = A\,1.67 \times 10^{-27}$ Kg. Setting $R = 1$ m, volume $= (4/3)\pi\,R^3 = (4/3)\pi\,(1.2 \times 10^{-15})^3\,A$ m^3 with mass $A\,1.67 \times 10^{-27}$ kg. The density is 2.307×10^{17} kg/m^3. The densities of neutron stars are estimated as two or three times this value. These values are seen to be much greater than density of the solar core, given as 1.62×10^5 kg/m^3.

Large nuclei like ^{238}U have several isotopes, corresponding to different neutron numbers, N, for the same Z. The chemical identity is controlled by Z, which sets the number of electrons that will collect around that nucleus.

This is shown by the Uranium alpha-decay series ^{238}U $\rightarrow$ ^{236}U $\rightarrow$ ^{234}U $\rightarrow$ ^{232}U $\rightarrow$ ^{230}U $\rightarrow$ ^{228}U, in each case by emitting an alpha particle. The decay lifetimes in this series range from about 6×10^9 y to about 300 s, all accurately predicted by the Gamow tunneling model. The plot of lifetime vs. $E^{-1/2}$ gives a straight line, consistent with our earlier discussion of Equation (4.24). In the simplified expression for $\gamma \approx \hbar^{-1}\,(2m_r\,E)^{1/2}$ $[\pi\,r_2/2 - 2(r_1\,r_2)^{1/2}]$, note that $r_2 = k_c\,90 \times 2\,e^2/E$, (for Uranium, $Z = 92$) and also that the $\pi\,r_2/2$ term in the square bracket dominates. Thus, γ is nearly proportional to $E^{-1/2}$. These considerations make clear that a plot of log lifetime vs. $1/(E)^{1/2}$ will be a straight line as shown. Similarly, the Thorium series ^{232}Th → ^{230}Th → ^{228}Th → ^{226}Th is also well fit. The value $r_2 = k_c$ 88 × 2 e²/E, (for Thorium, Z = 90) is slightly smaller than for Uranium, leading to systematically shorter decay times. Such decays, releasing energy are partly responsible for the high temperature in the core of the Earth.

Formally, all of these isotopes are *metastable*, trapping alpha particles at *positive* energies. In Fig. 1.5 and Fig. 4.3a, these alpha particle levels would be positive, ranging from 6.8 MeV to 4.3 MeV. In the decay process, the alpha particle is imagined as moving freely inside the nucleus, is treated as a square well potential, a spherical trap for alpha particles. The rate of decay is formally the product of a collision frequency against the wall and the Gamow tunneling probability, T. The intermediate barrier corresponds to the Coulomb potential k_c(Ze)(2e)/r, where r is the spacing between the two charges. The peak of the Coulomb potential for ^{238}U decay would be at contact between 90 protons inside, 2 protons outside at spacing set by the nuclear radii for A = 234 and A = 4. This barrier energy is:

$$V_B = k_C\,(90\ e)(2)/\left[R_o\left(4^{1/3} + 234^{1/3}\right)\right] = 27.8\ \text{MeV}.$$

In the Gamow tunneling probability the reduced mass is 4 × 234/(238) = 3.93, that is nearly 4 (proton masses).

The data in Fig. 4.6 are well fit by Gamow's application of Schrodinger's Equation. These fits absolutely require quantum mechanics, that we can see as central in our understanding of matter on the scale of nuclei and atoms. Such fits can only come from

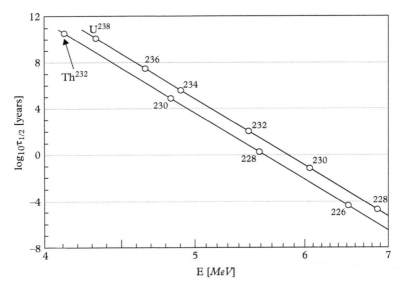

Fig. 4.6 *Alpha decay (^{4}He^{++} emission) by isotopes of Uranium and Thorium, fit to Gamow tunneling model. Plot of logarithm of lifetime against alpha decay vs 1/E$^{1/2}$ where E is the energy of the emitted alpha particle (Park, 1992)*

quantum theory, there is no way for alpha decay to occur in a classical theory. The fit is remarkable in that it accurately covers decay times from a billion years to a few minutes.

The square-well potential that has been useful in explaining alpha-particle decay is a simplification, we know that the only interaction is the short-range attraction between nearest neighbor nucleons. The outer edge of the well will be rounded off again on the scale of $\delta \approx R_o$ and the physical reason for the edge of the well is simply that outside the boundary there are no more particles to offer binding. The whole array is spherical in order to maximize the attractive energy, that can be restated as minimizing a (fictitious) repulsive surface energy. While the closest analogy is to a drop of water, where the intermolecular interactions are also short range, limited to nearest neighbors, the same argument, maximizing attractive (long-range) gravitational energy, explains the spherical shape of the Sun. The same approach of Eq. 4.36 can be applied to a metal like liquid mercury, where, however there is no accumulating Coulomb repulsion, since each electron is accompanied by a positive ion.

It seems that the motion of the strongly bound alpha particle, within this "liquid sphere" of nucleons, is reasonably free, and this situation of free motion of a particle in a dense medium also occurs in the motion of electrons in solids such as gold or silicon. We will come back later to reasons for free motion in these cases, such free motion is beneficial to the operation of devices including solar cells. In many cases in physics the motion of a single particle in a many-particle medium can be understood approximately by treating the effect of "other" particles by an effective potential or even as empty space with a boundary imposed. In physics it is productive to use the simplest picture that works.

5

Atoms, Molecules, Crystals and Semiconductor Devices

We found in Chapter 4 that the release of energy from the Sun starts when protons come together to form deuterons, a process that would not happen in classical physics. The wave property of particles of matter, as developed in the Schrodinger Equation, was needed, also, to explain the alpha-decay of Uranium and Thorium nuclei. We now extend Schrodinger's method to more familiar matter, in the form of atoms, molecules and semiconductors. The solar cell, that produces electrical energy from sunlight, in fact requires a sophisticated understanding of the semiconductor PN junction. So, we need to become expert in the application of Schrodinger's equation introduced in Chapter 4 to the cases of interest, including photovoltaic solar cells.

5.1   Atoms: Bohr's model of hydrogen

To begin, we describe a useful simple model of the atom, essentially an electron orbiting around a proton. Bohr made a semi-classical model of the atom that first explained the sharply defined energy levels, a puzzling feature not present in any classical model of atoms. These levels were suggested by the optical spectra, which were composed of sharp lines. Even though this model is incorrect in some respects, it easily leads to exact results for the orbit radius, energy levels, and the wavelengths of light absorption and emission of the one electron atom. It is well worth learning.

Bohr's model describes a single electron orbiting a massive nucleus of charge + Ze. Bohr knew that the nucleus of the atom was a tiny object, much smaller in size than the atom itself, containing positive charge Ze, with Z the atomic number, and e the electron charge, 1.6×10^{-19} C. The proton $m_p = 1.67 \times 10^{-27}$ kg is much more massive than the electron, $m_e = 9.1 \times 10^{-31}$ kg, thus M/m = 1835, so that nuclear motion can often be neglected. In the motion of two particles about a common center of mass, the relative motion can be corrected for the small motion of the heavier particle M by using the reduced mass $m_r = mM/(m + M)$ so $m_r \approx m_e (1 - m_e/m_p) = m_e (1 - 1/1835)$. The attractive Coulomb force, $F = k_c Ze^2/r^2$, where $k_c = (4\pi\varepsilon_o)^{-1} = 9 \times 10^9$ Nm^2/C^2, must match $m_e v^2/r$, that is the mass of the electron, $m_e = 9.1 \times 10^{-31}$kg, times the required acceleration to the

Physics and Technology of Sustainable Energy. E. L. Wolf, Oxford University Press (2018).
© E. L. Wolf. DOI: 10.1093/oso/9780198769804.001.0001

center, v^2/r. The total kinetic energy of the motion, $E = mv^2/2 - k_c Ze^2/r$, adds up to $-k_c Ze^2/2r$. This is true because the kinetic energy is always -0.5 times the (negative) potential energy in a circular orbit, as can be deduced from the mentioned force balance.

We thus find the relation between the total energy of the electron in the orbit, E, and the radius of the orbit, r, to be:

$$E = -k_C Ze^2/2r. \tag{5.1}$$

This classical relation predicts collapse (of atoms, of all matter): for small r the energy is increasingly favorable (negative). So, the classical electron would spiral in toward $r = 0$, giving off energy in the form of electromagnetic radiation. Fortunately, the non-zero value of Planck's constant, h, keeps this collapse from happening.

Bohr imposed an arbitrary quantum condition to stabilize his model of the atom. Bohr's postulate of 1913 was of the quantization of the angular momentum L of the electron of mass m circling the nucleus, in an orbit of radius r and speed v, as a multiple of Planck's constant $h = 6.6 \times 10^{-34}$ J-s, divided by 2π:

$$L = mvr = n\hbar = nh/2\pi. \tag{5.2}$$

Here n is the arbitrary integer quantum number $n = 1, 2\dots$. Note that Planck's constant, already described in Chapter 3, has the correct units, J-s, for angular momentum. This additional constraint leads exactly to basic properties of electrons in hydrogen and similar one-electron atoms:

$$E_n = -k_c Ze^2/2r_n, r_n = n^2 a_0/Z, \text{ where } a_0 = \hbar^2/mk_c e^2 = 0.053\,\text{nm}. \tag{5.3}$$

The energy of the electron in the nth orbit can thus be given as $-E_0 Z^2/n^2$, $n = 1, 2\dots$ where $Z = 1$ for hydrogen and

$$E_0 = m_r k_c^2 e^4/2\hbar^2 = 13.6\,\text{eV}. \tag{5.4}$$

All of the previously puzzling spectroscopic observations of sharply defined light emissions and absorptions of one-electron atoms were nicely explained by the simple quantum condition,

$$h\nu = hc/\lambda = E_0(1/n_1^2 - 1/n_2^2). \tag{5.5}$$

The energy of the light is exactly the difference of the energy of two electron states, n_1, n_2 of the atom. This was a breakthrough in the understanding of atoms, and explained sharp absorption lines as are seen in Fig. 1.4. For example, the hydrogen $n = 2$ to $n = 3$ transition absorbs red light at 656 nm, this is called the Balmer line, present as a dip in the light spectrum of the Sun. In evaluating the wavelength in (5.5) it is convenient to note that $hc = 1240$ eV-nm, since $hc = 6.6 \times 10^{-34}$ J-s $\times 3.0 \times 10^8$ m/s $[1/1.6 \times 10^{-19}$ J/eV$] = 1.2375 \times 10^{-6}$ eV-m ≈ 1240 eV-nm.

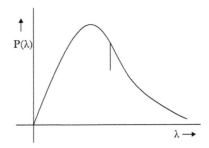

Fig. 5.1 *Sketch of solar spectrum vs wavelength showing hydrogen n = 2 to n = 3 (Balmer) absorption at 656 nm*

The wavelength of the n = 2 to n = 3 absorption is given by λ = ΔE/hc = 1240/ΔE = 1240/1.888 = 656 nm, is in the red part of the spectrum, see Fig. 5.1. As we know, the composition of the Sun is 66 percent H by volume, so certainly Hydrogen atoms are present in its atmosphere. The appearance of this sharp dip feature in the spectrum implies that the Sun's atmosphere contains some H atoms in the n = 2 state, with excitations ΔE of 3/4 × 13.6 eV = 10.2 eV. The probability of such an excitation of the atom is P = exp (−ΔE/k_BT). For T = 5973 K this is P = 2.5 × 10^{-9}. This suggests that in the gas of hydrogen atoms above the surface of the Sun, about 2.5 per billion will be able to absorb, and contribute to the observed dip.

Note that the reduced mass m_r enters the energy formula Eq. (5.4), and also in (5.5). In Chapter 2, we described deuterium as the first step in the energy releasing fusion process of the Sun. The deuteron D forms a slightly different version of "hydrogen" atom, the only difference is the mass of the nucleus, now 2 m_p for deuterium. We can find the small differences Δ in the energy and wavelength values for D vs H,

$$\Delta[h\nu] = \Delta[hc/\lambda] = \Delta[E_o](1/n_1^2 - 1/n_2^2), \qquad (5.6)$$

that arise from the small difference in the reduced mass Δ m_r ≈ m_e [(1 − m_e/2m_p) − (1 − m_e/m_p)] = m_e (1 + 1/3670). This makes the energy for D larger by about 2.7 × 10^{-4} and correspondingly makes the wavelengths smaller by the same factor. For the Balmer line, 656.5 nm will be shifted by −0.178 nm to appear at 656.3 nm. This is observable, because the spectral lines are sharply defined, and is how deuterium was discovered (on the Sun). We will see later that in "muonic hydrogen", where the electron is replaced by a *muon*, an electron-like particle whose mass is 206.8 m_e, so that the reduced mass with the proton is m_r ≈ 186 m_e, that the binding energy is about 186 E_o ≈ 2528 eV. If this "muonic" atom, further, is made using deuterium, and then a molecule between two such atoms is formed, we will see later that the Deuterons actually come close enough to undergo fusion at a modest rate.

The Bohr model, that does not incorporate the basic wavelike aspect of microscopic matter, fails to correctly predict some aspects of the motion and location of electrons. (It is found that the idea of an electron orbit, in the strict planetary sense, is no longer correct, in nanophysics. The Uncertainty Principle mentioned in Chapter 2 indicates that the position and motion of a particle cannot be known simultaneously.)

In spite of this, the Bohr model values for the electron energies $E_n = -E_o Z^2/n^2$, spectral line wavelengths, and the characteristic atomic size, $a_o = \hbar^2/mk_c e^2 = 0.053$ nm, are all exactly preserved in the fully correct treatment based on nanophysics, to be described.

5.2   Charge motion in periodic potential

Semiconductors and their electrical conduction are important in photovoltaic cells and are strongly influenced by the wave properties of electrons. Charge motion is a central issue in efficient solar cells. We discussed the idea of the mean free path Λ for protons in the Sun in Chapter 4, using formula (4.28), $\Lambda = 1/(n_s \sigma_{scatt}) = 1/n_s \pi r_2^2$, for a scattering center of radius r_2 present at density n_s. This formula is useful also in connection with the electrical conductivity, for which the conventional symbol is:

$$\sigma = 1/\rho, \tag{5.7}$$

where ρ is the resistivity, and with σ in units Siemens or $[\Omega m]^{-1}$.

The resistivity of pure crystalline metals tends to zero at low temperature, and the ratio of the resistivity at 300 K to that at 4.2 K is called the "residual resistance ratio". It can be as large as a million in a pure crystalline sample.

The large residual resistance ratio for extremely pure metals implies large mean free path Λ at low temperature. The formula for resistivity $\rho = 1/\sigma$ can be written in several forms:

$$\rho = 1/Ne\mu = 1/[Ne(e\tau/m)] = m_e/(Nee\tau) = m_e v_F/(Ne^2\Lambda). \tag{5.8}$$

Here N is the number of electrons freely moving, τ is the time between scattering events for a given electron, v_F is a characteristic electron speed and the *mobility* μ is defined as:

$$\mu = e\tau/m. \tag{5.9}$$

The mobility, whose units are m²/Vs, provides the *drift velocity*

$$v_D = \mu E \tag{5.10}$$

of the electron in an applied electric field E = V/L given in volts per meter. Looking at the final form in (5.8), we see that the only way ρ can increase by 10^4 to 10^6 going from 300 K to 4 K is if mean free path Λ increases, because other factors such as N and v_F are constant. The data for pure metals in Fig. 5.2 show that the interaction of electrons and metal ions is surprisingly weak: the resistivity goes to zero at low temperature. Let's

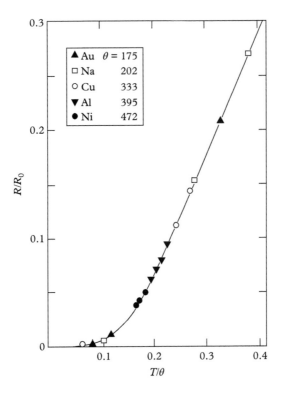

Fig. 5.2 *Resistivity of pure crystalline metals tends to zero at zero temperature, limited by impurities. The ratio of resistivity at 300 K to that at 4K, called the "residual resistance ratio" can be as large as a million in an extremely pure crystalline metal. The linear region of resistivity with temperature is a well-known effect of the thermal vibrations of the atoms, characterized by θ_D, the Debye temperature ("Electrons and Phonons" by J. M. Ziman, 1960) Fig. 11.5*

look to find the most important modification of free particle motion that would show up if we slowly increase a weak scattering interaction between it and an array of atoms.

5.3   Effects of periodicity: energy bands and gaps

As we learned in connection with Fig. 4.2, the Davisson–Germer experiment implies that a free electron of momentum $p = h/\lambda$ is described by a wave $\psi = e^{ikx} = e^{(i2\pi x)/\lambda}$. Suppose this wave, moving to the right, weakly scatters from atoms at spacing a, as sketched in Fig. 5.3.

Here $k = 2\pi/\lambda$ is the definition of *wave number*, the number of radians advance in phase per meter, corresponding to 2π per wavelength, λ.

We see that the condition for strong coherent back reflection (*Bragg scattering*) is:

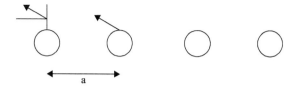

Fig. 5.3 *Sketch of weak scattering of an electron wave by atoms at spacing a*

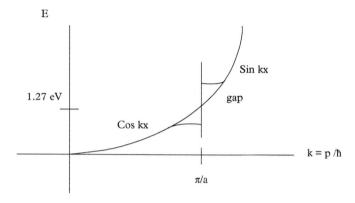

Fig. 5.4 *Sketch of E(k) for electrons influenced by weak scattering from atoms at spacing a. The background curve corresponds to the free particle condition $E = (k\hbar)^2/2\,m$. The discontinuities at $k = \pm n\pi/a$ arise by coherent back-scattering, see text. The energy 1.27 eV corresponds to $(k\hbar)^2/2\,m$ at $k = \pm \pi/a$, using the lattice constant for crystalline Silicon a = 0.543 nm, with m = m_e. This Figure shows the origin of electron energy bands and gaps in a periodic lattice*

$$2a = n\lambda = n\,2\pi/k, \tag{5.11}$$

so that the wavenumber condition for back scattering is: $k = n\pi/a$, where n is any integer.

Sketched in Fig. 5.4 is the curve $E = \frac{1}{2}\,(\hbar^2 k^2/m)$ vs k, that can be regarded as the momentum $p/\hbar$. The curve is modified from that for a free particle, near $\pm n\pi/a$ because the scattering produces linear combinations, $\psi = e^{ikx} + e^{-ikx} = (1/2)$ cos kx or $e^{ikx} - e^{-ikx} = (i/2)$ sin kx. These combinations are standing waves, and the energy is reduced for the coskx vs the sinkx choice. For Si, where a = 0.543 nm, $k = \pi/a = \pi/0.543$ nm $= 5.78 \times 10^9$ m^{-1} and $\hbar^2 k^2/2m = 1.27$ eV, a value similar to measured values. Near $k = \pi/a$, the back-scattered wave is as strong as the forward wave: the result is a standing wave. It is generally true, for Schrodinger's Equation or any linear differential equation, that if one has two solutions, such as exp($\pm ikx$), then linear combinations of these, such as sinkx and coskx are equally valid, with specific choices to be based on physical reasoning. (It is also generally true that the wavefunction can be a complex quantity, such as $\psi = $ exp(ikx), because the measured probability of finding the particle is the product of ψ with its complex conjugate, $P = \psi^*\psi$, that is a positive number.) In this case, for k just below π/a, the coskx form is more stable, because the electron spends more time near the ions located at $x = 0$, $x = na$. The probability density $P = (\text{coskx})^2$ peaks at ion locations. For the sinkx combination for k, larger than π/a, the peaks of $P(x)$ lie between the ions, see Fig. 5.5.

The Kronig–Penney potential $V(x)$ model for perfect conduction along a row of atoms is sketched in Fig. 5.6.

This model assumes a linear array of N atoms spaced by a along the x-axis, $0 < x < Na = L$. The 1-D potential $U(x)$ is a square-wave with period a:

$$U(x) = 0, \quad 0 < x < (a - w),$$
$$V_0, \quad (a - w) < x < a. \tag{5.12}$$

The model potential $U(x) = 0$ except for periodic barriers of height V_0 and width w.

The solutions $\Psi = U_k (x) e^{ikx}$ are compatible with the 1-D Schrodinger Equation introduced in Section 4.2

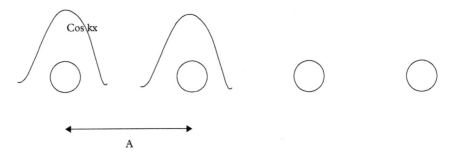

Fig. 5.5   *Sketch of standing wave probability $P = (coskx)^2$ expected at k just below the Bragg point π/a, the cos(kx) combination of solutions concentrates charge near the positive Ions, stabilizing the state*

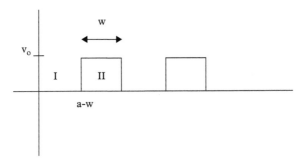

Fig. 5.6   *Sketch of square wave potential $V(x)$ assumed by Kronig and Penney. The band-determining condition is obtained by matching solutions of Types I and II at the boundary, $x = a$-w. A useful simplification is to approximate the square wave potential by repulsive delta function potentials at $x = a$-w/2, preserving the barrier potential area wV_0 as $w \rightarrow 0$*

$$(-\hbar^2 2m)d^2\Psi/dx^2 + [U(x) - E]\ \Psi = 0, \text{with } \Psi = U_k(x)e^{ikx} \tag{5.13}$$

with periodic $U(x)$ of Eq. (5.13), *only* if the following condition is satisfied:

$$\text{coska} = \beta(\sin qa/qa) + \cos qa = R(E), \tag{5.14a}$$

where $\beta = V_0wma/\hbar^2$ and $q = (2\ mE)^{1/2}/\hbar$. (This simplified form of Eq. (5.14) is actually obtained in a limiting process where the potential barriers are simultaneously made higher and narrower, preserving the value β. This can be described as N δ-functions of strength β.) The parameter β is a dimensionless measure of the strength of the periodic variation.

We can test Eq. 5.14a by examining what happens in the simple cases of vanishing and extremely strong potential barriers V.

One can see that Eq. (5.14a) in the zero-barrier limit, $\beta = 0$, the condition coska = cosqa leads to $k = (2mE)^{1/2}/\hbar$, that recovers the free electron result, $E = \hbar^2k^2/2m$.

Next, if β becomes arbitrarily large, the only way the term $\beta(\sin qa/qa)$ can remain finite, as the equation requires, is for sinqa to become zero. This requires qa = $n\pi$, or $a(2mE)^{1/2}/\hbar = n\pi$, which leads to $E = n^2h^2/8ma^2$ (see Eq. 4.15). These are the levels for a 1-D square well of width a (recall that in the limiting process the barrier width w goes to zero, so that each atom will occupy a potential well of width a, the atomic spacing).

The new interesting effects of band formation occur for finite values of β. Fig. 5.7 gives a sketch of the right-hand side, R(E), of Eq. (5.14a), vs. Ka (qa in the text). Solutions of this equation are only possible when R(E) is between −1 and +1, the range

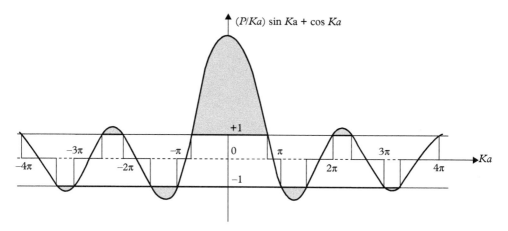

Fig. 5.7 *Kronig–Penney model, plot of ordinate R(E) vs. abscissa Ka, for $\beta = 4.712$. Shaded areas in this plot denote ranges of Ka that do not allow traveling wave states. Traveling wave solutions occur only when ordinate R(E) has magnitude unity or less. The bottom of the conduction band E_o corresponds to R(E) = 1, the first crossing, and the band-edge k a = π corresponds to R(E) = −1. In this plot dark areas correspond to band gaps. In the text the symbol $\beta \pi$ is used for P. Ka in this figure corresponds to qa in the text*

of the coska term on the left. Solutions for E, limited to these regions, correspond to allowed energy bands. Note that allowed solutions are possible for $-1 <$ coska < 1, that corresponds to $-\pi <$ ka $< \pi$. More generally, boundaries of the allowed bands are at k = $(\pm)n\pi/a$, n = 1, 2, 3....It is conventional to collect the allowed bands as shown in Fig. 5.8 into the range between $(\pm)\pi/a$ in k-space, called the Brillouin zone.

The result of this analysis, then, is that in the allowed bands *no scattering occurs*, as long as the potential is periodic! This can explain why the resistivity approaches zero at low temperature for a pure crystalline metal like Au. It also explains the "band structure" of semiconductors that are important in their application in photovoltaic devices. This effect has an analogue in electrical circuitry, where a periodic "transmission line" of lumped inductor and capacitor elements can have a pass band and a stop band. So, the Kronig–Penney model explains the observation as in Fig. 3.2, of very long mean free path for electrons in pure metals, and also in semiconductors.

Each band $(-\pi/a < k < \pi/a)$ contains N atoms and can accommodate 2N electrons. In gold, one finds a half-filled band (metal) because Au releases only one electron. On the other hand, an **even** number of electrons per atom *fills* one or more bands, leading to an *insulator* because electrons cannot respond by changing k in an E field. Si has valence 4, will have *two filled bands*, is thus a "band insulator". It conducts electricity only because of thermally induced jumps of electrons from the valence band to the conduction band.

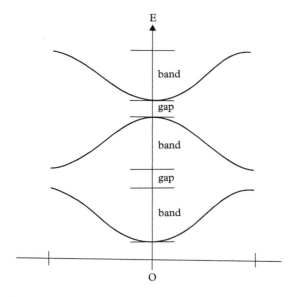

Fig. 5.8 *Schematic of bands E vs. k in a periodic potential, based on Kronig-Penney model. The bands are restricted in k to values less than π/a. Energy gaps occurring at k = $\pm$ (π/a) are also physically understood on the basis of Bragg reflections at k = $\pm(\pi/a)$. Each band accommodates exactly 2N electrons, so that 1 electron per atom gives a half-filled band and a metal, while 2 electrons per atom gives a filled band, an insulator*

5.4   Properties of a metal: electrons in an empty box (I)

Some essential properties of a metal such as sodium or gold can be obtained from a simple 3-D model. This idealized picture is of electrons confined in a box of side L, with infinite potential outside. More realistically the potential barrier is the "work function" ϕ, with typical values between 3 and 5 electron volts. This is a classic case of the situation mentioned at the end of Chapter 4 where the effects of "other" particles can be summarized in a useful way by a potential, or just by a confining boundary condition.

The Schrodinger Equation can be solved inside the box (see Eq. 5.1) because of the boundary condition $\psi = 0$ at the walls, and the solutions inside an empty cube are easier to deal with than the solutions inside an empty sphere. This $\psi = 0$ boundary condition is still useful for the finite barrier provided by ϕ, the metallic work function, whose measured value is 4.83 eV for gold.

However, as we will see, the main parameter describing a metal is E_F the *Fermi energy*, that is determined only by the density of electrons, independent of the work function value.

The wavefunction (see Eq. (4.16) in Chapter 4) is easily extended to three dimensions:

$$\psi = (2/L)^{3/2} \sin(n_x \pi x/L) \sin(n_y \pi y/L) \sin(n_z \pi z/L). \tag{5.14}$$

While these were presented as bound states, they are equally valid as linear combinations of traveling waves. This is true since $\sin kx = (e^{ikx} - e^{-ikx})/i2$, and we can consider these states to be superpositions of oppositely directed traveling waves $\psi_+ = \exp(ikx)$ and $\psi_- = \exp(-ikx)$. Here the moving waves, $\psi_\pm$, are more fundamental for a description of conduction processes.

We simulate a metal by adding electrons into the states defined by (4.1a). The important quantum aspect of this situation is Pauli's exclusion principle, such that only one electron of specified spin can occupy a state. For a given choice of n_x, n_y, n_z only two electrons, one of spin up and one spin down, can be accommodated. If we add a large number of electrons to the box, the quantum numbers and energies of the successively filled states will be given by:

$$E_n = \left[h^2/8m_e L^2 \right] \left(n_x^2 + n_y^2 + n_z^2 \right). \tag{5.15a}$$

We need to know how the highest filled energy changes as we add electrons. To learn this, it is convenient to rewrite equation (5.15a) as:

$$E_n = E_o r^2, \tag{5.15}$$

where $E_o = [h^2/8mL^2]$, as an aid to counting the number of states filled up to an energy E, in connection with Fig. 5.9. In coordinates labeled by integers n_x, n_y, and n_z, constant energy surfaces are spherical, and 2 electron states occupy a unit volume. Since the states are labeled by positive integers, only one octant of a sphere is involved.

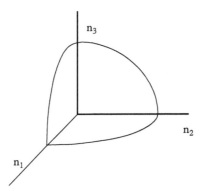

Fig. 5.9 *Positive octant showing spherical surface of constant energy. The number of states is twice the volume of this octant*

The number N of states out to radius r (energy E_n) is:

$$N = (2)(1/8)(4\pi r^3/3) = \pi r^3/3 = (\pi/3)(E/E_o)^{3/2}, \tag{5.16}$$

for a box of side L. This is equivalent to:

$$E_F = (h^2/8m)(3N/\pi L^3)^{2/3} = (h^2/8m)(3N/\pi V)^{2/3}. \tag{5.17}$$

Setting $N/V = n$, the number of states/m^3, some algebra gives:

$$dn/dE = g(E) = (3n/2)E^{1/2}E_F^{-3/2} = c' E^{1/2}, \tag{5.18}$$

as the density of electron states per unit energy and per unit volume at energy E with c' a constant. This is the preferred form of the formula, and shows the characteristic dependence on $E^{1/2}$. We can use this to find the average kinetic energy at $T = 0$ as:

$$E_{av} = n^{-1}\int_0^{Ef} E\,g(E)\,dE = 3/5\,E_F.$$

This is quite different from the classical value, $(3/2)\, k_B\, T$, and a changed form of the P(V) gas-law relation, from an "ideal gas" to a "degenerate gas" is a consequence that we will return to.

The Fermi velocity v_F is defined by $mv_F^2/2 = E_F$, and T_F is defined by $k_B T_F = E_F$. The metallic Fermi velocity and Fermi temperature exceed their thermal counterparts. The reason for this is that the boundary conditions prescribe allowed states, limited to 2 electrons per state by the Pauli principle, raising the energy and velocity, as particles are added. A test is to compare the de Broglie wavelength of the electron to the interatomic spacing $n^{-1/3}$. Taking the highest energy state E = E_F (for n = 5.9 × 10^{28} m^{-3}, E_F = $(h^2/8m)$ $(3n/\pi)^{2/3}$ = 5.53 eV) we find λ = h/p = h/$(2m_e E)^{1/2}$ = 6.6 × 10^{-34}/(2 × 9.1 × 10^{31} × 5.53 × 1.6 × 10^{-19})$^{1/2}$ = 0.52 nm. The spacing between atoms is $n^{-1/3}$ = (5.9 × 10^{28})$^{-1/3}$ = 0.257 nm. So, on this criterion, electrons at the Fermi energy in gold behave predominantly as waves rather than classical particles.

We will return to a more realistic discussion of electrical conduction in an "empty box" metal after we have presented a more accurate description of electronic shells in atoms.

As an application of this simple development, we digress from the properties of a metal to estimate the pressure of the dense electron gas in the core of the Sun, to extend the discussion in Chapter 4. While we found that the protons in the Sun act as classical particles, we now ask if the free electrons in the Sun act as quantum particles, or whether they are also classical in their behavior. If the electrons are following the quantum description, their equation of state is modified from the ideal gas form $P = RT/V$ (where R, the gas constant, is $N_A k_B$, the product of Avogadro's number and the Boltzmann constant), to:

$$P = 2NE_F/5V = 0.4\,n\,E_F. \tag{5.19}$$

(Following Chandrasekhar, 1983, this formula comes from two statements, $(PV) = 2/3\,N\,E_{av}$, that is a general ideal gas result, and, for the Fermi gas, $E_{av} = 3/5\,E_F$.)

In the new formula, as expected, the Fermi energy, $E_F = (h^2/8m_e)\,(3n/\pi)^{2/3}$ (5.17) replaces the thermal energy $k_B T$.

To return to the condition of electrons in the Sun's core, recall that $n = N_p + 2\,N_{He} = 6.026 \times 10^{31}$ m^{-3} (see Eq. 4.2) we find, first, $E_F = (h^2/8m_e)\,(3n/\pi)^{2/3} = [(6.6 \times 10^{-34})^2/(8 \times 9.11 \times 10^{-31})]\,(3 \times 6.026 \times 10^{31}/\pi)^{2/3} = 8.90 \times 10^{-17}$ J $= 556$ eV.

Since this number is less than the thermal energy $k_B T = 1293$ eV, we can assume that the electrons at the center of the Sun actually behave in a nearly classical fashion, and formula (5.19) is not called for.

So, to estimate the pressure in the Sun's core, we can use the classical relation $(PV) = 2/3\,N\,E_{av}$. N is the total particle density, protons, helium and electrons, that sums to:

$(6.026 + 3.106 + 1.46) \times 10^{31}$ m^{-3} $= 10.59 \times 10^{31}$ m^{-3}. The pressure then is $10.59 \times 10^{31} \times 1293$ eV $\times 1.6 \times 10^{-19}$ Pa $= 2.19 \times 10^{16}$ Pa $= 217$ G Bar, where 1 Bar $= 101$ kPa. This is close to the value 232×10^9 Bar (Bahcall et al, 2001) for the total hydrostatic pressure at the core of the Sun. The total hydrostatic pressure is the sum of pressures from electrons, protons, He ions and radiation. The radiation pressure is smaller, and is estimated as 0.126 GBar for 15 million K. So, these numbers are in good agreement, and the electrons in the Sun behave approximately classically.

To return to the properties of a system of electrons at zero temperature, states below E_F are filled and states above E_F are empty. At non-zero temperatures the occupation (probability that the state contains one electron) is given by the Fermi–Dirac occupation function,

$$f_{FD} = \left[\exp\left(\{E - E_F\}/k_B T\right) + 1\right]^{-1}. \tag{5.20}$$

The energy width of transition of f_{FD} from 1 to 0 is about $2k_B T$.

Some of these features are sketched in Fig. 5.10, for the three dimensional case, using notation $n(E) = f_{FD}(E)$ and $N(E) = g(E)$, note the characteristic $E^{1/2}$ of the upper two curves.

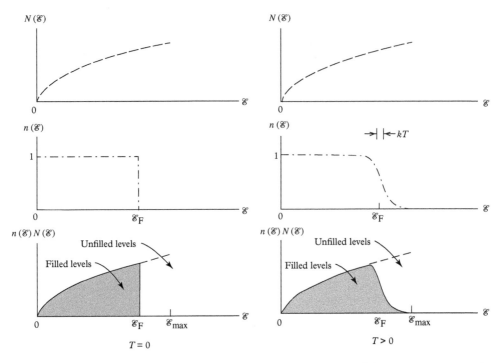

Fig. 5.10 *Density of states g(E) and occupation f(E) at temperature T = 0 (left) and T non-zero (right) in 3D case*

5.5 Schrodinger approach to atoms, molecules, and the covalent bond

The Schrodinger Equation introduced in Chapter 4, together with the laws of electricity and magnetism, are capable of describing details of atoms, molecules and solids, and their interaction with photons. We need to better understand these essential methods, to allow us to extend the approach to semiconductors, PN junctions and solar cells. We have seen that the Bohr model of the atom gives the correct energies, and allows an initial understanding of the magnetic and optical properties of one-electron atoms. But a more thorough approach, available through Schrodinger's equation in spherical polar coordinates, is necessary to incorporate the wave aspects of the electron, and to understand the nature of covalent bonding.

5.5.1 Hydrogenic wavefunctions

The atom is basically spherical, since the potential energy U of the electron in the electric field of the nucleus depends only on the radius r. The Schrodinger Equation (4.9) is easily extended to 3-D x, y, z coordinates (see Eq. 4.16a in connection with solutions in the box of side L) but is more complicated when expressed in spherical polar coordinates (4.19). Using these coordinates, for a spherically symmetric potential U(r), one finds, where θ, and φ, respectively are the polar and azimuthal angles:

$$\frac{-\hbar^2}{2m}\frac{1}{r^2}\frac{\partial}{\partial r}\left(\frac{r^2\partial\psi}{\partial r}\right) + \frac{\hbar^2}{2mr^2}\left[\frac{1}{\sin\theta}\frac{\partial}{\partial\theta}\left(\sin\theta\frac{\partial\psi}{\partial\theta}\right) + \frac{1}{\sin^2\theta}\frac{\partial^2\psi}{\partial\phi^2}\right] + U(r)\psi = E\,\psi. \quad (5.21)$$

The Schrodinger Equation is applied to the hydrogen atom, and any one-electron atom with nuclear charge Z, by choosing $U = -k_C\,Ze^2/r$, where k_C is the Coulomb constant. It is found, because of the spherical symmetry, that the equation separates into three equations, in variables r, θ, and φ, by setting:

$$\psi = R(r)f(\theta)g(\phi). \quad (5.22)$$

The solutions are conventionally described as the quantum states $\Psi_{n,l,m}$, specified by quantum numbers n, l, m.

The principal quantum number n, setting the energy, is associated with the solutions for the radial wavefunction,

$$R_{n,l}(r) = (r/a_o)^l \exp(-r/na_o)\mathcal{L}_{n,l}(r/a_o). \quad (5.23)$$

Here $\mathcal{L}_{n,l}(r/a_o)$ is a Laguerre polynomial in $\rho = r/a_o$, and the radial function has $n-l-1$ nodes. The parameter a_o is identical to its value in the Bohr model, but it no longer signifies the exact radius of an orbit. The energies of the electron states of the one electron atom, $E_n = -Z^2E_o/n^2$ (where $E_o = 13.6$ eV, and Z is the nuclear charge) are unchanged from the Bohr model. The energy can still be expressed as $E_n = -kZe^2/2r_n$, where $r_n = n^2a_o/Z$, and $a_o = 0.0529$ nm is the Bohr radius, as we found in Eqs. (5.1 – 5.4).

The lowest energy wavefunctions $\Psi_{n,l,m}$ of the one-electron atom are listed in Table 5.1. As before, to represent the hydrogen atom, set Z = 1. In common chemical usage, the letters s, p, d, and f correspond to angular momentum values 0, 1, 2, and 3. An "s-state" is spherically symmetric and has no angular momentum.

$$\Psi_{100} = (Z^{3/2}/\sqrt{\pi}\,a^{3/2})\exp(-Zr/a_o) \quad (5.24)$$

represents the lowest energy ground state that we earlier found had energy –13.6 eV.

The probability P(r) of finding the electron at radius r is:

$$P(r) = 4\pi r^2\Psi^2_{100}, \quad (5.25)$$

that is a smooth function with a maximum at $r = a_o/Z$. This is not an orbit of radius a_o, but a spherical probability cloud where the electron's most probable radius is a_o. There is no angular momentum associated with this wavefunction.

Table 5.1 *One-electron wavefunctions in real form.*

Wavefunction designation	Wavefunction name, Real form	Equation for real form of wavefunction*, where $\rho = Zr/a_o$ and $C_1 = Z^{3/2}\sqrt{\pi}$
Ψ_{100}	1s	$C_1\, e^{-\rho}$
Ψ_{200}	2s	$C_2\, (2 - \rho)\, e^{-\rho/2}$
$\Psi_{21,\cos\phi}$	$2p_x$	$C_2\, \rho \sin\theta \cos\phi\; e^{-\rho/2}$
$\Psi_{21,\sin\phi}$	$2p_y$	$C_2\, \rho \sin\theta \sin\phi\; e^{-\rho/2}$
Ψ_{210}	$2p_z$	$C_2\, \rho \cos\theta\; e^{-\rho/2}$
Ψ_{300}	3s	$C_3\, (27 - 18\rho + 2\rho^2)\; e^{-\rho/3}$
$\Psi_{31,\cos\phi}$	$3p_x$	$C_3\, (6\rho - \rho^2)\, \sin\theta \cos\phi\; e^{-\rho/3}$
$\Psi_{31,\sin\phi}$	$3p_y$	$C_3\, (6\rho - \rho^2)\, \sin\theta \sin\phi\; e^{-\rho/3}$
Ψ_{310}	$3p_z$	$C_3\, (6\rho - \rho^2)\, \cos\theta\; e^{-\rho/3}$
Ψ_{320}	$3d_{z^2}$	$C_4\, \rho^2\, (3\cos^2\theta - 1)\; e^{-\rho/3}$
$\Psi_{32,\cos\phi}$	$3d_{xz}$	$C_5\, \rho^2 \sin\theta \cos\theta \cos\phi\; e^{-\rho/3}$
$\Psi_{32,\sin\phi}$	$3d_{yz}$	$C_5\, \rho^2 \sin\theta \cos\theta \sin\phi\; e^{-\rho/3}$
$\Psi_{32,\cos2\phi}$	$3d_{x^2-y^2}$	$C_6\, \rho^2 \sin^2\theta \cos2\phi\; e^{-\rho/3}$
$\Psi_{32,\sin2\phi}$	$3d_{xy}$	$C_6\, \rho^2 \sin^2\theta \sin2\phi\; e^{-\rho/3}$

* $C_2 = C_1/4\sqrt{2}$, $C_3 = 2C_1/81\sqrt{3}$, $C_4 = C_3/2$, $C_5 = \sqrt{6}C_4$, $C_6 = C_5/2$. Pilar, F. J. (1990). *Elementary Quantum Chemistry*, 2nd Ed. (Dover: McGraw-Hill), p. 125.

This new solution represents a correction in concept, and in some numerical values, from the results of the Bohr model. Note that Ψ_{100} is real, as opposed to complex, and therefore the electron in this state has no orbital angular momentum. Both of these features correct errors of the Bohr model.

The n = 2 wavefunctions start with Ψ_{200} and Ψ_{210}, that has one node in r, but is spherically symmetric. The first anisotropic (non-spherical) wavefunctions are:

$$\Psi_{21,\pm1} = R(r)f(\theta)g(\phi) = C_2\rho\sin\theta\; e^{-\rho/2} \exp(\pm i\phi), \tag{5.26}$$

where $\rho = Zr/a_o$ and where Z, the number of positive nuclear charges, is 1 for hydrogen.

These are the first two wavefunctions to exhibit orbital angular momentum, here $\pm\hbar$ along the z-axis. Generally,

$$g(\phi) = \exp(\pm im\phi), \tag{5.27}$$

where m, known as the magnetic quantum number, represents the projection of the orbital angular momentum vector of the electron along the z-direction, in units of $\hbar$.

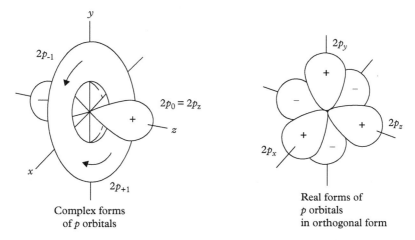

Complex forms
of *p* orbitals

Real forms of
p orbitals
in orthogonal form

Fig. 5.11 *2p (n = 2, l = 1) wave-functions in schematic form. Left panel, complex forms carry angular momentum. Right panel, linear combinations have the same energy, now assume aspect of bonds*

The orbital angular momentum **L** of the electron motion is described by the quantum numbers *l* and *m*.

The orbital angular momentum quantum number *l* has a restricted range of integer values:

$$l = 0, 1, 2..., n - 1. \tag{5.28}$$

This rule confirms that the ground state, n = 1, has zero angular momentum. In the literature the letters s, p, d, f, g, respectively, are often used to indicate $l = 0, 1, 2, 3,$ and 4. So a 2s wavefunction has n = 2 and $l = 0$, and the wavefunctions sketched in Fig. 5.11 are called the 2p wavefunctions, where p is chemical shorthand for $l = 1$.

The allowed values of the *magnetic quantum number m* depend upon both n and *l* according to the scheme,

$$m = -l, -l + 1, ..., (l - 1), l. \tag{5.29}$$

There are $2l + 1$ possibilities. Again, *m* represents the projection of the angular momentum vector along the z axis, in units of ℏ. For $l = 1$, for example, there are three values of *m*: −1, 0, and 1, and this is referred to as a "triplet state". In this situation the angular momentum vector has 3 distinct orientations with respect to the z-axis: polar angle θ = 45°, 90° and 135°. In this common notation, the n = 2 state (containing four distinct sets of quantum numbers) separates into a "singlet" (2s) and a "triplet" (2p).

For each electron there is also a spin angular momentum vector **S** with length $[s(s+1)]^{1/2}$ ℏ, where s = 1/2, and projection,

$$m_s = \pm(1/2)\hbar. \tag{5.30}$$

These strange rules, mathematically required to solve Schrodinger's Equation, are known to accurately describe the behavior of electrons in atoms. We can use these rules to enumerate the possible distinct quantum states for a given energy state, n.

Following these rules, one can see that the number of distinct quantum states for a given n is $2n^2$. Since the Pauli exclusion principle for electrons (and other Fermi particles) allows only one electron in each distinct quantum state, $2 n^2$ is also the number of electrons that can be accommodated in the nth electron shell of an atom. For n = 3 this gives 18, that is seen to be twice the number of entries in Table 5.1 for n = 3. The factor of two represents the "spin degeneracy", that is formally 2s + 1 evaluated for s = ½, see Fig. 5.12.

A further peculiarity of angular momentum is that the vector **L** has length L = $\sqrt{(l(l+1))}$ ħ and projection $L_z = m$ ħ. A similar situation occurs for the spin vector **S**, with magnitude S = $\sqrt{(s(s+1))}$ ħ and projection m_sħ. For a single electron $m_s = \pm1/2$. In cases where an electron has both orbital and spin angular momentum (for example, the electron in the n = 1 state of the one-electron atom has only S, no L), these two forms of angular momentum combine as **J** = **L** + **S**, and has a similar rule for its magnitude: J = $\sqrt{(j(j+1))}$ ħ. The rules are required by the solutions of the Schrodinger Equation.

The wavefunctions $\Psi_{21,\pm1} = C_2 \rho \sin\theta\ e^{-\rho/2} \exp(\pm i\phi)$ are the first two states having angular momentum. A polar plot of $\Psi_{21,\pm1}$ has a node along z, and resembles a doughnut flat in the x,y plane.

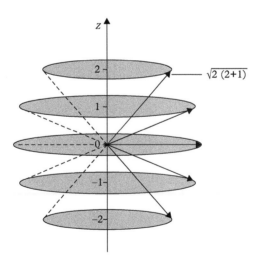

Fig. 5.12 *Five allowed orientations of angular momentum l = 2, length of vector and z-projections in units of ħ. Azimuthal angle φ is free to take any value*

Linear combinations of states are important in quantum mechanics. Here, the sum and difference of the $\Psi_{21,\pm1}$ states are also solutions to Schrodinger's Equation, for example:

$$\Psi_{211} + \Psi_{21-1} = C_2\,\rho\,\sin\theta\,e^{-\rho/2}\,[\exp(i\phi) + \exp(-i\phi)] = C_2\,\rho\,\sin\theta\,e^{-\rho/2}\,2\cos\phi. \qquad (5.31)$$

This is twice the $2p_x$ wavefunction in Table 5.1. This linear combination, Eq. (5.31), is exemplary of the real wavefunctions in Table 5.1, where linear combinations have canceled the angular momenta to provide a preferred direction for the wavefunction.

A polar plot of the $2p_x$ wavefunction (5.31), shows a node in the z direction from the $\sin\theta$ and a maximum along the x direction from the $\cos\phi$, so it is a bit like a dumbbell at the origin oriented along the x axis. Similarly, the $2p_y$ resembles a dumbbell at the origin oriented along the y-axis.

These real wavefunctions, where the $\exp(im\phi)$ factors have been combined to form $\sin\phi$ and $\cos\phi$, are more suitable for constructing bonds between atoms in molecules or in solids, than are the equally valid (complex) angular momentum wavefunctions. The complex wavefunctions that carry the $\exp(im\phi)$ factors are essential for describing orbital magnetic moments as occur in iron and similar atoms. The electrons that carry orbital magnetic moments usually lie in inner shells of their atoms.

5.5.2 Chemical Table of the Elements

The rules governing the one-electron atom wavefunction $\Psi_{n,l,m,m}$ and the Pauli exclusion principle, stating that only one electron can be accommodated in a completely described quantum state, are the basis for Chemical Table of the Elements. The number of electron states per atom is simply Z the nuclear charge. As we saw at the end of Chapter 4, the maximum Z for any nucleus is set by Coulomb repulsion among the Z protons.

As we have seen, the rules allow $2n^2$ distinct states for each value of the principal quantum number, n. There are several notations to describe the *filled atomic shells*. The "K shell" of an atom comprises the two electrons of n = 1 ($1s^2$), followed by the "L shell" with n = 2 ($2s^2\,2p^6$) (Ne); and the "M shell" with n = 3 ($3s^2\,3p^6\,3d^{10}$)(Ar). These closed shells contain, respectively, 2, 8, and 18 electrons. Completely filled electron cores occur at Z = 2 (He), Z = 10 (Ne), Z = 18 (Ar), Z = 36 (Kr), Z = 54 (Xe) and Z = 86 (Rn).

5.5.3 More about metals: electrons in an empty box (II)

We come back to our description of metals, first to remark that forming the "empty box", i.e., the binding of the metal atoms to make the container for the electrons, can be imagined to start with the fundamental van der Waals short range inter-atomic attraction U. Briefly, any two atoms at radius r attract via $U = k_c^2\,c_{vdw}/r^6$, where c_{vdw} is proportional to the product $Z_1\,Z_2$ of the number of orbitals in each atom and to a mean ionization energy, and k_c is the Coulomb constant.

Once the "empty box" is formed of real atoms, we can discuss how the electrons are bound inside it. To address the first question, we can apply Eq. (4.35) to the binding of a group of metal atoms, if we replace the nuclear binding energy $U_o \rightarrow U_{coh}$, by the *cohesive energy* U_{coh} per atom for the metal. For gold U_{coh} is listed as 3.81 eV per atom, while for the rare gas atoms from Ne to Rn the cohesive energy per atom ranges from 0.02 to 0.2 eV/atom (Kittel, 1986, p. 55).

The value 3.81 eV/atom greatly exceeds the values for rare gases, that are in the range expected for the van der Waals attraction. The rare gases form liquids or weakly bound solids only at very low temperatures, but of course metals are strongly bound, many with high melting points. The cohesive energy is the energy needed to remove one neutral atom from the metal, that does not involve ionization of any atom.

Since the only difference between the rare gas like radon, and a monovalent metal like gold is one valence electron on the outside of a filled rare gas electron shell, this one valence electron greatly strengthens the binding.

The "surface energy" correction term (4.35) again means only that outside the metal there are no more atoms to bind a surface atom. This term describes why a liquid metal like Hg will minimize its surface area to form a spherical drop. The collection of atoms in a metal becomes more strongly bound if the outer valence electrons "delocalize" to go into extended states similar to Eq. (4.16a), which have lower kinetic energies than the bound atomic valence states. These electrons are now filling the whole box.

The outer valence electron localized states are similar in concept to the hydrogenic 1s state (Section 5.5.1). The valence electron for gold has principal quantum number $n = 6$, and is called a "6s" state, to indicate a spherical state of zero angular momentum. These states oscillate rapidly varying with radius (following Eq. 5.23 we state that the number of radial nodes is $n - l - 1$, thus 5 for the 6s state), and the sharp $d\psi/dx$ variations lead to large kinetic energy. This part of the binding energy of a metal comes from the reduction in electron kinetic energy related to the delocalization of its wavefunctions. The smooth electron states extended away from the atomic cores, definitely have lower $d\psi/dx = p/\hbar$ than localized atomic states, that have rapidly varying wavefunctions (large $d\psi/dx$, i.e. momentum, and hence large kinetic energy), as one can see by looking at Eqs. 4.4 and 4.9. This clearly implies a reduction of kinetic energy $p^2/2m$, and this contributes to the strength of the *metallic bond*. A more detailed estimate coming to the same conclusion is given by (Kittel, 1986), p. 235.

That the starting point for an electron in the metallic "empty box" is a free particle, rather than an atomic state, is suggested by the fact that the atomic density exceeds the Mott critical value, mentioned in Chapter 4 following Eq. 4.2 (Mott, 1973).

Mott predicted that hydrogen atoms (see Section 5.1) when packed together at a density,

$$N_{Mott} \geq (0.3/a_H)^3 \qquad (5.32)$$

revert to an ionized, free electron state. While Mott did not discuss the physics in much detail, the lower kinetic energy of the delocalized electrons certainly is a part of this transition. (A second aspect is electron screening.)

In Mott's formula a_H is the expected Bohr radius in the situation (we will see later that this value is affected by the principal quantum number n, a possible effective mass parameter and by the permittivity). For hydrogen atoms in vacuum $N_{Mott} = (0.3/0.0529 \text{ nm})^3 =$ 18.2×10^{28} m^{-3}. This value is much less than the proton density in the core of the Sun, $N_p = 7.14 \times 10^{31}$ m^{-3}, so that hydrogen there is certainly ionized. However, before comparing to the density of atoms in gold, 5.9×10^{28} m^{-3}, we note that the gold valence electron has quantum number n = 6, and therefore (see Eq. 5.3) the Bohr radius is n^2 $a_o = 6^2 a_o = 1.9$ nm. So, the relevant Mott concentration is reduced to $N_{Mott}{}^3 [0.3/(36 \times a_o)]^3 = 3.91 \times 10^{24}$ m^{-3}. This allows us to proceed with our assumption of the valence electrons enclosed in a box of side L, imagined to be built of gold atoms, are correctly assumed to be in free particle states such as those in Eq. (4.16), but more accurately, those of Eq. 5.13.

We assume that the positive charge arising from the gold ions neutralizes the negative charge of the electrons, and proceed, assuming the potential seen by the electrons is U = 0 (compared to U = $\phi \approx \infty$ outside). In more common usage (see Fig. 5.13), the work function is measured above the Fermi energy, so that the minimum energy for an electron in the interior, relative to the outside vacuum energy, is:

$$U_o = -(E_F + \phi).$$

(a)

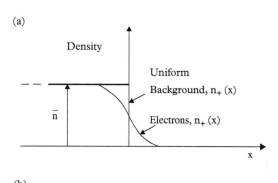

(b)

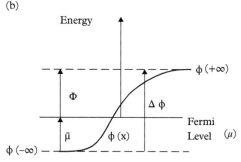

Fig. 5.13 *Schematic representation of (a) density distributions at a metal surface and (b) various energies relevant to a study of the work function (Lang and Kohn, 1971)*

Returning to the work function value, it is reasonable that the work function exceeds the cohesive energy U_{coh} because the former involves separating charge, ionizing an atom. In a metal, the work function barrier arises from an electric dipole layer. The electrons can tunnel slightly outside the perimeter of the metal ions, as we have discussed in connection with Gamow's probability and Fig. 1.6, putting negative charge outside the metal, which will then be compensated by positive charge on the inner side of the metal-vacuum boundary. This generates an electric dipole layer, that leads to a jump in electric potential, the work function barrier (see Fig. 5.13).

Careful calculations of the work function for a wide range of metals have been reported by Lang and Kohn, 1971, whose sketch of the surface dipole barrier is shown in Fig. 5.13. Their theory applied to gold gives work function values 3.5 eV, 3.65 eV, and 3.80 eV, respectively, for 110, 100 and 111 surfaces of the face centered cubic crystal. (The 111 surface is the body diagonal, 110 the face diagonal and 100 is the cube face surface, see Fig. 5.15).

We now turn to diatomic molecules, starting with H_2, as an introduction to atmospheric diatomic oxygen and nitrogen. These are held together by "electron exchange", an effect that is purely quantum in its nature, although the final result is an electrostatic attraction. The covalent bond can involve one, two or more electrons. The underlying effects are shown in the simplest cases, the one- electron bond in H_2^+, and the "covalent" bond in H_2. The idea of exchange or hopping between sites is brought into view by consideration of molecular bonding.

5.5.4 Hydrogen molecule ion H_2^+

The physics of the hydrogen molecular ion, the simplest (one-electron) covalent bond, is inherently quantum-mechanical. For the system of one electron and two protons at large spacing there are two obvious wavefunctions: $\psi_a(x_1)$ and $\psi_b(x_2)$, that represent, respectively, the electron on the left proton and on the right proton. For large spacing these states will be long-lived, but for smaller spacing they will be unstable. An electron starting in $\psi_a(x_1)$, say, will tunnel to $\psi_b(x_2)$, at a frequency f.

To find the ground state we make use of the idea that in quantum mechanics a linear combination of allowed solutions is also a solution. A general solution is:

$$\Psi = A\psi_a(x_1) + B\,\psi_b(x_2), \tag{5.33}$$

where $A^2 + B^2 = 1$.

The linear combinations that are stable in time are the symmetric and anti-symmetric,

$$\Psi_S = 2^{-1/2}[\psi_a(x_1) + \psi_b(x_2)], \quad \Psi_A = 2^{-1/2}[\psi_a(x_1) - \psi_b(x_2)]. \tag{5.34}$$

These states are stable in time, because the electron is equally present on right and left, and the tunneling instability no longer occurs.

It is easy to understand that the symmetric combination Ψ_S has a lower energy than Ψ_A, because the probability of finding the electron at the midpoint is non-zero, while that probability is zero in Ψ_A. The midpoint is an energetically favorable location for the electron, because it sees attraction from both protons. The energy difference between the symmetric and anti-symmetric cases is:

$$\Delta E = hf, \tag{5.35}$$

where f is the tunneling frequency of an electron started on one side for tunneling to the other side. The value ΔE is about twice the binding energy (Pauling and Wilson, 1935, pp. 326–36), that is 2.65 eV for H_2^+. So, the tunneling rate is 1.28×10^{15}/s, corresponding to a residence time 0.778 fs.

The exchange of an electron between the two sites is also referred to as "hopping" or "resonance" and a more detailed treatment will give us the "hopping integral" that determines the rate. (A modification of this treatment will be applied to a scheme for inducing D + D fusion.)

Consider two protons, (at sites a and b, assume they are massive and fixed) a distance R apart, with one electron. If R is large and we can neglect interaction of the electron with the second proton, then, following the discussion of Pauling and Wilson, 1935 we have:

$$[-(\hbar^2/2m)\nabla^2 + U(r)]\psi = \mathcal{H}\psi = E\psi \tag{5.36}$$

with r the electron position. This gives solutions, with no interactions, $\psi = \psi_a(x_1)$ and $\psi = \psi_b(x_2)$ at energy $E = E_o$.

The interactions of the electron and the first proton with the "second" proton, $-k_c e^2(1/r_{a,2}+1/R)$ are now considered. The attractive interaction, primarily occurring when the electron is between the two protons, and is attracted to both nuclear sites, stabilizes H_2^+.

We can write the interaction as:

$$\mathcal{H}_{int} = k_C e^2 [1/R - 1/r_{a,2}]. \tag{5.37}$$

where the first term is the repulsion between the two protons spaced by R. Following Pauling's treatment, one finds:

$$E - E_o = (k_C e^2/Da_o) + (J + K)/(1 + \Delta) \text{ for } \Psi_S \tag{5.38}$$

$$E - E_o = (k_C e^2/Da_o) + (J - K)/(1 - \Delta), \text{ for } \Psi_A, \text{ where} \tag{5.39}$$

$$K = \iint \psi_b^\star (x_2)[-ke^2(1/r_{a,2})]\psi_a(x_1)d^3x_1 d^3x_2 = -(ke^2/a_o)e^{-D}(1 + D) \tag{5.40}$$

$$J = \iint \psi_a^\star (x_1)[-ke^2(1/r_{a,2})]\psi_a(x_1)d^3x_1 d^3x_2 = -(ke^2/a_o)[-D^{-1} + e^{-2D}(1 + D^{-1})] \tag{5.41}$$

$$\Delta = \int\int \psi_b^\star (x_2) \psi_a (x_1) d^3x_1 d^3x_2 = e^{-D}(1 + D + D^2/3), \text{ where } D = R/a_0. \quad (5.42)$$

K is known as the resonance or exchange or hopping integral, and measures the rate at which an electron on one site moves to the nearest neighbor site. One sees that its dependence on spacing is essentially $e^{-D} = e^{-R/a}$, as one would expect for a tunneling process, and that the basic energy (the prefactor of the exponential term) is $-(k_c e^2/a_0) = -2E_0 = -27.2$ eV. In these equations k_c is the Coulomb constant 9×10^9.

The energy E of the symmetric case is shown in (5.38). The major negative term is K, and this term changes sign in (3.39), the anti-symmetric case. So, the difference in energy between the symmetric and anti-symmetric cases is about 2 K (see Eq. (5.40)), which amounts to about 2×2.65 eV = 5.3 eV for H^+_2. The predicted equilibrium spacing is 2.4 a_0 for this molecule.

The energy can be expanded as a function of $D = R/a_0$ and has a minimum at 2.4 a_0. The energy vs spacing, near the minimum, can be expressed as:

$$E(D-2.4) = E(2.4) + dE/dD(D-2.4) + \tfrac{1}{2} \ d^2E/dD^2 (D-2.4)^2 + ... \quad (5.43)$$

Since the slope is zero at the minimum, that minimum is locally parabolic, that is the basis for simple harmonic motion, and one sees that the spring constant is $K_{spring} = d^2E/dD^2 = E''$. The oscillator frequency then is:

$\omega = (E''/m_{red})^{1/2}$. The value of $E''(2.4) = 0.1257 \ E_0/a^2$ (Griffiths, 2005), that gives $\omega = (E''/m_{red})^{1/2} = (0.1257 \ 13.6 \ 1.6 \ 10^{-19}/\tfrac{1}{2} \ 1.67 \ 10^{-27})^{1/2}/0.0529$ nm $= 3.42 \times 10^{14}$ /s. This corresponds to a zero-point energy $\hbar\omega/2 = 0.113$ eV.

5.5.5   Diatomic molecules: hydrogen H_2

The atmosphere is largely composed of diatomic molecules N_2 and O_2. Molecular hydrogen will serve as a prototype for these covalently bonded molecules. The covalent bond contains two electrons, that are identical particles of the class called Fermions. Before describing the hydrogen molecule, we need to think about wavefunctions for two particles $\psi(x_1, x_2)$, leading to a probability distribution $P(x_1, x_2) = \psi^*(x_1, x_2)\psi(x_1, x_2)$ where $\star$ represents the complex conjugate. This is the probability of finding particle 1 at location x_1 and particle 2 at location x_2. In systems with two or more electrons, their indistinguishable nature plays an important role. If two electrons are present in a system, the probability distributions $P(x_1, x_2)$ and $P(x_2, x_1)$ must be identical.

5.5.5.1   *Symmetric and anti-symmetric two-particle wavefunctions*

No observable change can occur from exchanging the two electrons. That is,

$$P(x_1, x_2) = P(x_2, x_1) = \left| \psi_{n,m}(x_1, x_2) \right|^2, \quad (5.44)$$

from it follows that either:

$$\psi_{n,m}(x_2, x_1) = \psi_{n,m}(x_1, x_2) \text{ (Symmetric case), or} \tag{5.45}$$

$$\psi_{n,m}(x_2, x_1) = -\psi_{n,m}(x_1, x_2) \text{ (Antisymmetric case).} \tag{5.46}$$

If we apply this idea to the two non-interacting electrons in, for example, a 1-D trap, with the wavefunction,

$$\psi_{n,m}(x_1, x_2) = A^2 \sin(n\pi x_1 / L) \sin(m\pi x_2 / L) = \psi_n(x_1) \psi_m(x_2),$$

we find that this particular 2-particle wavefunction is neither symmetric nor anti-symmetric.

However, the combinations,

$$\psi_S(1,2) = [\psi_n(x_1)\psi_m(x_2) + \psi_n(x_2)\psi_m(x_1)] / \sqrt{2}, \text{ and} \tag{5.47}$$

$$\psi_A(1,2) = [\psi_n(x_1)\psi_m(x_2) - \psi_n(x_2)\psi_m(x_1)] / \sqrt{2}, \tag{5.48}$$

respectively, are correctly symmetric and anti-symmetric.

The antisymmetric combination ψ_A, Eq. (5.46), is found to apply to electrons, and to other spin = 1/2 particles, including protons and neutrons, that are called *fermions*. By looking at ψ_A in the case m = n, one finds $\psi_A = 0$. Thus the wavefunction, and hence the probability, for two fermions in exactly the same state, is zero. This is a statement of the *Pauli exclusion principle*: only one Fermi particle can occupy a completely speci-fied quantum state.

For other particles, notably photons of electromagnetic radiation, the symmetric combination $\psi_S(1, 2)$, Eq. (5.47), is found to occur in nature. These particles are char-acterized by integer spin. Macroscopically large numbers of photons can have exactly the same quantum state, and this is important in the functioning of lasers. Photons, alpha particles, and deuterons (with nuclear spin I = 1) are examples of Bose particles, or *bosons*.

5.5.5.2 *Orbital and spin components of wavefunction*

To complete the description of the electron state, one must add its spin projection: $m_s = \pm 1/2$. It is useful to separate the space part $\phi(x)$ and the spin part χ of the wavefunction, as:

$$\psi = \phi(x)\chi. \tag{5.49}$$

For a single electron $\chi = \uparrow$ (for $m_s = 1/2$) or $\chi = \downarrow$ (for $m_s = -1/2$). For two electrons there are two categories, S = 1 (parallel spins) or S = 0 (anti-parallel spins). While the S = 0 case allows only $m_s = 0$, the S = 1 case has three possibilities, $m_s = 1, 0, -1$, referred to as a "spin triplet".

A good notation for the spin state is $\chi_{S,m}$, where the spin triplet states are:

$$\chi_{1,1} = \uparrow_1 \uparrow_2, \chi_{1,-1} = \downarrow_1 \downarrow_2, \text{ and } \chi_{1,0} = \uparrow_1 \downarrow_2 + \downarrow_1 \uparrow_2. \text{ (spin triplet)} \tag{5.50}$$

For the singlet spin state, $S = 0$, one has:

$$\chi_{0,0} = \uparrow_1 \downarrow_2 - \downarrow_1 \uparrow_2. \text{ (spin singlet)} \tag{5.51}$$

Inspection of these makes clear that the spin triplet ($S = 1$) is symmetric on exchange, and the spin singlet ($S = 0$) is anti-symmetric on exchange of the two electrons.

Since the *complete* wavefunction (for fermions like electrons) must be *anti-symmetric* for exchange of the two electrons, this can be achieved in two ways:

$$\psi_A(1, 2) = \phi_{sym}(1, 2)\chi_{anti}(1, 2) = \phi_{sym}(1, 2) \quad S = 0 \text{ (spin singlet)} \tag{5.52}$$

$$\psi_A(1, 2) = \phi_{anti}(1, 2)\chi_{sym}(1, 2) = \phi_{anti}(1, 2) \quad S = 0 \text{ (spin triplet)} \tag{5.53}$$

The structure of the orbital or space wavefunctions $\phi_{sym,anti}(1, 2)$ here is identical to those shown previously for $\psi_{S,A}(1, 2)$, Eqs. (5.47, 5.48).

5.5.5.3   Back to the hydrogen molecule

Consider two protons, (labeled a, and b, assume they are massive and fixed) a distance R apart, with two electrons. If R is large and we can neglect interaction between the two atoms, then, following the discussion of Tanner, 1995 pp. 186–9,

$$[-(\hbar^2/2m (\nabla_1^2 + \nabla_1^2) + U(r_1) + U(r_2)]\psi = (\mathcal{H}_1 + \mathcal{H}_2)\psi = E\psi \tag{5.54}$$

where r_1, r_2 represent the space coordinates of electrons 1 and 2. Solutions to this problem, with no interactions, can be $\psi = \psi_a(x_1)\psi_b(x_2)$ or $\psi = \psi_a(x_2)\psi_b(x_1)$ (with $\psi_{a,b}$ the wavefunction centered at proton a, b) and the energy in either case is $E = E_a + E_b$.

The interaction between the two atoms is the main focus, of course. The interactions are basically of two types. First, the repulsive interaction $k_C e^2/r_{1,2}$, with $r_{1,2}$ the spacing between the two electrons. Secondly, the attractive interactions of each electron with the "second" proton, $-k_C e^2(1/r_{a,2} + 1/r_{b,1})$. The latter attractive interactions, primarily occurring when the electron is in the region between the two protons, and can derive binding from both nuclear sites at once, stabilize the hydrogen molecule.

Altogether we can write the inter-atom interaction as:

$$\mathcal{H}_{int} = k_C e^2 [1/R + 1/r_{1,2} - 1/r_{a,2} - 1/r_{b,1}]. \tag{5.55}$$

To get the expectation value of the interaction energy the integration Eq. (5.56), must extend over all six relevant position variables q. (Spin variables are not acted on by the interaction.)

$$<E_{int}> = \int \psi^* \mathcal{H}_{int}\psi dq. \tag{5.56}$$

The appropriate wavefunctions have to be overall anti-symmetric. Thus, following Eqs. (5.52, 5.53), the symmetric $\phi_{sym}(1, 2)$ orbital for the anti-symmetric $S = 0$ (singlet) spin state, and the anti-symmetric $\phi_{anti}(1, 2)$ orbital for the symmetric $S = 1$ (triplet) spin state.

The interaction energies are:

$$<E_{int}> = A^2(K_{1,2} + J_{1,2})\, S = 0 \text{ (spin singlet)} \tag{5.57}$$

$$<E_{int}> = B^2(K_{1,2} - J_{1,2})\, S = 1 \text{ (spin triplet), where} \tag{5.58}$$

$$K_{1,2}(R) = \int\int \phi_a^\star(x_1)\, \phi_b^\star(x_2) \mathcal{H}_{int}\, \phi_b(x_2)\, \phi_a(x_1) d^3x_1 d^3x_2 \tag{5.59}$$

$$J_{1,2}(R) = \int\int \phi_a^\star(x_1)\, \phi_b^\star(x_2) \mathcal{H}_{int}\, \phi_a(x_2)\, \phi_b(x_1) d^3x_1 d^3x_2 \tag{5.60}$$

The physical system will choose, for each spacing R, the state providing the most negative of the two interaction energies. For the hydrogen molecule, the exchange integral $J_{1,2}$ is negative, so that the covalent bonding occurs when the spins are anti-parallel, in the spin singlet case.

The qualitative understanding of the result is that in the spin singlet case the orbital wavefunction is symmetric, allowing more electron charge to locate halfway between the protons, where their electrostatic energy is most favorable. A big change in electrostatic energy (about $2J_{1,2}$) is linked (through the exchange symmetry requirement) to the relative orientation of the magnetic moments of the two electrons. This effect can be summarized in what is called the exchange interaction:

$$\mathcal{H}_e = -2J_e\, S_1 \cdot S_2. \tag{5.61}$$

In the case of the hydrogen molecule J is negative, giving a negative *bonding* interaction for *anti-parallel* spins. The parallel spin configuration is repulsive or *anti-bonding*. The difference in energy between the bonding and anti-bonding states is about 9 eV for the hydrogen molecule at its equilibrium spacing, R = 0.074 nm. The bonding energy is 4.7 eV, see Fig. 5.14.

This is a huge effect, ostensibly a magnetic effect, but actually a combination of fundamental symmetry and electrostatics. The covalent bond is what makes much of matter stick together. Other forms of bonding are ionic as in NaCl and van der Waals, as in liquid rare gases. Organic molecules are on the whole bonded together in a covalent fashion.

The covalent bond as described here is a short-range effect, because it is controlled by overlap of the exponentially decaying wave functions from each nucleus. This is true even though the underlying Coulomb force is a long-range effect, proportional to $1/r^2$.

In other cases, J is negative, making the energy favorable arrangement one of *parallel spins*. This can lead to *ferromagnetism*, a cooperative state of matter in which huge numbers of magnetic moments are all locked parallel, leading to a macroscopic magnetization, M. It is important to recognize that the driving force for the spin alignment is electrostatic, in the exchange interaction, represented by Eq. (5.61).

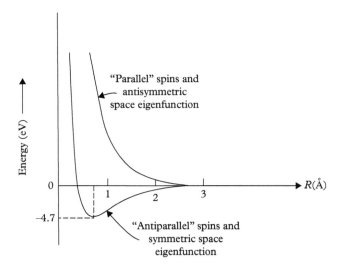

Fig. 5.14 *Energy curves for bonding and anti-bonding states of the hydrogen molecule. The bonding state requires anti-parallel spins. The equilibrium separation is 0.074 nm*

5.6   Tetrahedral bonding in silicon and related semiconductors

The most important linear combinations of angular momentum wavefunctions (hybrids) cases are sp^3, sp^2, and sp hybrids. sp^3 hybrids describe tetrahedral bonds at 109.5° angles in methane, CH_4 and diamond. This scheme also describes bonding in the important semiconductors Si and GaAs, see Fig. 5.15.

In these materials, four outer electrons (2 each from 3s and 3p orbitals in Si and GaAs, and 2 each from 4s and 4p in Ge), are stabilized into 4 tetrahedral covalent bond orbitals that point from each atom to its 4 nearest neighbor atoms, located at apices of a tetrahedron.

The hybridization effect is essential to understanding molecules and solids. To get a better understanding, recall (see Table 5.1) that wavefunctions $\Psi_{21,\pm1} = C_2 \, \rho \sin\theta \, e^{-\rho/2}$ $\exp(\pm i\phi)$ are the first two states having angular momentum, where $\rho = r/a_0$ and ϕ is the angle in the x, y plane. A polar plot of $\Psi_{21,\pm1}$ has a node along z, and resembles a dough-nut flat on the x, y plane. But these wavefunctions can be combined to make equivalent wavefunctions that point in particular directions.

The sum and difference of these wave functions are also solutions to Schrodinger's Equation, examples are:

$$\Psi_{211} + \Psi_{21-1} = C_2 \, \rho \sin\theta \, e^{-\rho/2} \, [\exp(i\phi) + \exp(-i\phi)]$$
$$= C_2 \, \rho \sin\theta \, e^{-\rho/2} \, 2 \cos\phi = 2 \, 2p_x \tag{5.62}$$

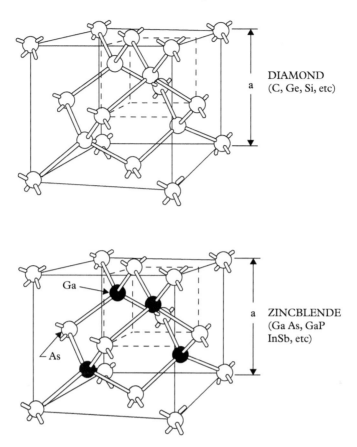

Fig. 5.15 *Diamond and zincblende crystal structures. Each atom is covalently bonded to four nearest neighbors in tetrahedral directions. The directed bonds are linear combinations of s and p orbitals (See Table 5.1), and analogous to directed orbitals sketched in Fig. 5.11. Specifically, the linear combinations are 2s and 2p³ for diamond (as in CH₄) and 3s and 3p³ for Si. There are 4 valence electrons per atom, leaving a band structure with filled bands, thus insulating except for thermal excitations*

$$\Psi_{211} - \Psi_{21-1} = C_2\, \rho\, \sin\theta\ e^{-\rho/2}\,[\exp(i\phi) - \exp(-i\phi)]$$
$$= C_2(i)\, \rho\, \sin\theta\ e^{-\rho/2}\, 2\sin\phi = 2\ 2p_y \tag{5.63}$$

and,

$$\Psi_{210} = C_2\, \rho\, \cos\theta\ e^{-\rho/2} = 2p_z. \tag{5.64}$$

These linear combinations, $2p_x$, $2p_y$ and $2p_z$, are equivalent to the original solutions that are eigenstates of angular momentum. They are also a bit like unit vectors **i**, **j**, and **k** that point along the x, y and z directions (see Table 5.1 and Fig. 5.11).

These details are more important than one might expect. The same pattern, the same angular dependences for the p- and d- wavefunctions, are duplicated for higher values of principal quantum number n beyond n = 2.

5.6.1   Wavefunctions for directed or covalent bonds

We can use these functions in turn to form other hybrid states. For example, the sp^3 set of wavefunctions, that point to the corners of a tetrahedron, occur in molecular CCl_4, and in crystalline germanium, silicon and diamond.

A tetrahedron has four vertices at equal radius from the center, and has both 3-fold (through each vertex) and two-fold (through face-centers) rotation axes thru the origin (cube center). If we think of a cube of side L = 2, then four diagonally related corners, out of the 8 corners of the cube, are vertices of a tetrahedron. These points have radius $\sqrt{3}$ from the center of the cube of side 2, and the spacing between adjacent vertices is $2\sqrt{2}$, which defines the bond angle β. The law of cosines ($a^2 = b^2 + c^2 - 2bc \cos\beta$) then gives $\cos\beta = -1/3$, or β = 109.47° for the tetrahedral bond angle. This illustrates a method that be applied to other bonding geometries.

To represent a linear combination of directed wavefunctions more simply, call it:
r = l **i** + m **j** + n **k,** where l, m, and n are direction cosines, which give the projection of **r** along each of the axes. To find the combinations that form the tetrahedron, imagine vectors from the center of a cube of edge 2, to four diagonally related corners, and represent these vectors by l, m, n values. Taking the origin at the center of the cube, a suitable set of linear combinations of p_x, p_y and p_z wavefunctions (see Table 5.1) is easily seen to be represented (by l,m,n values) as (1,–1,1), (-1,1,1), (1,1,–1), and (–1,–1,–1).

5.6.1.1   Bond angle

A useful formula for the angle Θ between two lines that are described respectively, by l,m,n values, is:

$$\cos\Theta = \left(l_1 l_2 + m_1 m_2 + n_1 n_2\right) / \left[\left(l^2_1 + m^2_1 + n^2_1\right)^{1/2}\left(l^2_2 + m^2_2 + n^2_2\right)^{1/2}\right]. \quad (5.65)$$

If we apply this to find the angle, e.g., between radius vectors (1,–1,1) and (–1,1,1), we have:

$$\cos\Theta = (-1-1+1)/[(3)^{1/2}(3)^{1/2}] = -1/3. \quad \text{So } \Theta = 109.47°,$$

to verify the tetrahedral bonding angle, we found for the diamond structures that include silicon. Again, the directed wavefunctions p_x, p_y and p_z have been represented by unit vectors **i, j,** and **k.** We have also identified a directed wavefunction with a covalent bond, which will be further discussed.

Table 5.1 includes linear combinations of 3-D wavefunctions that have characteristic shapes, with lobes pointing in particular directions. These wavefunctions also can participate in hybrid bonding.

5.6.2   Band structures, donor and acceptor impurities

On the other hand, the theory predicts that a filled band will lead to no conduction, i.e., to an insulator, because no empty states are available to allow motion of the electrons in response to an electric field. Such a situation is approximately present in the pure semiconductors Si, GaAs, and Ge.

The 4 electrons per atom (8 electrons per unit cell) that fill the covalent bonds of the diamond-like structure, completely fill the lowest 2 "valence" bands (because of the spin degeneracy, mentioned previously), leaving the third band empty. In concept this would correspond to the first two bands in Fig. 5.8 as being completely filled, and thus supporting no electrical conduction at zero temperature. Referring to Fig. 5.8 one sees that at k = 0, just above the second band, there is a forbidden gap, E_g. There are no states allowing conduction until the bottom of band 3, that is about 1 eV higher in these materials. For this reason, at least at low temperatures, pure samples of these materials do not conduct electricity.

Electrical conductivity at low temperature and room temperature in these materials is accomplished by "doping"; substitution for the 4-valent Si or Ge atoms either acceptor atoms of valence 3 or donor atoms of valence 5. In the case of 5-valent donor atoms like P, As, or Sb, 4 electrons are incorporated into tetrahedral bonding and the extra electron becomes a free electron at the bottom of the next empty band. In useful cases, the number of free electrons, n, is close to the number of donor atoms, N_D. This is termed an N-type semiconductor. In the case of Boron, Aluminum, and other 3-valent dopant atoms, one of the tetrahedral bonding states is unfilled, creating a "hole". A hole acts like a positive charge carrier, that moves when an electron from an adjacent bond jumps into the vacant position. Electrical conductivity by holes is dominant in a "P-type semiconductor".

The band structures for Si and GaAs are sketched in Fig. 5.16. These results are calculated from approximations to Schrodinger's Equation using more realistic 3-D forms for the potential energy U(x,y,z). The curves shown have been verified over a period of years by various experiments.

In these semiconductors, the charge carriers of importance are either electrons at a minimum in a nearly empty conduction band, or holes at the top of a nearly filled valence band. In either case, the mobility $\mu = e\tau/m^*$, such that the drift velocity $\mathbf{v_d} = \mu\mathbf{E}$, is an important performance parameter. A high mobility is desirable as increasing the frequency response of a device such as a transistor. A useful quantity, that can be accurately predicted from the band theory, and measured by a cyclotron resonance experiment, is the effective mass, m^*. This parameter is related to the inverse of the curvature of the energy band. The curvature, $\partial^2 E/\partial k^2$ can be calculated, and the formula, simply related to $E = \hbar^2 k^2/2m^*$, is:

$$m^{\star} = \hbar^2/(\partial^2 E/\partial k^2). \tag{5.66}$$

In looking at the energy bands for Si and GaAs in Fig. 5.16, one can see that generally the curvature is higher in the conduction band than in the valence band, meaning that the effective mass is smaller and the mobility therefore higher for electrons than for holes. Secondly, comparing Si and GaAs, the curvature in the conduction band

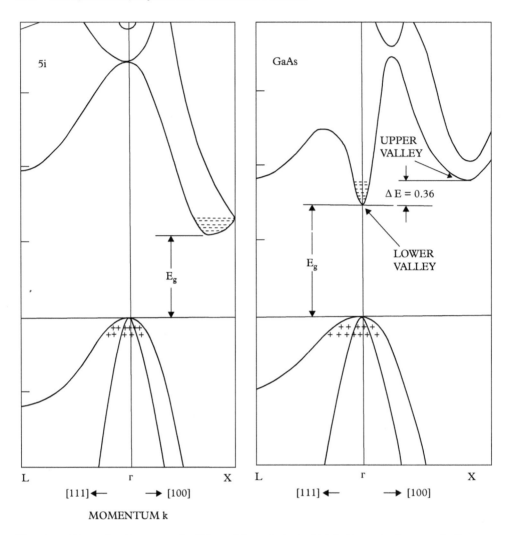

Fig. 5.16 *Energy band structures for Silicon (left) and GaAs (right). Energy is shown vertically, measured from the top of the valence band, and k horizontally. The horizontal line marks the top of the filled "valence" bands; in pure samples the upper bands are empty except for thermal excitations (indicated by +
and – symbols, for holes and electrons, respectively). The zero of momentum is indicated as "Γ", and separate sketches are given for E vs k in (111) left and (100) right directions. GaAs (right panel) has a "direct bandgap" because the minimum in the conduction band and the maximum in the valence band occur at the same value of wavevector k. This makes it easier for electrons and holes to recombine to produce light*

minimum is higher in the latter case, leading to a higher mobility for electrons in GaAs than in Si. A further aspect is that the conduction band minimum in Si is shifted from k = 0, that has important effects especially on the absorption of photons by Si, and similar semiconductors having an "indirect" bandgap. Parameters describing several important semiconductors are collected in Table 5.2.

Table 5.2 *Energy gaps and other electronic parameters of important semiconductors.*

Semi-conductor	Band Gap (eV)		Mobility at 300°K (cm²/volt sec)		Effective Mass m^*/m_o		Permittivity ε
	300°K	0°K	Electrons	Holes	Electrons	Holes	
C	5.47	5.51	1800	1600	0.2	0.25	5.5
Ge	0.66	0.75	3900	1900	$m_l^* = 1.6$ $m_t^* = 0.082$ $m_t^* = 0.19$	$m_{lh}^* = 0.044$ $m_{hh}^* = 0.28$	16
Si	1.12	1.16	1500	600	$m_l^* = 0.92$ $m_{DOS}^* = 0.32$	$m_{lh}^* = 0.16$ $m_{hh}^* = 0.5$	11.8
Grey Tin		~ 0.08					
AlSb	1.63	1.75	200	420	0.3	0.4	11
GaN	3.4				0.2	0.8	8.9
GaSb	0.67	0.80	4000	1400	0.047	0.5	15
GaAs	1.43	1.52	8500	400	0.068	0.5	10.9
GaP	2.24	2.40	110	75	0.5	0.5	10
InSb	0.16	0.26	78000	750	0.013	0.6	17
InAs	0.33	0.46	33000	460	0.02	0.41	14.5
InP	1.29	1.34	4600	150	0.07	0.4	14

The energy gaps of semiconductors range from about 0.2 eV to about 3.5 eV, as indicated in Table 5.2. Carbon in diamond form, listed at 5.5 eV, is normally considered an insulator. For devices that operate at room temperature a gap of at least 1eV is needed to keep the number of thermally excited carriers sufficiently low. GaN, with bandgap 3.4 eV, from a transport point of view is an insulator but from the point of view of making an injection laser is more reasonably regarded as a semiconductor. For Si with ellipsoidal electron energy surfaces, the density of states mass is $m^*_{DOS} = (m_l^* \, m_t^* \, m_t^*)^{1/3} = 0.322$ (Greene, 1989).

A final topic relates to the origin of resistivity. Conceptually, the electron states that we have been dealing with are perfectly conducting, in the sense that an electron in such a state maintains a velocity $v = \hbar k/m$. A perfectly periodic potential gives a perfect conductor. Indeed, very pure samples of GaAs, especially epitaxial films, measured at low temperatures, give mobility values of millions, in units of cm²/volt-sec, and mean free paths many thousands of atomic spacings. The meaning of τ then in the expression for the mobility is the lifetime of an electron in a particular k state. The cause of limited state lifetime τ in pure metals and semiconductors at room temperature is loss of perfect periodicity as a consequence of thermal vibrations of the atoms on their lattice positions. Calculations of this effect in metals, for example, lead to the observed linear dependence of the resistivity ρ on the absolute temperature T.

5.6.2.1   *Hydrogenic donors and excitons in semiconductors, direct and indirect bandgaps*

Pure semiconductors have filled valence bands and empty conduction bands and thus have only small thermally activated electrical conductivity, depending on the size of the energy gap. As mentioned previously, larger electrical conductivity is accomplished by "doping"; substitution for the 4-valent Si or Ge atoms either acceptor atoms of valence 3 or donor atoms of valence 5.

In the case of 5-valent *donor* atoms like P, As, or Sb, 4 electrons are incorporated into to tetrahedral bonding and the extra electron, that cannot be accommodated in the already filled valence band, must occupy a state at the bottom of the next empty band, that can be close to the donor ion, in terms of its position. This free electron is attracted to the donor impurity site by the Coulomb force, and the same physics that was described for the Bohr model should apply. However, in the semiconductor medium, the Coulomb force is reduced by the permittivity, or relative dielectric constant, ε, sometimes called κ. Referring to Table 5.2, values of ε are large, 11.8 and 16, respectively, for Si and Ge.

A second important consideration for the motion of the electron around the donor ion is the effective mass that it acquires because of the band curvature in the semiconductor conduction band. These are again large corrections, $m^\star/m$ for electrons is about 0.2 for Si and about 0.1 for Ge. So the Bohr model can usefully be scaled by the change in dielectric constant and also by the change of electron mass to $m^\star$.

The energy and Bohr radius are, from Eqs. (5.1–5.3), $E_n = -k_C Ze^2/2r_n$, and $r_n = n^2 a_0/Z$, where $a_0 = \hbar^2/mk_C e^2 = 0.053$ nm. Consider the radius first, and notice that its equation contains both k_C, the Coulomb constant (representing the Coulomb energy, proportional to $1/\varepsilon$) and the mass. So, the scaled Bohr radius will be:

$$a_0^\star = a_0(\varepsilon m/m^\star).\qquad(5.67)$$

Similarly considering the energy, $E_n = -k_C Ze^2/2r_n = -E_0 Z^2/n^2$, n = 1, 2, ..., where, $E_0 = m k_C {}^2 e^4/2\,\hbar^2$, it is evident that E scales as $m^\star/(m\,\varepsilon^2)$ in the dielectric medium:

$$E_n^\star = E_n[m^\star/(m\,\varepsilon^2)].\qquad(5.68)$$

For donors in Si, we find $a^\star = 59\ a_0 = 3.13$ nm, and $E_0^\star = 0.0014\ E_0 = 0.0195$ eV. The large scaled Bohr radius is an indication that the continuum approximation is reasonable, and the small binding energy means that most of the electrons coming from donors in Si at room temperature escape the impurity site and move freely in the conduction band. An entirely analogous situation occurs with the holes circling acceptor sites.

An exciton is a bound state of an electron and hole created by light absorption. The important point in calculating the behavior of the exciton is to consider the reduced mass of the electron and hole as they circle about the center of mass. The exciton is an analogue of positronium, a hydrogen-like atom formed by a positron and an electron. Exciton diffusion plays a role in energy transport in solar cells made of organic compounds.

5.6.3 Carrier concentrations in semiconductors

The objectives of this section are, first, to explain a standard method for finding N_e and N_h, the densities of conduction electrons and holes, respectively, in a semiconductor at temperature T. Secondly, an important special case occurs when the numbers of donors or acceptors becomes large, and the semiconductor becomes "metallic".

At high doping, these systems behave like metals, with the Fermi energy actually lying above the conduction band edge in the N^+ case, or below the valence band edge, in the P^+ case. These separate rules apply to the so-called N^+ regions used in making electrical contacts. In these "metallic" or "degenerate" cases the number of mobile carriers remains large even at extremely low temperature, as in a metal. (In the usual semiconductor case, the carrier density goes to zero at low temperature, making the material effectively insulating.)

The essential data needed to find N_e and N_h, are:

a) The Fermi energy E_F, the energy at which the available states have 0.5 occupation probability, i.e., $f_{FD}(E) = 0.5$ (see Eq. (5.20));

b) the size of the bandgap energy E_g;

c) the temperature T; and

d) the concentrations of donor and acceptor impurities, N_D and N_A, respectively.

Smaller effects on the carrier densities come from the values of the effective masses, see Eq. (5.66) for electrons and holes. For accurate numerical work, note that for silicon, with indirect band gap, the electron energy surfaces are ellipsoidal and the correct density of states mass for each valley is $m^\star_{DOS} = (m_l{}^\star m_t{}^\star m_t{}^\star)^{1/3} = 0.32$. (Greene, 1989). For Si the six equivalent conduction band minima mean that to calculate the electron density N_e the N_c should be multiplied by six, or alternatively the six can be absorbed into a new effective mass of the form $6^{2/3} (m_l{}^\star m_t{}^\star m_t{}^\star)^{1/3}$.

In addition, it is important to recognize that the semiconductor as a whole must be *electrically neutral*, because each constituent atom, including the added donor and acceptor atoms, is an electrically neutral system.

To find the expected number of electrons in the conduction band, requires integrating the density of states, multiplied by the occupation probability, over the energy range in the band. In a useful approximation, valid when the Fermi energy is several multiples of $k_B T$ below the conduction band edge, a standard result is derived, which simplifies the task of finding carrier densities N_e and N_h.

In formulas for carrier concentrations, the energy E is always measured from the top of the valence band. The standard formulas assume that the energies of electrons in the conduction band (and of holes in the valence band) vary with wavevector k as $E = \hbar^2 k^2/2m^\star$, where $\hbar = h/2\pi$, with h Planck's Constant, and $m^\star$ the effective mass, as given by (5.66).

In the formula for the energy E, m* must be given in units of kg, while the common shorthand notation, as in Table 5.2, is to quote m* as a dimensionless number, in which case the reader must multiply that number by the electron mass m_e, in kg, before entering it into a calculation. (The same applies to the mass of a hole m_h*, recalling that the motion of a hole is really the motion of an adjacent electron falling into the vacant bond position). The bandgap, E_g, is measured from the top of the valence band $E_V = 0$, so that $E_C = E_g$.

The density of electron states (per unit energy per unit volume) g(E) in a bulk semiconductor is:

$$g(E) = C_e\,(E - E_C)^{1/2},\ C_e = 4\pi(2m_e^\star)^{3/2}/h^3. \tag{5.69}$$

This formula (and the similar one for the density of hole states) is closely related to (5.17), that was derived for a three-dimensional metal. Note also that the effective density of states mass $m^\star_{DOS} = (m_l^\star m_t^\star m_t^\star)^{1/3}$ is appropriate for a single conduction band in silicon.

Similarly, the density of hole states in a bulk semiconductor is given by:

$$g(E) = C_h\,(-E)^{1/2},\ C_h = 4\pi(2m_h^\star)^{3/2}/h^3. \tag{5.70}$$

It is evident that this formula is valid for E < 0, corresponding to states lying below the top of the valence band. The probability of occupation of an electron state is $f_{FD}(E)$ (5.20) and the probability that the state is occupied by a hole is $1 - f_{FD}(E)$. The shallow donor electron has a binding energy E_D. Thus, an electron at the bottom of the conduction band has energy E_g and an electron occupying a shallow donor site has energy $E = E_g - E_D$. The density of shallow donor atoms is designated N_D.

If the Fermi energy E_F is located toward the middle of the energy gap, or at least a few k_BT below the conduction band edge, the exponential term in the denominator of (5.20) exceeds unity, and the occupation probability is adequately given by $f_{FD}(E) \approx \exp[-(E - E_F)/k_BT]$. Thus, for the total number of electrons we can calculate,

$$N_e = \int_{Ec} C_e\,(E - E_C)^{1/2} \exp\left[-(E - E_F)/k_BT\right]dE$$
$$= C_e \exp\left[-(E_g - E_F)/k_BT\right]\int_0^\infty x^{1/2}e^{-x}dx. \tag{5.71}$$

To get the second form of the integral change variables to E' = E − E_C, and then x = E'/k_BT, recalling that $E_C = E_g$. Move the upper limit on × to infinity, that is safe since the exponential factor falls to zero rapidly at large x, and then the value of the integral is $\pi^{1/2}/2$. Making use of the definition $C_e = 4\pi\,(2m_e^\star)^{3/2}/h^3$, we find:

$$N_C = 2(2\pi m_e^\star\,k_BT/h^2)^{3/2}. \tag{5.72}$$

This is the effective number of states in the (single) band at temperature T, per unit volume.

The number of electrons per unit volume in equilibrium at temperature T is then given as:

$$N_e = N_C \exp\{-(E_G - E_F)/k_B T\}. \tag{5.73}$$

(It is still necessary to know E_F in order to get a numerical answer: always in such problems the central need is to find the value of E_F. In the case of a pure "intrinsic" material, where the number of electrons equals the number of holes, the Fermi level to a first approximation lies at the center of the gap, halfway between the filled and the empty states. As a first step in a pure sample one often will assume $E_F = E_g/2$.)

In the same way, the effective valence band density of states for holes is:

$$N_V = 2\left(2\pi m_h^\star k_B T/h^2\right)^{3/2}. \tag{5.74}$$

The number of holes per unit volume in equilibrium at temperature T is given as:

$$N_h = N_V \exp(-E_F/k_B T). \tag{5.75}$$

In a pure semiconductor the number of holes must equal the number of electrons. By forcing the equality,

$N_e = N_C \exp\{-(E_G - E_F)/k_B T\} = N_h = N_V \exp\{-E_F)/k_B T\}$, one can solve for E_F:

$$E_F = E_g/2 + \tfrac{3}{4} kT \ln\left(m_h^\star/m_e^\star\right) \quad \text{(for the pure sample)}. \tag{5.76}$$

Finally, consider the algebraic form of the product,

$$N_e N_h = N_c N_v \exp(-E_G/k_B T). \tag{5.77}$$

This product, that is called N_i^2 (or n_i^2) is seen to be *independent of E_F*, meaning that it is unchanged by doping (impurity levels).

As an implication: if we know that there are a large number of donors N_D of small binding energy $E_D \ll k_B T$, so that it is reasonable to assume that all electrons have left the donor sites to become free electrons (i.e., $N_e = N_D$) then we find $N_h = N_i^2/N_D$, that can be small for large N_D.

In this case, since there are electrons in large numbers near E_g, (at the donor level or the conduction band edge, measured from the top of the valence band) the Fermi energy (where occupation probability is ½) must be near the conduction band edge.

This implies, in the heavily electron-doped case, that the probability of occupation of a hole state is extremely small, $\sim \exp(-E_G/k_B T)$. These holes (in the N-doped case) are called "minority carriers, of concentration p_n", and we see that they are small in number and have the strong temperature dependence $\exp(-E_G/k_B T)$. This is the reason for the strong temperature dependence in the reverse current of the PN junction. This can be used to make a thermometer, called a "thermistor".

5.6.3.1 *The degenerate metallic semiconductor*

Now consider an extreme case, that will lead to an understanding of a heavily doped, metallic region of semiconductor, that is used to contact solar cells, for example.

Consider pure InAs, for which Table 5.2 indicates a bandgap of 0.33 eV, electron mass, $m_e^*/m_e = 0.02$, hole mass $m_h^*/m_e = 0.41$ and permittivity $\kappa = 14.5$.

First we estimate N_e at 300 K, assuming for simplicity that the Fermi energy is at the center of the energy gap. We find $N_e = 2(2\pi m_e^* k_B T/h^2)^{3/2} \exp\{-(E_G - E_F)/k_B T\} = 7.165 \times 10^{22} \times 1.7 \times 10^{-3} = 1.22 \times 10^{20}$ m^{-3}, and the same number applies to holes for the pure sample if we neglect the shift of the Fermi energy with effective mass difference.

Next, consider *heavily N-doped InAs, with 10^{18} /cc = 1.0×10^{24} m^{-3} shallow donor impurities*, at 300 K. This is a large doping, greatly exceeding the thermal intrinsic number of electrons, 1.22×10^{20} m^{-3}. What fraction of these electrons will remain on the donor sites at 300 K?

We must calculate the value of the donor binding energy, E_D, taking effective mass $m^* = 0.02$ and permittivity, $\kappa = \varepsilon = 14.5$, and scale the energy from 13.6 eV for the hydrogen atom, and find 0.00129 eV. This is small compared to $k_B T$ at 300 K, that is 0.0259 eV. Thus, it is reasonable to assume that all of these electrons are in the conduction band! The same conclusion, of delocalized donor electrons, would be reached in use of the Mott transition criterion, as expressed in Eq. (5.32).

To estimate what range of energies in the conduction band will be filled by these electrons, we use the formula for the Fermi energy for a metal, (5.17), again using the effective mass $m^*/m = 0.02$, and taking N/V as 1.0×10^{24} m^{-3}. The result is $E_F = 0.181$ eV (measured up from the conduction band edge), or 0.511eV from the valence band edge. Formally this would be:

$$E_{FN} \approx E_G + (h^2/8m)(3N_D/\pi)^{2/3} \tag{5.78}$$

So, heavily doped InAs is really a metal, and we would expect that the number of free electrons will not change appreciably if this sample is cooled to a low temperature.

Finally, to complete discussion of heavily doped InAs, making use of the "mass action law" $N_e N_h = N_i^2$ at 300 K, we find that $N_h = (1.22 \times 10^{20})^2/10^{24} = 1.49 \times 10^{16}$/m^3 = 1.49×10^{10}/cc. This "minority concentration" of holes is much smaller than the electron concentration, 10^{18}/cc. The minority concentration will change rapidly with varying temperature, in this case according to $\exp(-0.33\text{eV}/k_B T)$.

5.7   The PN junction, diode I–V characteristic, photovoltaic cell

To begin thinking about a PN junction in Si, we represent the semiconductor by two horizontal lines, the conduction band edge at $E = E_c$ and the valence band edge $E_V = 0$, at a distance $E_G = 1.1$ eV. The Fermi energy is indicated by a dashed line, and for a pure Si sample it will be near the middle of the gap at 0.55 eV, as sketched in Fig. 5.17a.

To introduce a PN junction, imagine at the center of our picture of the two horizontal lines we make a transition to P-type material on the left and to N-type material on the right.

In the N-type region, using Eq. (5.73), we can see that the Fermi level is given by,

$$E_{FN} = E_C - k_B T \ln(N_C/N_e), \tag{5.79}$$

while on the P-type side we have,

$$E_{FP} = k_B T \ln(N_V/N_h) \tag{5.80}$$

In such an imagined *abrupt junction*, electrons will quickly flow from donor levels on the right to negatively ionize acceptor levels on the left until the bands have shifted to align the Fermi level as constant across the structure. The energy level shift will be eV_B, called the *band-bending*, and V_B the *built-in potential*. The shift will be:

$$E_C - k_B T [\ln(N_C/N_e) + \ln(N_V/N_h)] = eV_B. \tag{5.81}$$

This will have the effect of shifting the bands on the left upward and shifting the bands on the right down. The voltage V_B called the *built-in potential* is typically near 0.6 volts for Si, depending on doping levels. The width W over which the shift occurs is called the *depletion region width*:

$$W = [2\varepsilon\varepsilon_0 (V_B - V)(N_D + N_A)/e(N_D N_A)]^{1/2}, \tag{5.82}$$

where we have introduced a possible forward external bias V, seen to reduce the band shift. Note that permittivity is expressed as κ and also as ε. In practice, one side, let us assume the N-side, may be highly doped, even to produce a metallic situation as discussed before (5.78). In this case of a "one sided junction" the width formula simplifies (for N_D large) to,

$$W = [2\varepsilon\varepsilon_0 (V_B - V)/eN_A]^{1/2}. \tag{5.82a}$$

In the depletion region, on the N-side, for example, the charge density is $\rho = +N_D e$. Using the Poisson Equation for the electric potential V, we find the condition,

$$\nabla^2 (V) = \rho/\varepsilon\varepsilon_0 = +N_D e/\varepsilon\varepsilon_0. \tag{5.83}$$

For one dimension, this becomes $\partial^2 V/\partial x^2 = +N_D e/\varepsilon\varepsilon_0$, so that the electric field $E = \partial V/\partial x = +N_D e\, x/\varepsilon\varepsilon_0 + c$, with c an arbitrary constant. So, the electric field increases linearly from zero, at the outer edges of the depletion layer, reaches a peak at the junction center, where $E(x)$ must be continuous from right to left. The potential $V(x)$, sketched in Fig. 5.17c, is therefore quadratic in × in the depletion region, with an inflection point at the center of the junction.

We now understand the outline of the PN junction in Panel (c) of Fig. 5.17. This is the most important semiconductor device! How does it work?

It is important to understand the role of *minority carriers*, n_p electrons in P-region and p_n holes in N-region. (As we will see, these concentrations set the reverse current

(a)

(b)

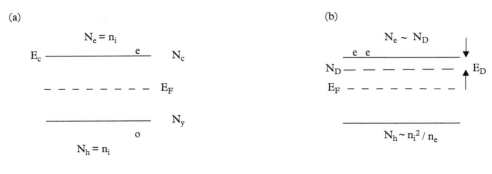

(c)

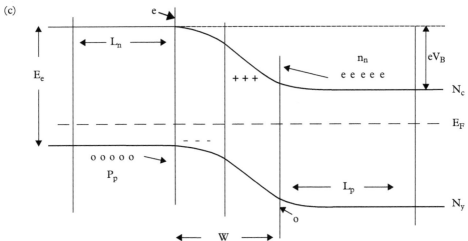

Fig. 5.17 *Sketches (a) intrinsic (pure) Si (b) N-Type doped Si and (c) PN junction. Electron energy vertically as in Fig. 5.15, but wave-vector $k = p/\hbar$, is now assumed to be fixed at the k location of the most important carriers, (usually $k = 0$, but displaced for electrons in Si), and x-axis denotes location in sample. It is seen (b) that Fermi energy moves from near center of gap in intrinsic material up toward conduction band edge E_c, with addition of donor impurities at concentration N_D. The concentration N_e of free electrons is given by Eq. 5.73, but the zero-order estimate is $N_e \approx N_D$. (The band density of states N_C is given by Eq. 5.72, similarly for N_V.) Panel (c) shows junction formation with energy shift eV_B, with minority carriers: n_p electrons in P-region and p_n holes in N-region, which set the reverse current density. The minority electrons in diffusion length L_n (c, upper left) fall down by eV_B (across the junction) in minority carrier lifetime τ, creating reverse current density $\mathcal{J}_{rev} \sim 2e\ (L_n n_p/\tau)$ (assuming holes behave similarly to electrons). Positive applied bias voltage V (not shown) raises bands on the right, as indicated by increasing probability P for electron on the right to increase kinetic energy toward barrier height and flow to the left: $P = exp[-e(V_B - V)/k_B T]$. Forward current in this device is electrons from right to left, forward bias reduces the band bending*

density). From Eq. (5.77) we see that these concentrations are small relative to the majority concentrations, that are essentially set by the levels of dopants, N_D on the N-side and N_A on the P-side. The numbers of minority carriers are set by the equilibrium Eqs. (5.72–5.77), but the minority carrier concentration can also be understood as a balance between a generation rate (a thermal bond-breaking rate) and a *minority carrier lifetime* τ. An electron in the upper left of Fig. 5.17 c lives only a short life-time τ before it encounters one of the many holes and recombines (annihilates), either by emitting a light photon or by giving off other forms of energy. The minority electrons within a diffusion length L_n (Fig. 5.17 c, upper left) can reach the junction electron field and cross to the other side in the minority carrier lifetime τ. Hence the sum of the two diffusion current densities, comprising the reverse current density, is given as:

$$J_{rev} = e\left(L_n n_p / \tau_n\right) + e\left(L_p p_n / \tau_p\right). \tag{5.84}$$

Since the diffusion lengths generally are larger than W, the minority carriers are in a field free region and diffuse randomly. Eq. (5.64) can be expressed in a slightly different form by making use of the minority carrier diffusion constant D, units m²/s. D is closely related to the minority carrier mobility μ, through the relation $\mu = eD/kT$, with units expressed as [m²/Vs]. The diffusion length is:

$L = (D\tau)^{1/2}$, the distance traversed in a random walk, so the reverse current can be expressed also as:

$$J_{rev} = e n_p \left(D_n / \tau_n\right)^{1/2} + e\, p_n \left(D_p / \tau_p\right)^{1/2} \tag{5.85}$$

and a further variation is possible using $D = k_B T \mu / e$. By use of the product rule (5.77), and setting $p_p \approx N_A$, etc., the approximate form (5.86) can be reached.

$$J_{rev} = J_o = e.n_i^2 \left[\frac{D_n}{N_A L_n} + \frac{D_h}{N_D L_h}\right]. \tag{5.86}$$

This makes explicit the strong temperature dependence of the reverse current, since we know (5.77) that: $n_i^2 = N_c N_v \exp(-E_G / k_B T)$ with strong exponential variation with T.

Since the reverse current density J_{rev} is an important parameter in the operation of solar cells, it is clear that their operating temperature is an important factor. The minority carrier diffusion length is important in solar cells. The absorbed light photons generate electron-hole pairs, and it is desirable to have the resulting minority carrier reach the junction field and transfer across to drive current in the external circuit, before recombining. In solar cell design, optimization is needed between the distance to fully absorb light (the absorption length) and the minority carrier diffusion length, to allow all of the photons to be converted, and then all of the resulting electron-hole pairs to diffuse to reach the junction field and cause external current flow.

The current density under applied bias V, from these considerations, can be expressed as:

$$J = J_{rev}\left[\exp\left(eV/k_B T\right) - 1\right].\qquad\qquad(5.87)$$

Positive applied bias voltage V (not shown in Fig. 5.17c) raises bands on the right, as indicated by probability P for electron on the right to attain kinetic energy equal to barrier height and flow to the left: $P = \exp[-e(V_B-V)/k_B T]$. Forward current in this device is electrons from right to left, forward bias reduces the band bending. Positive or forward bias for this device is defined as the bias direction that reduces the band shift, and shifts the electron conduction band (on the N-side, the right side) up to allow more transfer of electrons into the p-type region, from right to left, the energy shift being $e(V_b-V)$. So, forward bias, positive V, reduces the shift of the bands.

At strong forward bias, with applied potential $V = V_B$ the electrons will flow from right to left without any barrier at all from the n-type side to the p-type side. So, the forward I–V characteristics of the PN junction is governed by the factor $\exp[e(V-V_b)/k_B T]$. It is an exponentially rising characteristic vs. V, but extending only to $V \approx V_B$.

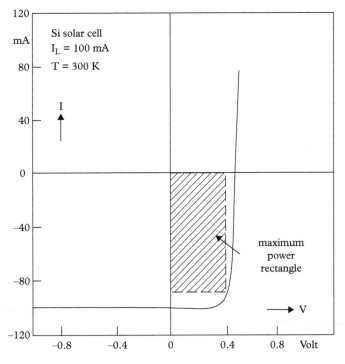

Fig. 5.18 *Current-voltage characteristics of a Silicon solar cell under illumination (M. B. Prince, J. Appl. Phys. **26**, 534 (1955))*

In reverse bias, the only current that flows is the described diffusion current that is independent of applied voltage. This current comes as thermally present electrons and thermally present holes diffuse and reach the junction electric field at the center of the junction.

If light falls on the PN junction, as sketched in Fig. 5.17c, the result will be generation of electron-hole pairs. This amounts to generating large numbers of minority carriers, and these photo-generated minority carriers will greatly enhance the reverse current. The resulting current density from absorbed light is described by $J = -J_L$ that would flow if the junction were short-circuited, with load resistance zero.

If the junction is open circuit, so the current is zero, then, $J = J_{rev} [\exp(eV/k_B T) - 1] - J_L = 0$, that implies a maximum open circuit voltage,

$$V_{oc} = (k_B T/e) \ln (J_L/J_{rev} + 1). \tag{5.88}$$

If the device is terminated in a load resistor the power density is:

$$P = JV = J_{rev} V[\exp(eV/k_B T) - 1] - J_L V \tag{5.89}$$

where the actual operating voltage V will depend on the choice of the load resistance, see Fig. 5.18.

5.8   Metals and plasmas

The fundamental properties of the plasma are independent motion of charges of opposite sign and charge neutrality, so semiconductors are plasmas. In a plasma each charge q is locally surrounded by its electric potential $V = k_c q/r$, but this potential will be felt by neighboring charges, that are mobile and will slightly relocate. Thus, a *shielding property* is inherent in a plasma. If a disturbance of potential, say V, appears, it will alter the local density of electrons to a perturbed value $n' = n \exp(eV/k_B T) \approx n (1 + eV/k_B T)$, assuming eV is small compared to $k_B T$. Recalling Eq. (5.83), where we can identify the relevant charge density as $e(n' - n)$ and find, since $n' - n \approx n \, eV/k_B T$,

$$\nabla^2 (V) = n \, e^2 V/k_B T \varepsilon_0 = V/\lambda_D^2 \tag{5.90}$$

where,

$$\lambda_D = (k_B T \varepsilon_0/ne^2)^{1/2} \tag{5.90a}$$

is the Debye length. This is important in plasmas and also in solution chemistry where ions are mobile and can shield surface potentials, for example.

To solve (5.83) in spherical geometry, as surrounding a point charge q, one finds:

$$V(r) = (k_c q/r) \exp(-r/\lambda_D) \tag{5.90b}$$

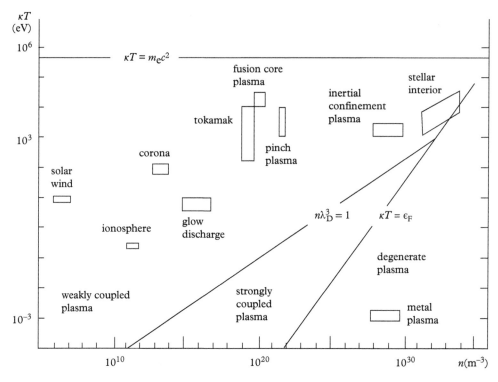

Fig. 5.19 *Survey of plasma domains in Temperature-density representation. Note that temperatures are represented in eV units. In this Figure the "metal plasma" region might be extended to higher temperatures, since some metals, e.g., W, Ta and Hf, are definitely Fermi gases up to 1000 K = 0.08 eV (K. Miyamoto "Plasma Physics and Controlled Nuclear Fusion", Springer, 2004) Fig. 1.1*

so that a potential disturbance is shielded away in a distance λ_D. A large value of the product $n\,\lambda_D^3$, denoted the "plasma parameter", indicates that the average Coulomb energy is small compared to kinetic energy, this is seen to apply in left portions of Fig. (5.19), to the Sun's interior and to tokamak plasma. If $n^{-1/3}$ is r, then in this regime $\lambda_D \gg r$, so that screening does not occur. The "strongly coupled plasma" region is $n\,\lambda_D^3$ small, screening length short compared to mean interparticle distance, that also means the Coulomb energy is large compared to kinetic energy, and this applies to metals.

6

Fusion Energy Technology on Earth

If it can be achieved, deuteron fusion would qualify as a renewable energy process given the large supply of deuterons in the sea. In this Chapter we build upon what we learned about fusion in the Sun with the hope of applying the same process on Earth.

A summary of fusion reactions of technological interest is given in Table 6.1. The reactions in Table 6.1 have much higher cross sections than the p-p reaction on the Sun as discussed in Chapter 4. The reactions involving deuterons D and tritons T (also represented as $^3H^+$) that have been observed in laboratories and may power a tokamak reactor to produce controlled fusion energy on Earth, are the main subject here.

In Table 6.1, Q is the kinetic energy release in the reaction, that is shared among the products depending on their masses. The cross section in units of cm^2 is that at the most favorable energy for the reaction to occur, that is given in the last column, in keV. (A typical variation of cross section with energy is sketched in Fig. 6.4.)

Fusion energy can be considered renewable if based on deuterium D, since the ocean contains a huge dilute reservoir of deuterons in forms of heavy water, molecules HDO and DDO. Further, the ocean contains large amounts of lithium, that can also be used as a fusion reactant (MacKay, 2009, p. 172).

The reactions shown in Table 6.1 have cross sections or probabilities about 25 orders of magnitude larger than the p–p reaction described in Chapter 4. These reactions take deuterons as starting material, avoiding the very slow p–p reaction to form deuterons. An interesting question posed by the set of reactions shown is—why should the DT reaction have a higher cross section than the DD reactions? It appears that this difference is related to an aspect of the quantum mechanical process of tunneling.

The primary thrust toward producing energy from fusion is the ITER tokamak reactor, under construction, with a toroidal DT plasma. International Thermonuclear Experimental Reactor (ITER)[1] is an 840 m^3 torus reactor being built in Cadamarche, France, by an international consortium. In a sense this is an attempt to scale the successful fusion conditions on the Sun to parameters, notably the lower pressure, that can be attained on Earth. The pressure has to be much lower, so scaling leads to a lower density of reactants, and the temperature can be raised to compensate. The temperature is easier to control in a laboratory plasma and can exceed the temperature in the Sun's core.

---

[1] http://www.iter.org/

Physics and Technology of Sustainable Energy. E. L. Wolf, Oxford University Press (2018).
© E. L. Wolf. DOI: 10.1093/oso/9780198769804.001.0001

Table 6.1 *Basic fusion reactions for technology.*[2]

Reaction	Q (MeV)	σ_{max} (barn)[a]	Energy Peak (keV)
D+T → α + n	17.6	5	64
D+D → T + p	4.0	0.10	1250
D+D → ^{3}He + n	3.3	0.11	1750
D+D → α + γ	23.9		
D+^{3}He → α + p	18.4	0.9	250
T+T → α + 2n	11.3	0.16	1000
P+^{6}Li → α + ^{3}He	4.0	0.22	1500
P+^{7}Li → 2α	17.4		
P+ ^{11}B → 3α	8.7	1.2	550

[a] $1 \text{ barn} = 10^{-24} \text{cm}^2 = 10^{-28} \text{m}^2 = (10 \text{ fm})^2$

The ITER approach is based upon magnetic confinement, to keep the 100 million K plasma from being cooled by, and damaging, the containing walls. Before we discuss the tokamak approach, we describe two small-scale demonstrations of deuterium fusion.

The first is based on field ionization of deuterium gas D_2 in a compact reaction chamber (Naranjo et al, 2005).

The resulting deuterons were accelerated to 80 to 115 keV and produced (small numbers of) fusion neutrons (the second reaction in Table 6.1) as they collided with the target, a layer of ErD_2, a solid with a large density of deuterons. This observation of fusion might suggest a reactor in which a solid surface of lithium or boron is bombarded with protons from a similar field ionization source. Such a direct solid state fusion reactor was discussed by (Ruggiero, 1992), who concluded that useful amounts of power would not be available. The problem is that the probability of fusion is low, because the D ions rapidly lose energy as they hit the Er ions, with their large numbers of orbiting electrons, in the target. Ionization of these electrons quickly slow the D ions to the point where fusion is impossible. In contrast, in the tokamak DT plasma the ions have a very long mean free path.

The second demonstration of fusion on a small laboratory scale (Strasser et al, 1996) is based on creating "muonic" deuterium molecular D_2^+ ions, which can decay by D-D fusion and muon release, as described by (Jackson, 1957), according to the 2nd and 3rd processes in Table 6.1. (A mu-meson, muon, or μ, is like an electron but 207 times

[2] Atzeni, S. and Meyer-Ter-Vehn, J. (2004). *The Physics of Inertial Fusion.* (Oxford: Oxford University Press) Table 1.2.

more massive.) It is unlikely that either of these methods can evolve toward useful release of energy. At the moment, the two demonstrate only that controlled fusion, as monitored by release of neutrons, can be achieved in modest laboratory conditions.

6.1   Deuterium fusion by field ionization

In Chapter 1 we estimated that the potential energy of two protons in contact, $U = k_c e^2/r$, is about 0.6 MeV, but that protons having thermal energies in the lower range of tens of keV in the Sun essentially accounted for its output of energy. As was explained in more detail at the end of Chap. 4, the particles are able to tunnel through the Coulomb barrier, as explained by Gamow, from distances much larger than the contact spacing. The value for two deuterons is similar but would formally be given from the nuclear radius formula $r = 1.07 A^{1/3}$ fm with $A = 2$ and 1 fm $= 10^{-15}$ m. So, the contact spacing is $2r_D = 2.14$ f $2^{1/3} = 2.7$ fm, so that $U = k_c e^2/2r_D = 533$ keV.

Naranjo et al (2005) achieved DD fusion in a small chamber filled with deuterium D_2 gas, using a 100 nm-radius tungsten W tip biased to a positive voltage 80–110 kV. The electric field near the tip is large enough to break up the deuterium molecules and to ionize the resulting deuterium atoms, producing D^+ ions and electrons. The D^+ ions are accelerated into an ErD_2 target and fusion neutrons were observed.

The apparatus is shown in Fig. 6.1. A $LiTaO_3$ "pyroelectric" crystal (center-left, in Fig. 6.1) produces a strong electric field between the crystal and the ground plane. A pyroelectric crystal, as the temperature is varied, gives a large surface charge density, σ (in C/m²) as electrons are pushed outward slightly from the crystal, and this means a large surface electric field $E = \sigma/2\varepsilon_0$. In this apparatus the voltages appear with change

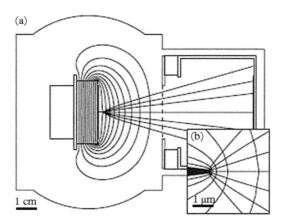

Fig. 6.1 *Sketch of compact field-ionizing neutron generator operating in D_2 gas. The inset shows the 100 nm radius W tip which, when biased to 80 kV by the pyroelectric crystal, ionizes nearby deuterium gas to produce free deuterons. The free deuterons D^+ are accelerated into the ErD_2 target on the right, providing a high density of deuterons for the reaction (Naranjo et al, 2005) Fig. 1a,b*

of temperature of the pyroelectric crystal. Once D^+ ions (free deuterons) are formed, they are carried into the target at energy 80 keV. The pyroelectric crystal, similar in nature to a ferroelectric crystal, is a clever way to achieve a high voltage in a small laboratory apparatus.

To generate ions, much larger localized ionizing electric fields E are generated near a tip of radius 100 nm (see inset in Fig. 6.1). The electric field E, at the surface of a metal sphere of radius a at voltage V is V/a, which in this case gives $E(a) = 80 \times 10^3/100 \times 10^{-9} = 0.8 \times 10^{12}$ V/m = 0.8 keV/nm at radius a = 100 nm. The field falls off as $E = Va/r^2$ for r larger than a.

The authors indicate that E = 25 V/nm is sufficient for 100 percent field ionization, thus generating electrons and D^+ ions from D_2 gas. Thus, ionization rapidly occurs out a radius $R_{ion} = 100 \times [0.8 \times 10^3/25]^{1/2} = 0.565$ μ from the tip, and the area of the ionizing hemisphere facing the target is $2\pi R_{ion}^2$. The rate of ion formation is the product of this area times nv/4, that is the rate, units 1/m^2 sec, at which gas molecules cross unit area. Here n is the density of deuterium molecules, of thermal velocity v, with another factor of 2 because each molecule releases two deuterons. The ion current I_i is then:

$$I_i = 2\pi R_{ion}^2 (nv/4) 2e. \tag{6.1}$$

The ambient gas D_2 is dilute, such that the mean free path λ exceeds the dimension of the chamber, so that the D^+ ions fall down the potential gradient without collisions and impact the (deuterated) target at 80 keV energy or more. This energy is enough to drive the nuclear fusion reaction:

$$D^+ + D^+ \rightarrow {}^3He^{2+} (820 \text{ keV}) + n(2.45 \text{ MeV}), \text{ (where D is the Deuteron,}$$
the nucleus of deuterium, 2H).

In the compact device shown in Fig. 6.1, no external high voltages are needed. In this case, heating the LiTaO$_3$ crystal from 240 K to 265 K, using a 2 W heater, is stated to increase the surface charge density σ by 0.0037 Cm^{-2}. This should correspond to a surface electric field $E = \sigma/2\varepsilon_0 = 3.7 \times 10^{-3}/(2 \times 8.85 \times 10^{-12}) = 0.209$ V/nm. The authors state that in the device geometry (see Fig. 6.1) this gives a potential of 100 kV. The observations, summarized in Fig. 6.2, confirm D-D fusion, based on the observation of neutrons.

Referring to Fig. 6.2, we see in (a) that a linear increase of temperature with time from 240 K to 280 K leads to the pyroelectric crystal's potential rising from zero to about 80 kV at t = 230 sec. Panel (b) shows onset of x-rays that come as electrons released from the ionization events fall onto the positively charged copper plate encasing the tantalate pyroelectric crystal. The x-ray energies, observed up to about 100 keV can only come from a tip potential of the same order, that produces a local electric field sufficient to strip electrons from the deuterium gas. Panel (c) records the ionic current, presumably the sum of electron current into the copper and positive ion current into the right-hand electrode, adding to 4 nA maximum. This number can be checked by elementary considerations involving the radius and surface area around the tip leading

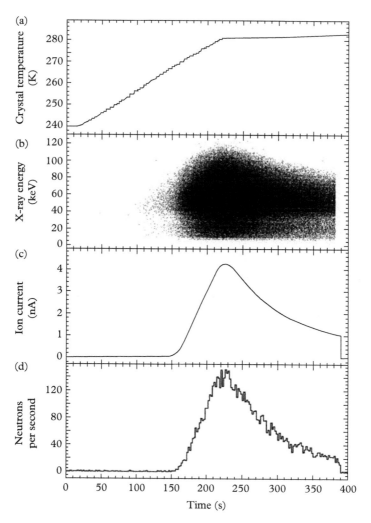

Fig. 6.2 *Documentation of one cycle of neutron generation in the compact device (Naranjo et al, 2005) Fig. 2*

to certain ionization, the density of the gas at the stated pressure, and the number of deuterium molecules in random gas diffusion that would cross that surface per unit time, indicated in Eq. 6.1. Finally, Panel (d) shows the measured number of neutrons per second. The satisfactory coincidence of the peaking of the several indicators at about 230 s. makes it clear that fusion has occurred.

Let us look at these results from the simplified theoretical model developed in Chapter 4, leading to Eq. 4.23. The tunneling transmission probability of the D through the Coulomb barrier (see Fig. 4.3) at 40 keV is now estimated as:

$$T = \exp(-2\gamma) \tag{4.23}$$

where, approximately,

$$\gamma = (2mE)^{1/2}/\hbar \left[(\pi/2)\, r_2 - 2(r_2 r_1)^{1/2} \right] \tag{4.26}$$

and r_2 is the classical turning point of the D. This is determined by the relation:

$$40\ \text{keV} = k_C\, e^2/r_2,$$

in the center of mass frame of the two deuterons, with accelerator voltage 80 kV, so that $r_2 = 36$ fm. We have already seen that $r_1 = 2.7$ fm. Using the reduced mass $m_R = 1.67 \times 10^{-27}$ kg, we find:

$$\gamma = \left[(2mE)^{1/2}/\hbar \right]\left[(\pi/2)\, r_2 - 2(r_2 r_1)^{1/2} \right] = 1.62.\ \text{Therefore}$$
$$T_{\text{Gamow}} = \exp(-2\gamma) = \exp(-3.239) = 0.0392.$$

Naranjo et al (2005), measured 130 neutrons/s at peak, when the accelerating voltage was actually 115 kV, rather than 80 kV. To adjust the tunneling probability T_{Gamow} value to 115 keV, note (text near Eq. 4.24) that the energy E dependence of γ is approximately $E^{-1/2}$. This gives tunneling probability $T_{\text{Gamow}} = \exp[-(E_G/E)^{1/2}]$.

For the D–D reaction in the compact chamber this would imply $3.239 = (E_G/E)^{1/2}$ at $E = 40$ keV, so that $E_G = 419.6$ keV. At (115/2) keV, then, $T_{\text{Gamow}} = 0.067$. (We neglect the small change in r_2.)

From our simplified approach in Chapter 4, following Eq. (4.28) the cross-section for fusion is, for $r_2 = 36$ fm, and $T_{\text{Gamow}} = 0.067$,

$$\sigma = \pi r_2^2 T_{\text{Gamow}} = 2.728 \times 10^{-28}\ \text{m}^2 = 2.728\ \text{barns}. \tag{6.2}$$

This cross-section value, for an energy near 100 keV, exceeds the maximum value quoted in Table 6.1, that is 0.11×10^{-28} m^2 at an energy of 1750 keV. We can attribute this discrepancy to a reaction probability $T = 0.04$, having to do with the details of the nuclear reaction, recalling that an analogous factor $T = 8.0 \times 10^{-24}$ was found for the p–p reaction on the Sun.

From a more general point of view, it may be useful to look at the expected energy dependence of the cross-section shown in Fig. 6.3, in the right-most curve.

Fig. 6.3 shows the dominant energy-dependent functions for nuclear reactions between charged particles in a thermalized plasma, as compared to a fixed energy beam as in the compact device of Fig. 6.1. While both the energy distribution function (Maxwell–Boltzmann) and the quantum mechanical tunneling function through the Coulomb barrier are small for the overlap region, the convolution of the two functions results in a peak (the Gamow Peak) near the energy E_o, giving a sufficient probability to allow a significant number of reactions to occur. The energy of the Gamow Peak is

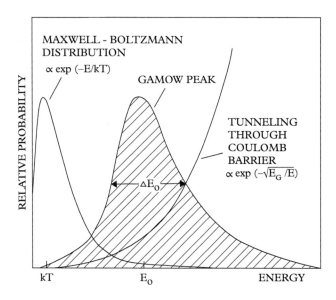

Fig. 6.3 *Right curve represents tunneling probability T_{Gamow} (Eq. 4.20) as a function of center of mass energy. The compact fusion device provides D at a fixed energy ~ 40 keV, rather than a thermal Maxwell–Boltzmann distribution of energies as occurs on the Sun or in a Tokamak reactor. The Gamow peak represents the optimal overlap of the energy distribution and the tunneling probability curves (Rolfs and Rodney, 1988) Fig. 4.6, p. 159*

generally larger than $k_B T$. In the compact device of Fig. 6.1 the energy is fixed by the accelerating potential, so the thermal distribution of energies in Fig. 6.4 are replaced by a single energy.

Returning to the results (Fig. 6.2) from the compact fusion device, we can find the experimental probability P_f per incoming D, of fusion. The result in the experiment, since the measured 4.4 nA corresponds to a deuteron flux of $4.4 \times 10^{-9}/(1.6 \times 10^{-19}) = 2.75 \times 10^{10}$ D/sec, is:

$$P_f = 130/(2.75 \times 10^{10}) = 4.72 \times 10^{-9} \tag{6.3}$$

at pyroelectric crystal accelerating voltage 115 kV.

We can estimate the density of deuterons in the ErD_2 target as $n_D = 6 \times 10^{28}$ m^{-3}. The mean free path for fusion then is $\Lambda = 1/n_D \sigma = 0.061$ m. However, the deuterons are rapidly slowed, and do not in fact penetrate more than a few atomic layers. The energy loss $dE/dx = -\alpha$ that comes from electron excitation as the energetic incoming ion proceeds into the target is on the order of 200 MeV/mm. We can estimate the distance of penetration as $x \sim 0.5 E_{in}/\alpha$. Taking $E_{in} = 115$ keV, we find $x = 2.8 \times 10^{-7}$. If fusion occurs certainly for 0.061 m, then in distance x the chance of fusion is:

$$P_f(x) = x/0.061 = 4.7 \times 10^{-6}. \tag{6.3a}$$

This value is larger than observed. The predicted efficiency from an energy release point of view is (3.5 MeV/115 keV) × 4.7 × 10^{-6} = 1.43 × 10^{-4}. These numbers are approximately applicable to the deuteron experiment, but predict a higher yield than was observed. From these results, confirmed by Naranjo et al (2005), there seems not much hope for making a practical fusion reactor from an energetic beam hitting a solid or liquid target, because of the rapid energy loss, as was earlier predicted by Ruggiero (1992).

6.1.1   Electric field ionization of deuterium

Ionizing an atom, as in the compact apparatus of Fig. 6.1, is done with an electric field, and the question is how large a field is required? The same question arises in a tokamak device where it is necessary to ionize the deuterium gas to form a plasma. So, the rate of "field ionization", governed by an electron tunneling process, is important.

The electronic properties of deuterium are the same as hydrogen H, that differs only by the additional (uncharged) neutron mass in the nucleus. Thus, the simplest estimate of the required electric field E to ionize deuterium might be the field of the proton at the Bohr radius in H, namely E = $k_C e/a_0^2$ = 514.5 V/nm.

This is an overestimate: a much smaller electric field, E ≈ 25 V/nm, quickly (~ 10^{-15} s) removes the electron by a process of tunneling.

In an electric field E, the potential energy of the electron has a term U(x) = –eEx, so that at a spacing $x^\star$ = E_0/eE, the electron will have same energy as in its ground state, and can tunnel out. Here E_0 is the electronic binding energy, 13.6 eV for deuterium, as well as for hydrogen. How quickly will this happen?

The electron can tunnel through the barrier that extends from x = a_0 to $x^\star$. In detail this is a difficult problem to solve, but a simplified treatment is possible. The earliest estimate of the lifetime of H against field ionization was given by Oppenheimer (1928). Oppenheimer's notable estimate, is that for H in a field of 1000 V/m the lifetime τ is $(10^{10})^{10}$ s.

We can use a simplified, one-dimensional, model to estimate the E-field ionization rate of H. We will find that for E = 25 V/nm, the field ionization lifetime of hydrogen and deuterium is on the order of 10^{-15} s, while, at E = 2.5 V/nm, τ is extremely long, estimated as 6 × 10^{33} s or 1.9 × 10^{26} y.

The ground state with E = 0 is spherically symmetric, depending only on r. With an electric field E_x = E, approximate the situation as depending only on x: electron potential energy:

$$U(x) = -k_C e^2/x - eEx, \qquad (6.4)$$

where k_C = 9 × 10^9 Nm^2/C^2, with E in V/m. Assume that the ground state energy is unchanged by the electric field, and remains –E_0 = –13.6eV.

The electron barrier potential energy is:

$$V_B(x) = U(x) - E_0. \qquad (6.4a)$$

For electric field $E = 0$, the electron is assumed to have energy $U = -E_o$ at $x = a_o = 0.053$ nm, the Bohr radius. When $E > 0$, there is a second location, near $x_2 = E_o/eE$, where U is again E_o. The problem is to find the time τ to tunnel from $x = a_o$ to $x = x_2 = E_o/eE$ through the potential barrier $U(x) - E_o$. For small field E, the maximum height of the barrier seen by the electron is $\approx E_o = 13.6$ eV, and the width t of the tunnel barrier is essentially $\Delta x = t = x_2 = E_o/eE$.

If the barrier potential $U(x)$ were V_B (a square barrier independent of x), the tunneling transmission probability T would be:

$$T = \exp\left[-2(2mV_B)^{1/2}\,\Delta x/\hbar\right] \tag{6.5}$$

for barrier width Δx and $\hbar = h/2\pi$, with h Planck's Constant.

A more accurate approach would replace $(V_B)^{1/2}\Delta x$ by $\int[U(x) - E_o]^{1/2}\,dx$, this is known as the Wentzel–Kramers–Brillouin (WKB) approximation, and was also used in the Gamow tunneling probability, see Eq. 4.22.

We will simplify further, and calculate based on the *average value* of the barrier: approximate the barrier function $U(x) - E_o$ by half its maximum value, $\frac{1}{2}V_{BMAX}$, over the width Δx.

The escape rate, $f_{escape} = 1/\tau$, with τ the desired ionization lifetime, is $T\,f_{approach}$. The orbital frequency of the electron will be taken from Bohr's rule for the angular momentum $L = mvr = h/2\pi$, so, $f_{approach} = h/[(2\pi)^2 ma_o^2] = 6.5 \times 10^{15}$ s^{-1}. The working formula $f_{escape} = 1/\tau$ is:

$$f_{escape} = f_{approach}\,\exp\left[-2(mV_{BMAX})^{1/2}\,\Delta x/\hbar\right]. \tag{6.6}$$

To find Δx one solves the quadratic $-E_o = -ke^2/x - eEx$, (E is the electric field):

$$\Delta x = \left[(E_o/e)^2 - 4k_C\,eE\right]^{1/2}/E. \tag{6.7}$$

(This limits the applicability of the approximation to $E < 32.1$ V/nm, where $\Delta x = 0$.)

So, we have the width of the barrier, and now want to find the location of its peak and the peak value.

The equation for U is $U(x) = -k_C e^2 x^{-1} - eEx$. Taking the derivative with respect to x, $dU/dx = U'$, we set $U' = 0$ to find the barrier peak. Thus,

$$U' = k_C e^2 x^{-2} - eE = 0 \text{ at:}$$

$$x' = (k_C e/E)^{1/2}, \tag{6.8}$$

that locates the peak of the barrier. Plugging this value, x', into the expression for U, we find the peak value of U, U_{max}, to be:

$$U_{max} = -k_C e^2 (E/k_C e)^{1/2} - eE(k_C e/E)^{1/2} = -2e(k_C Ee)^{1/2}. \tag{6.9}$$

Thus, the peak height of the barrier is:

$$V_{BMAX} = E_o - 2e \left(k_C eE\right)^{1/2}.$$ (6.9a)

Next, we evaluate f_{escape} for E = 25 V/nm, using:

$$f_{escape} = f_{approach} \exp[-2(2mV_{BMAX})^{1/2} \Delta x/\hbar].$$ (6.6)

taking the average barrier as $V_{BMAX}/2$. To evaluate the square bracket term one finds, at 25 V/nm,

$$[-2(mV_{BMAX}/2)^{1/2} \Delta x/\hbar] = -2.35.$$ (6.10)

The escape frequency is then, $f_{escape} = 6.5 \times 10^{15} \exp(-2.35)$ s^{-1} = 6.2×10^{14} s^{-1}, and the ionization time is:

$$\tau = 1.61 \times 10^{-15} \text{s}, \quad \text{(for E = 25 V/nm)}.$$ (6.11)

If we repeat the analysis of f_{escape} from the working formula for E = 2.5 V/nm, we find that the square bracket term is:
−118.8. The lifetime is about

$$\tau = 6 \times 10^{33} \text{s} \quad \left(\text{for E = 2.5 V/nm}\right)$$ (6.12)

or about 1.9×10^{26} y. The simple tunneling model predicts a sharp cutoff in the ionization rate for electric fields falling from 25 V/nm to 2.5 V/nm, as observed.

Tunneling rates are sensitively dependent on the parameters related to the barrier. This is observed experimentally as well as in model calculations like the one we have given.

6.2   Deuterium fusion in muonic deuterium molecule ions D_2^+ (μ)

This section is based on Eqs. 5.33–5.42 in Chapter 5. The hydrogen molecule ion has the same defining equations as the molecular ion based on two deuterons and one muon, that we now describe.

The muon has charge −e and mass 207m_e. Although its lifetime is only 2.2 μsec, the properties of the corresponding "muonic" hydrogen and deuterium atoms are well known. The reduced mass m_r = mM/(m+M) enters the Bohr radius in the denominator, and therefore multiplies the Bohr energy. Focusing on the "Dμ" atom formed by a muon and a deuteron D, the reduced mass is m_r = 207 × 2 × 1836/[207 + 2(1836)] = 196 m_e. The equivalent Bohr radius is 270 fm and the binding energy is 196 × 13.6 eV = 2.666 keV. The Dμ atom is smaller by a factor 196 than hydrogen and has a binding energy 196 times larger. If a beam of muons passes through a gas of hydrogen atoms or hydrogen molecules the electrons will be ejected with 2.66 keV energy and the muonic atoms will form. In dense hydrogen exposed to muons in this way it is observed that muonic DDμ⁺ molecule-ions form, for example, by the reaction,

$$\mu + DeeD \rightarrow DD\mu^+ + 2e.$$

(6.13)

It is experimentally known that the lifetime of the DDμ⁺ for fusion of the D particles is about 1.5 ns, based on detecting the 8.45 MeV neutrons that are emitted by the fusion (see text following Eq. (6.1)) It is also known that during the 2200 ns lifetime of a single muon in a dense deuterium gas several hundred cycles of molecule formation and fusion can occur.

The D-D-μ ion (DDμ⁺) can be understood by scaling the properties of the H-H-e (H$_2^+$) ion as described in Eqs. 5.36–5.39. The scaling of the Bohr radius and Bohr energy lead to an equilibrium spacing R = r$_e$ = 2.4a$_0$/196 = 648 fm. The binding energy of the scaled ion is 519.4 eV. (We will see that this binding energy is reduced by about half by the positive confinement energy of the vibrational motion.)

Following the discussion of Eq (5.38a), we equate the energy E with the potential sketched in Fig. 6.4. First estimate the vibrational frequency ω = (E"/m$_{red}$)$^{1/2}$, where the curvature E" = d²E/dr². We scale the curvature quantity,

$$d^2E(r_e)/dr^2 = E"(r_e) = 0.1257 \; E_o/a^2$$

(6.14)

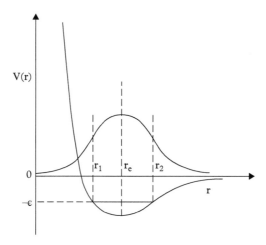

Fig. 6.4 *Ground state potential V(r) and lowest vibrational wavefunction sketched for DDμ or DTμ ion. The large extent of the vibrational wavefunction suggest a chance for fusion, related to a nonzero probability of nuclear spacing near zero (Jackson, 1957) Fig. 1*

with Bohr energy and Bohr radius to get the DDμ oscillation frequency,

$$\omega = \left(E''/m_{red}\right)^{1/2} = \left(0.1257\ 13.6\ 196\ 1.6\ 10^{-19}/1.67\ 10^{-27}\right)^{1/2}/270$$
$$= 6.64 \times 10^{17}\,s^{-1}. \tag{6.14a}$$

This high frequency corresponds to a "zero-point" energy,

$$\hbar\omega/2 = 217.8\ eV, \tag{6.15}$$

roughly half the electronic binding energy, as suggested in Fig. 6.4. (The zero-point energy is a confinement energy, not present in the classical oscillator, that is implied by the uncertainty principle $\Delta p\ \Delta x \geq h/2\pi$ as applied to the momentum of the oscillator.) The corresponding spring constant:

$$K = m\omega^2 = 0.735 \times 10^9\,N/m. \tag{6.16}$$

In Fig. 6.4, it appears that $V(r)$ is zero near D-D spacing $r_e/2 = 324$ fm. If we assume a parabolic variation of V in the range:
$r_e/2 < r < r_e$, we can make an alternative rough estimate for the spring constant:

$$K = 1.58 \times 10^9\,N/m, \tag{6.16a}$$

that is not too different from (6.16), considering the approximate nature of the analysis.

We can now estimate the time for fusion T_f, using the properties of this molecular DDμ quantum mechanical oscillator.

The formal problem of the 1-D oscillator in Schrodinger's Equation gives:

$$E_n = \left(n + \tfrac{1}{2}\right)\hbar\omega, \tag{6.17}$$

where $\omega = (K/m)^{1/2}$ as previously and $n = 0, 1, 2 \ldots$.

We can find the zero-point vibrational width (see Fig. 6.4), dimension in meters,

$$\sigma = (\hbar/m\omega)^{1/2} \tag{6.18}$$

that enters the probability distribution for the $n = 0$ (zero point) vibration,

$$P_{n=0}(x) = \sigma^{-1}(1/\pi)^{1/2}\exp-(x/\sigma)^2. \tag{6.19}$$

where $x = r - r_e$. We can evaluate the vibrational width of motion

$$\sigma = (\hbar/m\omega)^{1/2} = \left[6.6 \times 10^{-34}/\left(2\pi 1.6710^{-27}6.64 \times 10^{17}\right)\right]^{1/2} = 307.8\ f. \tag{6.18a}$$

At this spacing, $r = r_e - \sigma = 340.2$ fm, in the ground-state motion of the oscillating masses, the probability is reduced to $1/e = 0.368$. This width is large, around half the equilibrium spacing, consistent with the observed DDμ decay by fusion of the deuteron particles.

As a first rough estimate of the fusion rate, we can simply evaluate the probability P'(0 – r$_1$) that the oscillator spacing is in the range 0 < r < r$_1$ (r$_1$ ≈ 2.7 fm, corresponding to contact between the two deuterons), and multiply by the oscillator frequency. However, since this assumes a parabolic potential rather than Coulomb repulsive potential, it will likely overestimate the fusion rate. The probability of D-D spacings in the range 0 to r$_1$ is the integral of P(x) over that range, that can be approximated as:

$$P'\left(0 - r_1\right) \approx r_1 \sigma^{-1} \left(1/\pi\right)^{1/2} \exp - \left[\left(r_e - r_1\right)/\sigma\right]^2 \tag{6.20}$$

= (2.7/307.8) 0.564 exp−[(648−2.7) /307.8)^2. The result is P'(0 – r$_1$) = 8.77 × 10^{-3} 0.564 0.0123 = 6.08 × 10^{-5}.

The Rate of fusion is then estimated as R = (ω/2π) P'(0 – r$_1$) = 6.43 × 10^{12} s^{-1}, corresponding to a lifetime,

$$T_f = 1.55 \times 10^{-13} \text{s}. \tag{6.21}$$

This is shorter than the observed lifetime, that is 1.5 ns.

While the full solution of Jackson (1957) is more advanced than we can present here, we suggest an approximation that improves on what we have just done.

The improved approximation is to replace the harmonic oscillator potential with the direct Coulomb potential, for the range between the classical turning point: R = r$_e$ – σ = 340.2 fm and the contact radius R = r$_1$. (In physical terms, this means that we neglect the muon charge density in that region, removing screening of the Coulomb potential.)

The corresponding Gamow transmission factor,

$$T_{\text{Gamow}} = \exp - 2\gamma \text{ where, for } r_2 = 340.2 \text{ fm, } E = 4.233 \text{ keV, } r_1 = 2.7 \text{ fm,} \tag{6.22}$$

$$\gamma = \left(2mE\right)^{1/2}/\hbar \left[\left(\pi/2\right)r_2 - 2\left(r_2 r_1\right)^{1/2}\right] = 6.78$$

and T$_{\text{Gamow}}$ = 1.28 × 10^{-6}. In this case the integration of the oscillator probability function is from 0 to r$_2$ = 340.2 fm and is about P' = 0.023. The resulting rate of fusion R = (ω/2π) P'(0 – r$_2$) T$_{\text{Gamow}}$ = 3.1 × 10^9, so:

$$T_f = 0.32 \times 10^{-9} \text{s}. \tag{4.23}$$

This is closer, but still below the observed value, that is 1.5 ns (Atzeni and Meyer-Ter-Vehn, 2004, p. 26) .

We attribute the difference to a reaction probability factor T = 0.21, that we can interpret as the fraction of the collisions of the two deuterons at the contact point that lead to fusion, an effect of the nuclear physics of the two particles. The estimate here is of the same sense but numerically somewhat larger than the value T = 0.04 we mentioned earlier.

The influence of the details of the nuclear reaction, suggested here, is supported by the well-known large difference between this lifetime and the fusion lifetime of the similar DTμ ion, $T_f = 7 \times 10^{-13}$ s. This represents an increase in the reaction rate R by a factor 21. (This is similar to the increase in the DT above the DD cross section in Table 6.1.) The Gamow factors are almost the same for the DD and the DT cases, so we need to learn more about the nuclear reactions.

From the point of view of our calculation, adapting the molecular properties from the DD case to the DT case will change in our estimate of T_f only slightly. This can be seen from the larger reduced mass entering the calculation of γ. For DT, the reduced mass is $m_r = (2 \times 3)/(2 + 3) = 1.2$, a 20 percent increase, that will increase γ and reduce the tunneling rate. The parameter r_1 however is increased from 2.7 fm to 2.89 fm, leading again to a small change, but in the opposite sense to that of the effective mass change. The resulting changes will be small, suggesting that details of the nuclear tunneling physics, neglected in our estimate, may be important.

Why is the DT fusion reaction easier to initiate than the DD reaction? The difference is that the DT system has a "resonant state" at energy 114 keV, as shown in data (Li et al, 2000) of Fig. 6.5.

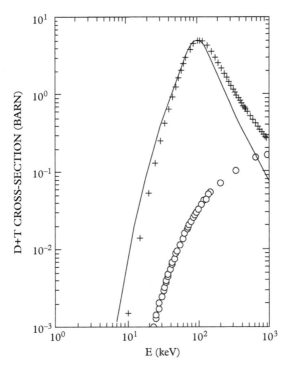

Fig. 6.5 *Measured cross-section for D-T reaction (solid curve) compared to resonant tunneling model (+) and measured values for DD reaction (circles) (Li et al, 2000) Fig. 2*

The solid curve in Fig. 6.5 is the measured cross-section for the D-T reaction. Not only is the cross-section much larger than the data for the D-D reaction (circles) but it has a clear peak, at a resonant energy 114 keV. The physical situation is as described by Fig. 4.3, imagining a D coming to collide with a T triton particle, except that there is a well-defined "resonant state" at positive energy E = 114 keV. This is a metastable state, that can decay either by back-reflection of the D, or by forming the ^{4}He + n final state. This final energy is negative (bound and stable), analogous to the –2.2 MeV marked in Fig. 4.3 denoting the D ground state.

There are, in effect, two tunnel barriers in this problem. One is the Coulomb barrier shown in Fig. 4.3 and the second effective barrier is between the resonant state and the final state. The large increase in cross section, approximately the inverse of the Gamow factor T$_{Gamow}$ = 0.067 found for 57.5 keV in connection with Eq. 6.2, namely 14.9, is expected for resonant tunneling, *when the incoming particle energy matches that of the resonant state.*

In the physics analogy, one has two barriers separated by a potential well. The interesting effect comes when the well has a bound state at the energy of the input tunneling particle. In this case the transmission probability (across the whole system) becomes 1, that corresponds to setting the Gamow factor to 1 as we mentioned. There is thus a strong peak in the transmission probability T(E) at the resonant energy E = E$_0$. In more detail, near the energy of the resonant state E$_0$:

$$T(E) = \Gamma^2 / \left[\Gamma^2 + (E - E_0)^2 \right]. \tag{6.24}$$

The (+) symbols shown in Fig. 6.5 as calculated by Li et al (2000) are in good agreement with the measurements and establish the resonance state interpretation of the large enhancement of the DT cross section over the DD cross section peaking at 114 keV. In (6.24), the energy width is of the form $\Gamma = \hbar/\tau$ where τ is the lifetime of the resonant state. The lifetime is the inverse of the rate of tunneling out of the state, $\omega = \omega_{attack} T_{Gamow}$, so for $\Gamma^2 \ll (E - E_0)^2$ we get:

$$T(E) \approx (\hbar\omega_{attack} T_{Gamow})^2 / E^2. \tag{6.24a}$$

This is not a familiar formula, but does indicate that the transmission factor doubly enters the rate as (T$_{Gamow}$)^2 away from the resonance, when the barrier width is effectively doubled.

6.2.1   Catalysis of DD fusion by mu mesons

Our discussion leaves no doubt that molecular ions DDμ^+ and DTμ^+ will be strongly bound until destroyed by fusion or by the decay of the mu meson, after 2200 ns. Fusion occurs in these ions on time scales about 1.5 ns for DD and 0.7 ps for DT. In both cases the muon is released, following a fusion event, and, after some delay, forms a new DDμ or DTμ ion. The role of the muon is as a catalyst.

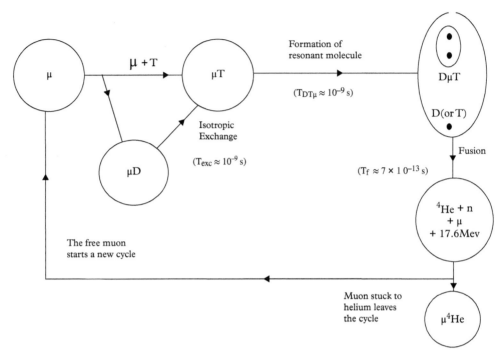

Fig. 6.6 *μ-catalysis cycle for the D T gas mixture, as irradiated with mu mesons. Mean reaction times are indicated for each process. After Atzeni, S. and Meyer-Ter-Vehn, J. (2004) Fig. 1.7*

The detailed cyclic process (Atzeni and Meyer-Ter-Vehn, 2004) for DT is shown in Fig. 6.6. A dense gas mixture of D and T is bombarded with muons, momentarily forming Dμ and Tμ atoms.

These atoms form molecules and molecular ions, as shown in Fig. 6.6. The DTμ molecule formed in a time about 1 ns, and promptly undergoes fusion in 0.7 ps. After fusion, the muon is released, and again available to catalyze further fusions. The cycle described takes place in about 5 ns, so in the muon lifetime about 440 fusion events could occur. However, as suggested in Fig. 6.6, a small fraction, about 0.006, of the fusion muons are not released, but remain attached to the helium fusion product. On this basis, the number of fusions per muon is estimated as 120. Experiments have shown up to 200 fusion events per muon.

This catalytic approach to fusion has been measured and modeled, but has not produced more fusion energy than needed to initiate the process.

In summary, Muon-catalyzed fusion (μCF) is a process allowing nuclear fusion to take place at temperatures significantly lower, even at room temperature, than the temperatures required for thermonuclear fusion. Although it can be produced reliably with the right equipment and has been much studied, it is believed that the poor energy balance will prevent it from ever becoming a practical power source. However, if muons ($μ^-$)

could be produced more efficiently, or if they could be used as catalysts more efficiently, the energy balance might improve enough for muon-catalyzed fusion to become a practical power source.

The observed time for fusion in DDμ is 1.5 ns, and shorter in DT μ, while the life-time of μ is 2200 ns. This would allow for 1400 cycles, emitting one neutron plus 3 MeV per cycle, but the best seen is only 200. This difference could make the process useful for energy production. At present, it seems that something else is limiting the neutron emission rate, maybe the time for a new DDμ or DTμ to form from the released μ, in the sample mixture of Deuterium, DeeD and related DeeT molecules. This is perhaps limited by diffusion.

6.3   The tokamak fusion reactor

The essential elements of a simplified tokamak reactor are sketched in Fig. 6.7. Here a toroidal vacuum chamber contains positive ions and electrons. A solenoid runs through the hole of the "doughnut" and its magnetic field is ramped to produce an EMF around the loop.

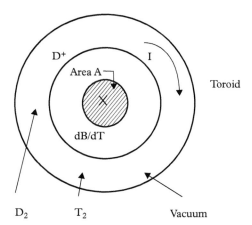

Fig. 6.7 *Essential elements of simplified Tokamak reactor, top view. A toroidal vacuum chamber is filled with Deuterium and Tritium gas at about $10^{20}/m^3$ density. Heating is applied to form a neutral high conductivity plasma of Deuterons, Tritons and electrons, from an initial charge of deuterium and tritium gases as shown. A nearly uniform toroidal magnetic field B, not shown, is of order 5 Tesla, produced by current through a superconducting coil wound around the toroid. The model for this is the uniform field $B = \mu_0 N I$, with I the current through a long solenoid of N turns per meter. The particles are guided by the field, following helical orbits that avoid the walls. The plasma current I is induced by a change of magnetic flux, A dB/dt, through the hole of the doughnut, where A is the cross-sectional area of the central vertical solenoid. The toroidal device can be regarded at first as a solenoidal coil of radius a, whose length is $2\pi R$, closed on itself to form a torus of major radius R and volume $V = 2\pi R\,\pi a^2$, with important corrections to be considered later*

6.3.1   Electrical heating of the plasma

Conventional electric plasma heating $P = I^2 R$ is provided by Faraday's Law induction of heating current from changing magnetic flux d/dt (BA) $= -\varepsilon = I R$. Fusion occurs at temperatures near 150 million K. When fusion occurs, released energy is collected from the energetic neutron emissions, in a surrounding blanket and heat exchanger (not shown, but see following description), while the charged reaction products are retained in the plasma by the magnetic confinement, and further heat the plasma.

We will model a simplified circular tokamak, assuming a pure deuteron-electron plasma at density $n = 10^{20}$ m^{-3}, at temperature $T = 150 \times 10^6$ K (ten times higher than the Sun's core).

Consider a torus of volume 489.5 m^3 (that we may take as 500 m^3) and this has major radius R = 6.2 m and minor radius a = 2 m. Think of this as a cylinder of length 2π 6.2 m and cross-sectional area π 2^2. Using the formula for the resistance R of a length L with cross-section A, R = ρL/A, the resistance around this torus, containing the plasma as described, is R = ρ 2π 6.2/π 2^2 = 3.1 ρ Ohms.

We have described in Chapter 4 the power density from fusion in the Sun, at some length. Our approach here will be to scale the analysis from the Sun to the situation of the model tokamak plasma. Scaling is an attractive approach with the assumption that a successful analysis allows extrapolation to a similar situation with modified parameters.

6.3.2   Scaling the fusion power density from that in the Sun

In Chapter 4, the power output from the Sun is analyzed from the point of view of proton-proton fusion, starting with Eqs. (4.23–4.29), leading to:

$$P = 0.5\ N_p^2\ <v\sigma>\ Q, \tag{4.34}$$

for the power per cubic meter generated, 313 W/m^3 in the core of the Sun, where Q is the energy release of the reaction.

Our program here is to scale this number to find the power density in a D-D reaction tokamak reactor, assuming deuterium density $N_D = 10^{20}$ m^{-3} (see Fig. 5.18), to be compared to the Sun's proton density $N_p = 3.11 \times 10^{31}$ m^{-3}. This, along with the change in the reaction probability factor T from 8 $\times$ 10^{-24} (text, following Eq. 4.32) to $T = 0.1$. Assume a factor of 10 increase in temperature T from 15 million K to 150 million K. Scaling must include the r_2 parameter, noting that the mass of the deuteron is twice that of the proton, and the temperature enters the velocity and also the cross-section for the geometric collision. Finally, if the tokamak has a volume 500 m^3, how much power does it release in fusion reactions (assume Q = 3.5 MeV for the DD reaction)?

The scaling ratio, P'/P, works out to be:

$$P'/P = \left(10^{20}/3.11 \times 10^{31}\right)^2 \left(1/10^{3/2}\right)\left(T'_{Gamow}/10^{-8}\right)\left(0.1/8 \times 10^{-24}\right)$$
$$(3.5\ \text{MeV}/26.2\ \text{MeV}) = 54.01.$$

where $T_{Gamow}' = 9.9 \times 10^{-4}$. (The second factor here comes explicitly from the change in temperature entering v and σ.) The value for T'_{Gamow} follows Eq. 4.24 with parameters $r_1 = 3$ fm, $r_2 = 111.3$ fm, $E = 12.93$ keV and $m_r = 1 \times m_p$.

So, the predicted tokamak power density, assuming pure deuterium fuel, is 16.9 kW/m³ and its power output is 8.45 MW.

6.3.3   Adapt DD plasma analysis to DT plasma as in ITER

To convert this result to a DT plasma, approximately, we consider two aspects. First is the factor 15 increase in the DT cross section over the DD cross section, and, second, the larger Q energy release, 17.6 MeV, vs. 3.5 MeV. Thus, if the plasma were a DT plasma, the projected Tokamak power output would be increased by 75, to,

$$Q = 634 \text{ MW.} \left(\text{DT plasma at 150 million K, 500 m}^3 \right). \tag{6.25}$$

This estimate is close to those projected for the ITER tokamak, which has a larger volume, 880 m³, by virtue of its larger, D shaped, rather than circular, cross section.

Let's continue to calculate properties of the assumed tokamak, starting with the energy U needed to heat the 500 m³ deuterium to 1.5×10^8 K, where we know it is 12.93 keV per particle. In thermal equilibrium the plasma contains equal numbers of ions (Deuterons plus Tritons) as electrons, and each has average energy 12.93 keV. Thus,

$$U = 2(N_D + N_T)k_B T \tag{6.25a}$$

$= 2 \times 10^{20} \times 500 \times 12.93 \times 10^3 \times 1.6 \times 10^{-19}$ J $= 206.8$ MJ $= 57.5$ kWh. The market cost of this energy, at 14c/kWh, the price of electricity in New York City, is $8.05. This is small, and the key is the assumption that only the gas is heated, the containment (to be described) allows the walls and the whole containment vessel to remain relatively cool.

We might next ask, how long can the reactor run at 634 MW before running low on fuel? For example, let's find $T_{1/2}$, when half the energy of the fusion reactions is consumed, assuming the DT reaction generating 634 MW:

$PT_{1/2} = 0.5 \times 10^{20} \times 500 \times 17.6 \times 10^6 \times 1.6 \times 10^{-19}$ J $= 7.04 \times 10^{10}$ J, so, $T_{1/2} = 7.04 \times 10^{10}$ J/$634 \times 10^6 = 111$ seconds, not quite 2 minutes.

So, the reactor will definitely need a continuous feed of DT mixture. (In the planned ITER device the tritium T generation is accomplished by feeding Li into system, that generates T by absorption of neutrons.)

We should confront an issue of how the fusion power is harvested. While the reaction of DT produces 17.6 MeV, this energy is shared between a neutron and an alpha-particle (first line in Table 6.1). In the frame of the fusion event, the momentum of the particles will add to zero, so $4V = -v$, with V and v, respectively, the speeds of the alpha (mass 4) and the neutron (mass 1). The kinetic energy of the neutron is ½ mv² and that of the alpha is ½ (4 m) (v/4)². On this basis the neutron gets 0.8 of the kinetic energy, which it carries out of the reaction zone into the containment structure. We will mention forthwith how the alpha particle is trapped into circling the magnetic field lines, and thus returns its heat energy back into the plasma. The ^{4}He is a waste product, termed "ash" in the literature of tokamaks. The neutrons are equally emitted in all spherical directions, and their motion is not at all impeded by the magnetic field.

Another aspect to be confronted is that we assume a uniform temperature distribution in the full torus. The major and minor radii, R and a, respectively, and volume should realistically be interpreted as the dimensions of the hot part of the plasma. Indeed, it is essential to establish a large radial temperature gradient in the cross-section of the torus to keep the walls cool, except for the 14.1 MeV fusion neutrons that carry the fusion energy out of the reactor.

Another energy cost is the toroidal magnetic field. To simplify we assume it is 5 Tesla uniformly in the cross section πa^2. The energy density of the 5T magnetic field B, (the correction to the field from the curvature of the torus and the toroidal current itself is addressed in Fig. 6.10)

$$U_B = B^2/2\mu_0 \tag{6.26}$$

$= 9.94 \times 10^6$ J/m^3, where $\mu_0 = 4\pi \times 10^{-7}$ N/A^2. Therefore, the energy of the whole uniform magnetic field is $U_B = 4.97$ GJ, with cost at $0.14 /kWh of $193.42. The pressure of the magnetic field on the walls is $u_B = 9.94 \times 10^6$ J/m$^3 = 9.94 \times 10^6$ N/m^2. Since a Pascal is 1 N/m^2, and one atmosphere (1 Bar) is 101 kPa, the pressure is 98.4 atmospheres. So, the containing vessel has to be very strong indeed.

The production of the field can be understood from the formula for the B field in a long solenoid, $B = \mu_0$ nI, where n is the number of turns of wire per meter and I is current. Once established, the field has no further energy cost because the current is maintained in a superconducting magnet.

The result of DT fusion as we have seen is an alpha particle of energy K = 0.2 × 17.6 MeV = 3.52 MeV = ½ $4m_p V^2$. The velocity of the alpha is thus V = $(2 K/4m_p)^{1/2}$ = 1.3×10^7 m/s. The basic force is:

$$\mathbf{F} = q\ \mathbf{V} \times \mathbf{B} = m\mathbf{a} \tag{6.27}$$

with m the mass. The acceleration a = V_t^2/R for a circular orbit of radius R, with V_t the component of the velocity transverse to the B field, gives us:

$$R = mV_t/qB = (2mK)^{1/2}/qB \tag{6.28}$$

for the radius R of the circular orbit, at kinetic energy K, around the B field lines. The path of the charged particle is a spiral, a helix whose orbital radius depends on the angle of the velocity vector relative to the magnetic field B. A particle moving perpendicular to the field has the largest orbital radius. The frequency of the orbital motion is:

$$\omega_L = qB/m, \tag{6.29}$$

called the cyclotron frequency. For the alpha-particle $^4\text{He}^{2+}$ released in the DT reaction at 5 Tesla we find for the maximum radius,

$$R = \left(4 \times 1.67 \times 10^{-27} \times 1.3 \times 10^7\right) / \left(2 \times 1.6 \times 10^{-19} \times 5\right) = 0.054 \text{ meters.} \qquad (6.30)$$

This means that the alpha particle, charge 2 e and mass 4 m_p, created in the plasma will at most move 0.054 meters toward the wall of the toroidal container. Only those alpha particles created within 5.4 cm of the wall, and moving specifically in the radial direction, will have a chance of colliding with the wall. Since the emission directions of the alphas are randomly distributed, an even smaller probability of wall collisions will occur. This is the basis of the confinement.

The cyclotron frequency for the electrons is of interest in regard to heating the plasma. It is

$$\omega_L = qB / m = 1.6 \times 10^{-19} \times 5 / 9.1 \times 10^{-31} \text{ radians} / \sec = 8.79 \times 10^{11}$$
$$\text{rad} / \sec = 140 \text{ GHz,} \qquad (6.31)$$

corresponding to radiation wavelength, $\lambda = c/f = 2.14$ mm. This corresponds to F- or D-band microwave radiation.

Electrons are in thermal equilibrium with K = 12.93 keV on average at the operating temperature. Thus $R = (2 \, mK)^{1/2}/qB = 76.7 \, \mu m$, and we see that thermal electrons are controlled even more tightly, to a smaller radius, and move closely along the toroidally oriented magnetic field lines. So, the charged particle system, closed on itself, is like an infinite 1-D system. The electron thermal velocity v in the plasma is of interest, and is $v = (k_B T/m_e)^{1/2} = 4.77 \times 10^7$ m/s, about 0.16 c. The motion of these electrons is not much affected by the theory of special relativity, because the relevant factor $[1- (v/c)^2]^{-1/2} = 1.012$ is close to 1, indicating Newton's Laws adequately describe the motion.

We want to learn about the electrical conductivity of the plasma, that is limited by the mean free path for scattering of the electrons by the ions. Assume the same parameter r_2 as was used in Chapter 4 (111.3f at 12930 eV) determines the collision cross section. So, for the electron mean free path we find:

$$\lambda_e = 1 / \left[N_D \pi \left(r_2\right)^2 \right] = 257 \text{ km (for electrons).} \qquad (6.32)$$

This is an amazingly long mean free path, indicating that the electrons orbit around the torus 6600 times between collisions. (We will see that there is a correction that reduces this by a factor 21.1.) The mean time between collisions, $\tau = \lambda_e/v = 5.38$ ms. Now we can find the mobility,

$$\mu = e\tau/m = 9.47 \times 10^8 \text{ m}^2/\text{Vs} \qquad (6.33)$$

and the electrical conductivity of the plasma,

$$\sigma = ne\mu = 1.52 \times 10^{10} \, (\Omega m)^{-1}. \qquad (6.34)$$

The resistivity of the plasma at 1.5×10^8 K and n $=10^{20}$ deuterons/m^3 is thus,

$$\rho = 1/\sigma = 6.6 \times 10^{-11} \Omega m. \qquad (6.35)$$

We can neglect the contribution of the ions to the conductivity because their mobility is at least 1835 times smaller due to the mass increase between proton and electron.

We can now find the Resistance R for the Torus plasma at its operating point as R = ρ 2π 6.2/π 2^2 = 3.1 ρ Ohms,

$$R = 0.204 \text{ n}\Omega. \qquad (6.36)$$

Suppose we want to heat this plasma at power 10 MW. Let us find the needed voltage V to make P = V^2/R = 10 MW. The needed voltage is V = (RP)$^{1/2}$, that will be (10^7 × 0.204 × 10^{-9})$^{1/2}$ = 0.045 Volts. This voltage is induced by the changing magnetic field through the hole of torus.

Using the Faraday Law, EMF = Voltage = –dΦ/dt, where Φ is magnetic flux in Webers, so that we need 0.045 Webers/sec. Suppose we have a superconducting coil with radius 1 meter, a bit bigger than used in a typical MRI Magnetic Resonance Imaging apparatus, cutting through the doughnut hole of the torus. Thus, we have

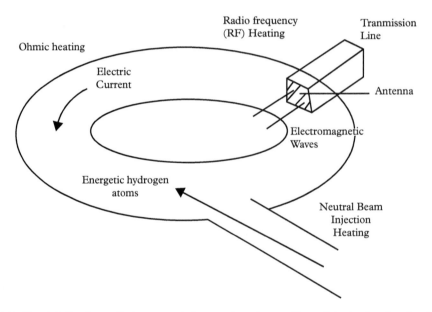

Fig. 6.8 *Sketch indicating radio frequency (microwave) and neutral beam injection heating, in addition to Ohmic heating of ionized gas mixture to reach the plasma operating temperature. Assuming the toroidal magnetic field 5T, the electron cyclotron resonance frequency was found in the text to be 140 GHz (after http://www-fusion-magnetique.cea.fr/gb/fusion/principes.htm) http://www-fusion-magnetique.cea.fr/gb/fusion/principes.htm)*

0.045 = π dB/dt, or dB/dt = 14.3 mT/sec, to provide 10 MW heating of the plasma. If the coil will support 5T (Tesla, a unit equal to 10 kilo Gauss) , then the time of the ramping can extend to 348 s. In this time the energy U = 10 MW 348 = 3.48 GJ, will be available. This is well beyond the 0.207 GJ needed to heat the plasma to 1.5×10^8 K.

One of the difficulties is that the induction heating needs a conductive plasma to start with. So, there are other methods used to start the plasma before the inductive heating. These are indicated in Fig. 6.8.

In looking at Fig. 6.9, note at the bottom that the "Primary fuels" are Li and D. The flow diagram indicates that Li is turned into T (tritium), some of which, along with D and ^{4}He, are pumped out at the end of each heating cycle. These residues undergo isotope separation, after which the ^{4}He is emitted as waste and the T and D are re-injected as fuel.

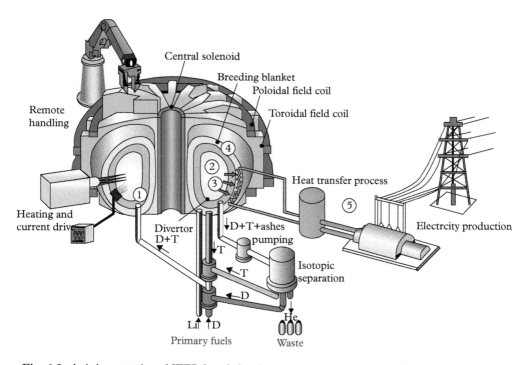

Fig. 6.9 *Artist's conception of ITER-based electric power plant. In the figure, (1) marks the injection point of DT mixture into the reaction chamber of "Dee" shaped cross section, where, due to confinement and various heating sources it goes into the plasma state and burns in fusion reactions (2). Location (3) marks the release of energy in the form of radiance and fast particles and helium ash. (4) Indicates the first wall, of the vacuum chamber, marked "breeding blanket" where fast particles are turned into heat but turn some particles into tritons T that enter the plasma. The first wall, breeding blanket and vacuum chamber are cooled by a heat extraction system used to produce steam and supply a conventional turbine and alternator electricity producing system (5) http://www-fusion-magnetique.cea.fr/gb/fusion/principes.htm*

6.3.4   Radiations from the plasma

The first role of the breeding blanket is to protect the vacuum vessel and magnets from neutron and gamma radiations. In its breeding function it produces from lithium the tritium T needed for continued fusion reactions. Finally, it converts neutron energy into heat and transfers that heat to superheated steam to drive the steam turbine and create electricity via a heat exchanger system. The breeding reaction is based on neutrons from the DT reaction:

$$^6\text{Li} + \text{n} \rightarrow {}^4\text{He} + \text{T} \tag{6.37}$$

$$^7\text{Li} + \text{n} \rightarrow {}^4\text{He} + \text{T} + \text{n}$$

In practice, the blanket can be maintained as a liquid, for example as a Pb + Li mixture, or as a solid ceramic, such as Li_4SiO_4, Li_2ZrO_3, or Li_2TiO_3. Figs. 6.9 and 6.10 indicate that there is a temperature distribution in the plasma: it is necessary that the first walls be maintained safely far below their melting temperatures. For reference, graphite and tungsten are viable up to the vicinity of 3000 K, but the plasma as we have assumed is 150 million K. The heat radial transfer by particle diffusion from the plasma toward the walls is inhibited by the magnetic confinement discussed previously. The plasma has no neutral particles, and all charged particles are magnetically confined to move principally in the toroidal direction, around the circle, and this greatly diminishes the ordinary heat flow by diffusion, in the radial direction.

Radiation transfer from the plasma to the walls is important, even though we ignored it in our simple estimate of the necessary heating power, Eq. 6.26. The DT plasma, containing only electrons and singly positive deuteron and triton positive ions, radiates electromagnetic energy. This power loss, however, is not well described by blackbody radiation formulas. The radiation from the plasma is known as *bremsstrahlung* (acceleration radiation) and can be estimated as follows.

The radiated power density P_{Brem} per cubic meter from *bremsstrahlung* is summarized by (Huba, 2006) for a plasma of equal numbers N_D of electrons and ions of charge Z_i. At thermal energy T (in eV), it is:

$$P_{\text{Brem}} = Z_i^2 N_i\, N_e\, T^{1/2} / \left(7.69 \times 10^{18}\right)^2 \text{ W/m}^3 \tag{6.38}$$

for N_i ions of charge Z_i and N_e electrons. If we evaluate this for the tokamak plasma, where $Z_i = 1$, with equal numbers of deuterons and electrons, we get:

$$P_{\text{Brem}} = \left(10^{20}\right)^2 (12930)^{1/2} / \left(7.69 \times 10^{18}\right)^2 \text{ W/m}^3 = 19.23 \text{ kW/m}^3. \tag{6.38a}$$

This radiation power density P_{Brem} is fortunately quite small compared with the fusion power density for the D-T case we are considering, that is 634 MW/500m³ = 1.27 MW/m³.

This formula (6.38) with factor Z_i^2 makes clear the danger posed by possible W contamination by tungsten walls. Since W has $Z = 74$, (with $Z_i^2 = 5476$), entering Eq. 6.38 if W ions should be eroded from the wall and appear in the plasma, these ions, even in small amounts, would cool the plasma. This makes graphite, with Z only 6, an attractive wall material.

It is not clear what fraction of this radiation might be reflected back by the walls into the plasma, because of its very short wavelength. This energy is more likely absorbed by the walls, making it part of the useful power output, but its nature makes it more likely to reduce the temperature of the plasma, than do the fusion events, which actually heat the plasma by thermalization of the trapped energetic-charged fusion products.

The distribution of this radiation peaks near $hf = 0.2\,T = 2586$ eV, corresponding to gamma ray photons of wavelength 0.48 nm, where T is the particle energy in eV. The frequency of this emitted acceleration radiation is far above the electron plasma frequency,

$$\omega_p = (N_D e^2/m_e \varepsilon_0)^{1/2}, \tag{6.39}$$

that works out to 5.63×10^{11} radians/s for this case, corresponding to 89.7 GHz. The blanket in the tokamak intercepts this radiation and turns it into heat, protecting the superconducting coils from the damage that the gamma rays might inflict.

For this DT plasma at 10^{20} m^{-3} the Debye screening length λ_D, discussed in conjunction with Eq. 5.71a, is 84.5 μm. The relation of the tokamak plasma to the Sun and other plasmas was summarized in Fig. 5.18 in Chapter 5.

It should be pointed out that the power density of the plasma is adjustable from the D, T density. Going from 10^{20} to 2×10^{20} m^{-3} will increase the power by a factor of 4. So, the difference between the full torus volume and the hot plasma volume, perhaps a factor of 2 in effective volume and power output, can easily be recouped by increasing the deuteron density.

6.3.5   Small-angle scattering correction, and comments

In our desire to make an opaque subject clearer to the non-expert we have focused on the essential parts of the nanophysics of fusion. The cross section for electron scattering from the ions, related to the resistivity of the plasma, it turns out, has been treated too simply, and needs a correction factor, that increases the plasma resistance (Miyamoto, 2004).

The correction factor, known as "$\ln(\Lambda)$" multiplies the strong-scattering cross section $\pi\,r_2^2$, which, when multiplied by T_{Gamow}, is the fusion cross section. In the scattering of electrons from ions, the cumulative effect of many small-angle scattering events make this substantial correction. This occurs because the Coulomb force has a long range. The correction does not apply to fusion, but only to the plasma resistance, because in that case the strong scattering is the only way to set up the fusion event, an accumulation of weak scattering events is of no use. The importance of the small scattering events is controlled by the Debye length,

$$\lambda_D = (k_B T \varepsilon_o / n e^2)^{1/2}, \tag{5.90a}$$

a measure of the maximum distance over which two particles in the plasma experience the Coulomb force. Beyond this distance, the small adjustments of positions of many electrons screen away the direct Coulomb force. In this case, the ratio of the Debye length to the classical turning point spacing r_2, that is called "b" in the plasma literature, is important. Thus (Miyamoto, 2004).

$$\ln(\Lambda) = \ln(2\lambda_D / b) = \int_b^{2\lambda D} dr / r. \tag{6.40}$$

Taking $\lambda_D = 84.5 \; \mu$ and $r_2 = 111.3 \; f$, we find $\ln(\Lambda) = 21.1$, for the DT tokamak plasma at 150 million K.

This means that the plasma resistivity and resistance values are too small by a factor 21.1, the electron mean free path reduced to 12.18 km, and the plasma resistivity becomes $3.0 \times 10^{-10} \; \Omega$-m. The voltage to produce the desired 10 MW heating rate has to be increased by a factor $(21.1)^{1/2} = 4.6$. So, the needed dB/dt in the previous example must be increased from 14.3 mT/s to 65.8 mT/s.

The design for ITER tokamak actually employs larger magnetic field values,[3] 11.5 T for the toroidal field, leading to a maximum stored magnetic energy 41 GJ, a volume of 840 m³ and a central solenoid field up to 13.5 T (see Fig. 6.10), that would limit the time of inductive heating at 10 MW to 13.5T/65.8 mT/s = 3.4 minutes, assuming the central solenoid radius of 1 m.

Beyond this, there are several technical issues that in practice are important. The stability of the plasma has been assumed, a simplification. The toroidal magnetic field is approximately constant across the cross section, but more accurately decreases from the inner to outer walls of the torus. The induced current in the torus produces the usual circling magnetic field around the current, that is termed a "poloidal" field. In fact, an additional magnet, the "poloidal magnet" produces a vertical magnetic field needed to keep the plasma from expanding outward in radius.

With reference to Fig. 6.9, we see that the high temperature part of the plasma occupies a cross section less than the full vacuum vessel cross section, πa^2, but let's assume a' ~ ≤ a is the radius of the current density when the Ohmic heating is applied. (The plasma is actually diffuse and somewhat free to move in the mechanical cross section, and could only roughly be described as a torus of major radius R' and minor radius a'.) The resistance of the plasma R_p (4.36) when corrected for the small angle scattering factor $\ln(\Lambda) = 21.1$ is raised from 0.204 nΩ to 4.30 nΩ. At the chosen heating power 10 MW, the plasma current I_p must satisfy:

---

[3] http://en.wikipedia.org/wiki/ITER

$$I_p^2 R_p = 10 \text{ MW} \tag{6.41}$$

that gives $I_p = 48.2$ MA.

One aspect of the plasma equilibrium may be suggested by estimating the magnetic field produced by the plasma current I_p itself. If the current were flowing in a straight solenoid of radius a, the magnetic field $B_p(a)$ at its radius can be found from Amperes Law,

$$\oint \boldsymbol{B} \cdot \boldsymbol{dl} = \mu_0 I_p. \tag{6.42}$$

In this situation then we estimate $B_p(a) = \mu_0 I_p/2\pi a = 4.82$ T, choosing a = 2 m. If we imagine bending the solenoid to fit the major radius R, then the field B will become smaller on the outside and bigger on the inside and the relevant factors will be approximately $(R\pm a)/R$. The difference in the plasma-current-induced magnetic fields B_p on the inside vs the outside is then $\Delta B \approx (2a/R) B_p(a)$. If a vertical field $B_z = (a/R) B_p(a)$ is added (to increase the field on the outside), the current I_p will tend to be stabilized, with total vertical fields equal on the inside and the outside. The required vertical field, on our rough analysis that neglects an outward force from the plasma pressure, is then,

$$B_z = \mu_0 I_p/2\pi R = 1.56 \text{ T}\left(\text{for } R = 6.2\text{m}, I_p = 48.2 \text{ MA}\right). \tag{6.43}$$

The vector sum of this circling poloidal field and the toroidal field is then a helical field, that the ions and electrons in the plasma will follow in a helical fashion, depending on I_p.

This situation is sketched in Fig. 6.10. On the left, the magnetic field lines shown on a cut through the torus are distorted, stronger on the inside and weaker on the outside, as estimated as ΔB previously. The uniform vertical field B_z is shown to strengthen the field on the outside of the torus. The "poloidal coil" system is used to shape the plasma, in particular to correct the tendency of the plasma to drift to larger radii. On the right,

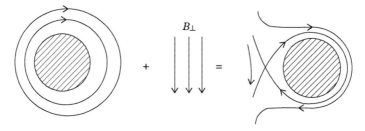

Fig. 6.10 *Poloidal magnetic field (on right side of torus) due to the combined toroidal plasma current (shown hatched) and externally applied vertical field. (Note that the field of the central solenoid is not part of this, it is almost wholly contained inside the solenoid.) A similar distortion of the magnetic field in the torus-cross section comes again from its curvature. A uniform winding of a coil around the torus leaves wires more closely spaced on the inside than on the outside, making the field again more intense on the inside of the torus. This sketch illustrates also how a crossing point of field lines can arise (Miyamoto, 2004) Fig. 6.3, p. 67*

the resultant corrected field is nearly symmetric as would occur for a straight solenoid. The detailed force balance has to include a component of the plasma pressure to the outside, that arises due to the curvature.

The difference ΔB is the origin of the tendency of the plasma to expand outward, that is corrected by the poloidal coil system. The poloidal coils stabilize the plasma and keep it away from the outer first wall. These are circular coils about the central axis, producing fields in the vertical direction. The vertical field B_z, estimated in Eq. (6.43) should be applied to maintain the plasma centered in the physical cross section of the torus. A more accurate analysis, that includes the plasma pressure, gives:

$$B_z = \left(\mu_0 I_p / 4\pi a\right) \left[\ln\left(8R/a\right) + p_{plasma} / p_{magnetic} + \delta\right], \tag{6.43a}$$

where the correction $\delta < 0$ is given by Miyamoto (2004). Here pressure,

$$p_{plasma} = 2N k_B T = 2 \times 10^{20} \times 12.93 \times 10^3 \times 1.6 \times 10^{-19}$$
$$= 0.414 \text{ MPa} = 4.1 \text{ Bar}, \tag{6.44}$$

and $p_{magnetic}$ is of the form $(B')^2/2\mu_0$. If we interpret $B' = \Delta B$ as defined previously, then $p_{magnetic} = (\Delta B)^2/2\mu_0 = 3.84$ MPa. It appears that our rough estimate is within a factor of two of the correct form (6.43a). A full analysis of the equilibrium is given in Section 6.3 of Miyamoto (2004).

The actual ITER torus cross-section is vertically elongated into a Dee shape. More realistic representations of this tokamak are given in Figs. 6.11 and 6.12. The plasma current itself, in the toroidal direction, produces its own magnetic field, now shown to be important. While we earlier suggested the magnetic field to be nearly uniform in the toroidal direction, with electrons and ions following those lines in helical orbits, in the working reactor the mega-ampere induced current produces its own poloidal field, circling the current itself, thus largely in the vertical direction. The resultant magnetic lines, in practice, vary vertically as much as they vary horizontally, and a parameter q, that might be 2 or 7, indicates how many times an orbit circles the horizontal torus circumference per total excursion in the vertical direction. Q is called the safety factor in the tokamak literature, as a large value of q, say 7, is associated with high stability of the plasma. In Figs. 6.11–12, the circles of transverse magnetic field in the transverse plane enclose succeeding fractions of the toroidal included current, for example said to be "Mega Amperes" in the MAST device, the "Mega Ampere Spherical Tokamak". The simplest form of the magnetic surfaces would be a set of nested toroidal surfaces. Magnetic field lines and current density lines are confined to these surfaces. The magnetic surfaces shown in these figures, circle the current flow, and may be labelled, starting from the core, the hottest- and highest-pressure region of the plasma, by the fraction of the total toroidal current they enclose. The 95 percent magnetic surface is near the separatrix, as suggested in the figure. Electron flow dominates the current that generates its own field, and enjoys the extremely long mean free path that was discussed earlier. But the ions then follow, in helical paths, the resultant magnetic field lines. At the outer part of the plasma, ions move to an important extent vertically, and are seen

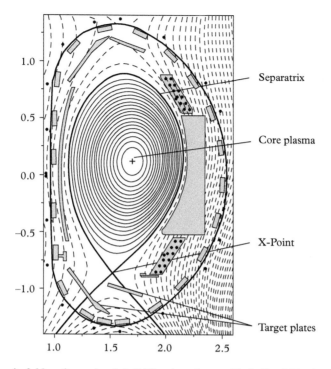

Separatrix

Core plasma

X-Point

Target plates

Fig. 6.11 *Magnetic field surfaces, closed (solid lines) and open (dashed) within the realistic Tokamak cross-section enclosed within the Dee shaped chamber wall. The denser magnetic surface spacing evident here to the right side of the "core plasma", results from the externally applied poloidal vertical magnetic field as suggested in Fig. 6.10. The core has the highest temperature, the helium ions there produced tend to migrate outward as they lose energy by heating the plasma. Once they reach the separatrix they move to the X-point and then exit the machine after giving much energy to the target plates (Ongena et al, 2016) Fig. 6b*

to follow the outermost magnetic field lines to exit the plasma. One essential new element in these figures is the "divertor" that serves as a particle and heat exhaust. The magnetic field structure is altered with the poloidal coil system to produce a crossing point that occurs in the lower part of the figure.

The shield or blanket is intended to reduce the energy of fusion neutrons to minimize their damage the tokamak structure, and, in the power reactor case, to transfer the fusion energy to hot steam to run a steam turbine. Elementary physics shows that the fraction of energy transferred in an elastic collision of an energetic mass m particle to a heavier particle of mass M at rest is $f = 4 \, mM/(m+M)^2$ that is 0.36 for m = 1 and M = 9 for beryllium.

A variety of undesirable plasma modes have been found to occur that reduce power. The underlying nanophysics of the fusion power generation is clear, but the engineering design of a practical reactor to avoid the plasma instabilities is difficult. Further, there are materials degradation problems. We have mentioned the power loss from the

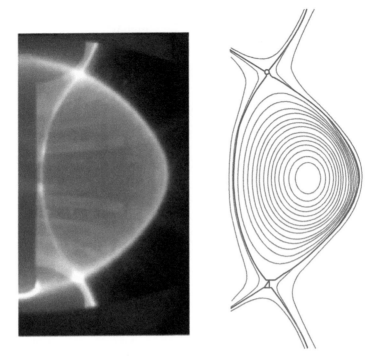

Fig. 6.12 *(Left) Plasma confinement evident from photograph. This is the Mega Ampere Spherical Tokamak (MAST) at Culham, England, with a small aspect ratio A = R/a = 0.9m/0.6 m = 1.5. The separatrix is shown to be a real effect in this image, leading particles to divertors at top and bottom, past the X-points of the magnetic field configuration. The overall plasma, defined by the outer separatrix, is nearly spherical; a small aspect ratio Tokamak is called a "spherical Tokamak". (Right). Magnetic field configuration in MAST Tokamak reactor corresponding to the plasma imaged on left. The central solenoid (not shown) runs vertically up the left side of the image, the inner boundary of the plasma is close to the central solenoid coil (Gusev et al, 2003) Fig. 15*

plasma if highly charged nuclei such as tungsten are eroded into the plasma. Wall erosion is also a maintenance problem, complicated by the fact that the first wall materials will become radioactive and require special handling.

An analysis of the availability in the sea of deuterium and lithium for a possible age of fusion-produced electric power has been given by MacKay (2009). Seawater contains 33 g of deuterium per ton, and this is a huge supply. Each gram of deuterium represents 100,000 kWh, and the sea contains 197 million tons of seawater per human (at 7.0 billion humans). Lithium is also available in sea water at 0.17 ppm, that translates to 3910 tons of lithium per person. If the lithium produces 2300 kWh per gram in the fusion reactors, and the energy usage per person is 105 kWh/day, then this energy source would last for over a million years. The success of the ITER reactor is not guaranteed, and even so it is not a power reactor but a research reactor. In the words of David MacKay (MacKay, 2009, p. 172).

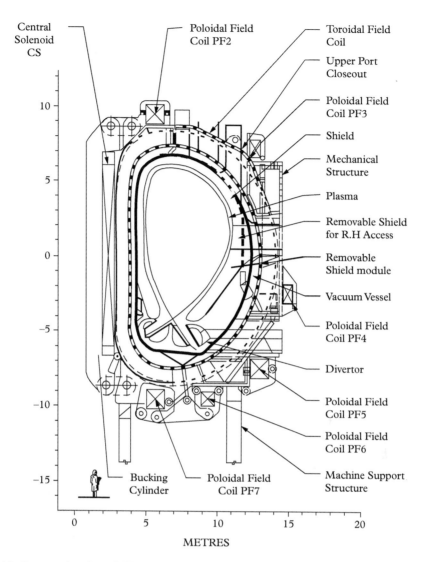

Central
Solenoid
CS

Poloidal Field
Coil PF2

Toroidal Field
Coil

Upper Port
Closeout

Poloidal Field
Coil PF3

Shield

Mechanical
Structure

Plasma

Removable Shield
for R.H Access

Removable
Shield module

Vacuum Vessel

Poloidal Field
Coil PF4

Divertor

Poloidal Field
Coil PF5

Poloidal Field
Coil PF6

Bucking
Cylinder

Poloidal Field
Coil PF7

Machine Support
Structure

METRES

Fig. 6.13 *Cross-section through ITER plan of 1995 showing the general plasma and divertor geometry. The 6 poloidal field coils, circular about the central vertical axis of the machine, and designed to shape the plasma, are labeled "PF2–PF7". These coils allow the vertical field structure to be varied to produce the field crossing points "X" shown in the previous Figure, and seen in this Figure as the crossing at the bottom of the innermost line labeled "plasma", the outer boundary of the central plasma, called the "separatrix" in the previous Figures. This is the boundary between closed magnetic surfaces and open magnetic surfaces. The magnetic lines below the crossing point lead ions downward to collide with "target plates" and thence to be pumped out of the machine. The "removable shield module", also called the "blanket" in other figures, is in the vacuum chamber outside the separatrix, and is designed to slow down the 14.1 MeV fusion neutrons to protect other components and to capture the primary energy of the fusion reactor. The blanket may contain water piping to extract heat. The "bucking cylinder" is a mechanical stabilizing element to counter forces on the central solenoid. The surface of the blanket is planned to be beryllium, with atomic mass 9, while the divertor plates are planned as tungsten (Dietz et al, 1995) Fig. 1*

"I think it is reckless to assume that the fusion problem will be cracked".

Even if the problem is "cracked", it is clear that a fusion reactor has to be large, a contributor to a power grid. While the fusion reactor seems quite safe, has no meltdown possibilities, and leaves few radioactive waste products, if indeed the engineering into a useful power reactor design succeeds, the end product will be very high technology only possible on a large scale, with high demands on an operator to make it work smoothly. The reader might look ahead to Fig. 10.15 that shows a working solar photovoltaic power plant. This is a view of the CdTe "Topaz" solar cell farm in San Luis Obispo County, California, completed in November 2014. There are 9 million fixed solar modules with capacity 0.55 GW, built by the American firm First Solar. This produces 550 MW of electricity directly when the Sun is shining, no need for steam turbines (that are not even part of the present ITER design, deferred to the following DEMO project). The solar cell facility has no moving parts. There is almost nothing that can go wrong with this facility, barring an episode of baseball-sized hail, unlikely in California. It requires minimal oversight and no long maintenance periods. Indeed, it is off at night, but existing electrolyzer technology could tap half the day's power to create hydrogen from water, with a corresponding array of fuel cells to return the hydrogen to electricity during the night hours. The Topaz facility cost $2.4 B, a figure that might be doubled by adding electrolyzers and fuel cells, or even lithium-ion or other batteries, to provide 24-hour power, but ITER has already cost $14 B. ITER cannot be built except by an initial fabrication facility on site, the coils are too large for railroad or truck. The comparison of the solar cell farm in Fig. 10.15 to Fig. 6.13 could lead one, on reflection, to conclude that ITER might more realistically be viewed as a physics project akin to the large hadron collider, rather than as a conceivably competitive approach to electric power.

7

More about the Atmosphere, Molecules, and their Interaction with Radiation

7.1 Introduction

In this chapter we provide a molecular basis for the absorption and emission from the atmosphere, first discussed in Chapter 3. This gives a better understanding of the solar spectrum as seen on Earth, that feeds photovoltaic devices as well as heating the Earth's surface, that in turn creates winds and waves that can be harvested. We then return to sophisticated measurement of the spectrum of outgoing infrared radiation, showing precisely how the greenhouse effect occurs.

The atmosphere plays two roles, as the medium of transmission of Sunlight to the Earth and as moderator of the Earth's radiation of infrared light back into the vacuum. The incoming radiation above the atmosphere is approximately a black-body spectrum at 6000 K (a detailed spectrum was shown in Fig. 1.4) and the emitted spectrum of the Earth is approximately a black-body spectrum near 250 K, as sketched in the upper panel (a) of Fig. 7.1. The areas under the two curves are equal, equating energy inflow and outflow, and allowing the Earth to maintain a constant temperature. The lower-panels (b) and (c), respectively show absorption spectra as measured at the Earth surface and at about 11 km.

All of the absorption features arise from molecules, and mostly from those that have three or more atoms. The main parts of the atmosphere, nitrogen and oxygen at 78 percent and 21 percent, respectively, are not important in the central wavelength region, although oxygen absorption is indicated at wavelengths less than 0.3 μm. At high-enough photon energy, the molecule can be dissociated or ionized, but this is not important in the energy balance between Earth and Sun. Diatomic oxygen and nitrogen are symmetrically bonded, much like hydrogen H_2 that was discussed in Section 5.5.5, and do not interact strongly with electromagnetic radiation in the visible range, incoming from the Sun.

Physics and Technology of Sustainable Energy. E. L. Wolf, Oxford University Press (2018).
© E. L. Wolf. DOI: 10.1093/oso/9780198769804.001.0001

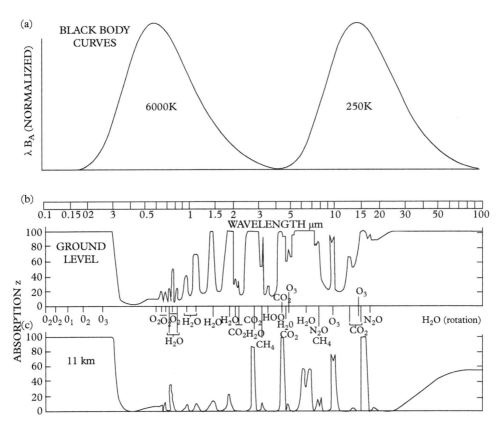

Fig. 7.1 *Atmospheric absorptions. (a) Black body curves for 6000 K and 250 K. (b) Atmospheric absorption spectrum for a solar beam reaching ground level. (c) The same for a beam reaching about 11 km, near what is called the temperate tropopause. Note in particular the ozone absorption at about 9.4 μm, that falls in the IR "window" 8–14 μm, and the CO$_2$ absorption at 14.85 μm (Goody and Yung, 1989) Fig. 1.1*

7.2   Electric dipole radiation

An electric dipole moment **p** is the primary means of molecule-radiation interaction, where **p** = d**e**, **d** being the vector spacing between two opposite charges e. In the ground state of the homonuclear diatomic molecule (e.g., N$_2$ as opposed to CO) the wavefunction is a symmetric (see Eq. (5.34)) linear combination of equal functions based on each atom and there is no electric dipole moment. But weaker effects are possible as we will see.

Radiation from an oscillating dipole moment can be described by a Poynting vector **S** $= 1/\mu_0$ **E × B** whose time average ⟨**S**⟩ is given by,

$$\langle \mathbf{S} \rangle = c^2\, Z_0 \pi^2\, |p|^2\, \sin^2\theta / (2\lambda^4 r^2) \quad \mathbf{r}^\wedge \tag{7.1}$$

Here r and θ describe the point of observation, along the unit vector $\mathbf{r}^\wedge$. The pattern of radiation is symmetric around the vector $\mathbf{p}$ and is maximum in the direction perpendicular to the dipole vector, $\theta = 90°$. Here $Z_0 = (\mu_0/\varepsilon_0)^{1/2}$ is the impedance of free space, 376.7 Ω, and $\lambda = c/f$ is the radiation wavelength at the oscillating dipole frequency f. The total radiated power from this system, from integrating the Poynting vector, is:

$$P = 4\ c^2 Z_0 \pi^3 \left|p\right|^2/(3\lambda^4) \tag{7.2}$$

The radiated power is proportional to $(1/\lambda)^4$ as was noted in connection with Rayleigh scattering that colors the sky blue, see Eq. (1.2).

Absorption of light by molecules occurs in an inverse fashion as the light interacts with the dipole moment.

7.3   Molecular structures and radiation interaction

Following our discussion of the simplest diatomic molecule H_2 in Chapter 5 we can look on H_2 as two masses connected by a spring, and expect it to have an oscillation frequency of the form $\omega = (K/m)^{1/2}$ where the spring constant K has units N/m and m is the reduced mass (4.25). The spring constant K is identified as the curvature of the energy-displacement relation for the two masses, as was discussed in Eqs. (6.14a–6.17). The potential and the allowed vibrational levels are indicated in Fig. 7.2. The simple harmonic oscillator rules apply for a strictly quadratic potential, and also for the lowest-energy states of a more realistic potential suggested in Fig. 7.2. The states of the simple harmonic oscillator, with energies,

$$E_n = (n + \tfrac{1}{2})\ \hbar\omega, \tag{6.17}$$

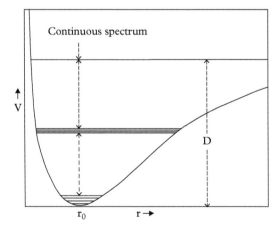

Fig. 7.2 *General molecular potential V(r) vs interatomic spacing r showing vibrational levels. D is the dissociation energy (Dennison, 1931) Fig. 1, p. 282*

are of the form:

$$\psi_n(y) = (m\omega/\pi\,\hbar)^{1/4}\,(2^n\,n!)^{-1/2}\,H_n(y)\,\exp(-y^2/2), \qquad (7.3)$$

where $y = r - r_e$, and $H_n(y)$ are Hermite polynomials. The first few of these are:

$$H_0(y) = 1,\; H_1(y) = 2y,\; H_2(y) = 4y^2 - 2,\; H_3(y) = 8y^3 - 12y. \qquad (7.4)$$

The probability distribution for the oscillator is $|\psi_n(y)|^2$ and approaches the classical distribution for large n. The selection rule specifically for the harmonic oscillator, but relaxed if the potential has terms beyond quadratic in the spacing, is:

$$\Delta n = \pm 1. \qquad (7.5)$$

When a dipole moment is present, absorption of light, since the oscillator is usually in its ground state, will appear at only one frequency. Again, if the potential has higher-order terms past quadratic, weak absorption may occur at $\Delta n = 2$, and this will be called an "overtone". (Absorption, as noted previously, will require an electric dipole moment, not present in molecules like H_2, O_2 and N_2.)

A second basic motion of the diatomic molecule is a rotation about its center of mass, that is associated with an energy $\frac{1}{2}\,I\,\omega^2$, with I the moment of inertia about the chosen axis. If the molecule has an electric dipole moment, this will allow absorption or emission of radiation. In quantum mechanics the energy states of the rotor are:

$$E_1 = (\hbar^2/2I)\,J(J+1), \qquad (7.6)$$

where J, an integer, governs the modulus of the angular momentum vector:

$$|\mathbf{L}| = \hbar(J(J+1))^{1/2}, \qquad (7.6a)$$

that has degeneracy $2J + 1$ as shown in Fig. 5.12. The symbol B is often used to represent $(\hbar^2/2I)$, that has units of energy. The selection rule for the rotor is:

$$\Delta J = \pm 1 \qquad (7.6b)$$

(Dennison, 1931) Fig. 1, p. 282.

The energy values and the selection rule lead to a series of nearly equally spaced absorption lines at frequency v in Hz with spacing $2B/h$:

$$v = (2B/h)\,(J+1). \qquad (7.7)$$

These are very low energies usually and the spectrum of rotational lines can look like a continuum.

7.3.1 The hydrogen molecule

If we take the hydrogen molecule as a starting point, looking at Fig. 5.14, and note that the measured vibrational frequency is 4161 cm^{-1} = 0.5195 eV = 2.387 μm. This also corresponds to ω = 7.91 × 10^{14}/s. If we use the basic oscillator formula ω = $(K/m)^{1/2}$, taking m as the reduced mass m_p/2 for the two protons, we find K =1047 N/m. Using the energy of a spring as E =1/2 Kx^2 to fit the binding energy curve of Fig. 5.14, if we set E = 4.7 eV = 7.52 × 10^{-19} J, we get x = 0.379 Angstrom. This is quite close to the curve shown. (The spring constant K is just the curvature d^2E/dx^2.)

The moment of inertia I is defined as:

$$I = \sum m_i r_i^2$$

(7.7a)

To estimate the rotational energy levels, we construct I = $\sum m_i r_i^2$ = 4.578 × 10^{-48} kg-m^2 for CO$_2$ (see Fig. 7.3), leading to the basic rotational energy B = $\hbar^2/2I$ = 1.205 × 10^{-21}J = 7.53 meV. This corresponds to an electromagnetic wavelength λ = hc/E = 164.7 μm, very far in the infrared, and off the scale to the right in Fig. 7.1, where the notation "rotation" appears attributed to water vapor near 100 μm. Hydrogen, and also molecular oxygen and nitrogen in the atmosphere all exhibit these vibrational and rotational motions but they do not interact with light because these majority molecules do not possess an electric dipole moment due to their symmetry.

7.3.2 Carbon dioxide, a linear triatomic molecule

Let us now consider an important molecule, carbon dioxide, that does develop an electric dipole moment in some of its vibrations, is important in the atmosphere, and is more complicated because it has three atoms. Fig. 7.3 shows this molecule and its vibrational "normal modes". "Normal modes" are linear combinations of atomic motions in the molecule that collectively act as single one-dimensional oscillators as described previously.

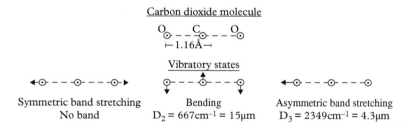

Fig. 7.3 *Molecular structure and vibrations for carbon dioxide. The energy of the symmetric vibration, not directly seen in infrared absorption, is 1337/cm. In the text the numbering of the modes 2 and 3 is interchanged. The molecule will possess also two modes of rotation about the central C atom (Peixoto and Oort, 1992) Fig. 6.5, p. 107*

This linear molecule with three masses is assumed to have a single spring constant k^1, and the equations for the linear motions of the masses at locations x_1, x_2 and x_3 are:

$$m_1 d^2x_1/dt^2 = -k(x_1 - x_2) \tag{7.8}$$

$$m_2 d^2x_2/dt^2 = -k(2x_2 - x_1 - x_3) \tag{7.9}$$

$$m_1 d^2x_3/dt^2 = -k(x_3 - x_2) \tag{7.10}$$

One assumes oscillatory motions for all three masses:

$$x_i(t) = x_i'e^{i\omega t}, \tag{7.11}$$

to find revised equations from which the frequencies can be deduced. Namely:

$$0 = -k(x_1' - x_2') + m_1\omega^2 x_1' \tag{7.12}$$

$$0 = -k(2x_2' - x_1' - x_3') + m_2\omega^2 x_2' \tag{7.13}$$

$$0 = -k(x_3' - x_2') + m_1\omega^2 x_3' \tag{7.14}$$

These can be compactly rewritten as a matrix equation:

$$\begin{pmatrix} m_1\omega^2 - k & k & 0 \\ k & m_2\omega^2 - 2k & k \\ 0 & k & m_1\omega^2 - k \end{pmatrix} \cdot \begin{pmatrix} x_1' \\ x_2' \\ x_3' \end{pmatrix} = \vec{0}. \tag{7.15}$$

For a nontrivial solution to the matrix equation the characteristic polynomial must vanish, following Bronner et al (2008), corresponding to:

$$(m_1\omega^2 - k)^2(m_2\omega^2 - 2k) - 2k^2(m_1\omega^2 - k) = 0. \tag{7.16}$$

Setting the common factor $(m_1\omega^2 - k)$ to zero gives the first eigenfrequency,

$$\omega_1 = (k/m_1)^{1/2}, \tag{7.17}$$

that can easily be identified from the Fig. 7.3 as the symmetric stretching vibration where only the ^{16}O masses m_1 are in motion. The overall symmetry of this stretching vibration indicates that no electric dipole is involved, and the mode will not appear in electromagnetic absorption or emission.

---

[1] Following Bronner et al 2008, http://users.physik.fuberlin.de/~essenber/Dateien/VersucheChemie/rotvib.pdf (accessed 4/27/2017)

The remaining equation,

$$\left(m_1 \, \omega^2 - k \right)\left(m_2 \, \omega^2 - 2k \right) = 2k^2, \tag{7.18}$$

when multiplied out, easily reduces to:

$$\omega^2 = k\left(2m_1 + m_2 \right)/m_1 \, m_2. \tag{7.19}$$

Thus, a second eigenfrequncy is given as:

$$\omega_2 = \left[k\left(2m_1 + m_2 \right)/m_1 \, m_2 \right]^{1/2} = \left(k/m_1 + 2k/m_2 \right)^{1/2}. \tag{7.20}$$

It is clear that this frequency is larger than the symmetric stretch eigenfrequency ω_1, and this is easily understood since the motion of the center mass C stretches two springs in this motion, rather than just one spring per mass in the symmetric motion. The factor of 2 enters as a consequence of the fact that the center of mass along × remains constant, with no external force, in the vibration motion. It is plausible that this asymmetric motion would move a center of negative charge relative to a center of positive charge, resulting in an oscillating dipole moment. So, this vibration indeed shows up in the absorptions attributed to carbon dioxide in the Earth's atmosphere.

In looking at Fig. 7.2 it is clear, in addition, that two perpendicular bending motions will occur, these are referred to as deformation modes. To understand these motions a new spring constant κ is defined, that gives the force resulting from an angular change, call it bending δ, in the O-C bond (of length L). Thus, an added Newton's law equation is needed:

$$F = -L\kappa\delta. \tag{7.21}$$

It turns out (Bronner et al 2008) that the corresponding frequency is:

$$\omega_3 = \left[\kappa \left(m_1 + 2m_2 \right)/m_1 \, m_2 \right]^{1/2}. \tag{7.22}$$

One might expect this frequency to be smaller than the frequencies that depend on stretching, rather than bending, of the covalent bonds.

To summarize this analysis, it follows that the spring constants for CO_2 vibrations can be expressed in terms of the infrared-active normal mode frequencies as:

$$k = m_1 m_2 \, \omega_2^2/\left(2m_1 + m_2 \right), \tag{7.23}$$

and,

$$\kappa = m_1 m_2 \, \omega_3^2/\left(m_1 + 2m_2 \right), \tag{7.24}$$

where m_1 and m_2, respectively, represent O and C atoms.

A wide-range laboratory measurement of the infrared absorption of CO_2 is shown in Fig. 7.4. The striking first point is that we see bands of absorptions, not sharp single absorptions as predicted by the linear normal mode model ignoring rotations. The good news is that two of the bands are indeed centered at the main energies predicted from the linear stretching mode and deformation mode analysis, 2349 cm^{-1} and 667 cm^{-1}. These bands are about 100 cm^{-1} wide. The most reasonable origin for the width, since the deformation mode energy is already 667 cm^{-1}, will be rotations of the molecule involving high-values of the rotational quantum number J. A second striking point is that there are unexpected strong absorptions near 3600 cm^{-1}, higher than the highest-frequency stretching vibration of the molecule. These are not at all predicted by the linear stretching analysis.

This spectrum shows two of the three vibrations mentioned, omitting, as expected, the symmetric vibration. However, it shows two additional peaks at high energy, 3609 and 3716 cm^{-1}, that cannot be attributed directly to individual vibration modes of the molecule. The asymmetric stretch vibration is the highest-frequency motion the system will exhibit, and it is agreed that the added lines must mean that multiples and linear combinations of individual absorption lines do occur, adding greatly the observed infrared spectra of molecules. These effects are attributed to departures of the interatomic forces from being strictly quadratic, as previously mentioned, and do not invalidate the Newton's Law analysis of the normal mode system. In addition, the lines are not simple peaks but have fine structure. The deformation peak at 667 cm^{-1} in particular has sidebands that probably are related to rotations of the molecule, beyond the prediction of the one-dimensional mechanical model described.

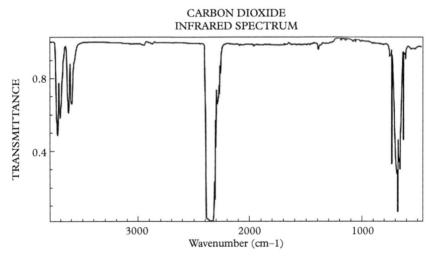

CARBON DIOXIDE
INFRARED SPECTRUM

Fig. 7.4 *Infrared spectrum for carbon dioxide. The energies from right to left are 667, 2349, 3609 and 3716 cm^{-1}. (The energy of the symmetric vibration, not directly seen in infrared absorption, but revealed by Raman spectroscopy, is 1337/cm.) http://webbook.nist.gov/cgi/cbook.cgi?Spec=C124389&Index=1& Type=IR*

Let us determine the spring constant k from the asymmetric stretch energy 2349 cm^{-1} corresponding to $\omega = 2\pi c/\lambda = (2\pi) \, 3 \times 10^{10} \times 2349 = 4.43 \times 10^{14}$ rad/s.

According to Eq. (7.23), taking m_1 and m_2, respectively, as ^{16}O and ^{12}C and $\omega_2 = 4.43 \times 10^{14}$ rad/s, we find:

$$k = \left[(16 \times 12)/(32 + 12) \right] \times 1.67 \times 10^{-27} \times \left(4.43 \times 10^{14} \right)^2 = 1430 \text{ N/m}.$$

As a check on the consistency one can also determine k from the reported frequency, 1337 cm^{-1}, of the symmetric vibration, and this gives:

$$k = 16 \times 1.67 \times 10^{-27} \left(2.5 \times 10^{14} \right)^2 = 1670 \text{ N/m}.$$

This is 17 percent higher, but close enough to leave the basic analysis intact, considering the width of the observed lines. That these values are larger than with the value 1047 N/m deduced previously for the hydrogen molecule, is reasonable as the CO_2 bonds are classified as double bonds, with four electrons, rather than a single covalent bond in hydrogen.

In the early work of Dennison (1940) p. 180, the extra lines at 3609 and 3716 cm^{-1} are attributed, respectively, to two units of the deformation vibration plus one unit of the asymmetric stretch (giving 3681 cm^{-1}), and one unit each of the symmetric and antisymmetric stretch motions (giving 3686 cm^{-1}). This single molecule is seen to be a complex system even if viewed only from its set of vibrations and rotations. According to Dennison, a total of 25 vibrational lines extending from 600 cm^{-1} to 12,775 cm^{-1} have been identified for CO_2. According to Dennison (1940) p. 180, this latter absorption is indexed as one unit of symmetric vibration plus 5 units of asymmetric distortion, adding to 13,072 cm^{-1} using our numbers.

An added complexity, first observed in the Raman spectrum of CO_2, as was pointed out by E. Fermi, relates to the unseen symmetric vibration located as 1337 cm^{-1}. This number is reached as the average of the energies of two strong Raman lines that are seen at 1285 and 1388 cm^{-1}. A single line might have been expected, rather than the pair.

A Raman line is observed when a molecule is raised to a high-energy state by virtually absorbing a high-energy photon, that creates momentarily in the molecule a distortion and electric dipole moment. In Rayleigh scattering, described in Eqs. (7.1–7.2) the same photon is re-emitted, but into the spatial pattern characteristic of electric dipole radiation. In Raman scattering, with much lower probability, an additional emission energy is observed, reduced by a characteristic energy that would in this case correspond to one of the vibrational excitations of the molecule. Careful measurements show that re-emission from CO_2, in addition to the elastic Rayleigh scattering, also occurs with characteristic losses in energy corresponding to 1285 and 1388 cm^{-1}. It was noted by Fermi that the average of these energies, 1337 cm^{-1} was close to twice the energy of the deformation vibration, $2 \times 667 = 1334$ cm^{-1}. Fermi predicted that that fact would cause an interaction between the two, with the result of nearly equal strong intensities of the two observed Raman lines, instead of an initially expected single Raman line at the symmetric stretch frequency.

The deformation peak centered at 667 cm^{-1} has fine structure. Carefully looking at the center of the peak, it appears that there are two lines at the center of the 667 cm^{-1} peak separated by about 25 cm^{-1}. The question is if that spacing could arise from the rotation of the linear CO_2 molecule about a line perpendicular running through the central C atom. The moment of inertia for that motion, taking the interatomic spacing as 1.16 Angstrom from Fig. 7.3, and the masses noted previously, is $I = 7.19 \times 10^{-46}$ kg-m^2.

The rotational energy $E_1 = (\hbar^2/2I)\, J(J+1) = B\, J(J+1)$ from (7.6), gives $B = 0.04795$ meV. In wavenumbers, since 1 meV is 8.064/cm, $B = 0.387$ cm^{-1} for the CO_2 molecule. Since the spacing of rotational lines is 2 B, to obtain a half width of about 50 cm^{-1} one would need $J = 50/[2 \times 0.387] \approx 65$.

We can estimate what values of J are thermally excited at 300 K by setting $k_B T = 26$ meV $= B\, J(J+1)$. This gives $J = 23$. Following Eq. (7.7), the energy would be $2B(J+1)$, so about $24 \times 2 \times 0.387$ cm^{-1} = 18.6 cm^{-1}. This is reasonable, so it looks like the width of the observed bands in the spectra simply arise from the molecule being thermally excited to a set of rotational states well beyond $J = 23$ at room temperature.

7.3.3   Vibration-rotation spectrum

Since the rotational states seem to account for the width of the observed bands, it is appropriate to look more carefully at the transition energies that are implied by Eq. (7.7). The line structure is suggested in Fig. 7.5.

In Fig. 7.5 only the first two vibrational levels are shown (see Fig. 7.2) with the rotational states added. All transitions shown start at vibrational state 0 and go to vibrational state 1. The transitions break into two types, on the right side of the diagram, where the J value increases by one unit, is the R branch, while on the left, where the J value in the transition decreases by one unit, is the P branch. (In some cases, not shown, transitions with $\Delta J = 0$ occur, and these are said to be on a Q branch.)

Returning to CO_2, a much higher-resolution infrared (IR) gas phase absorption spectrum near the asymmetric stretching frequency 2349/cm is shown in Fig. 7.6 (Castle, 2007). Most of the lines are indexed in detail, within the framework just described. The indexing is shown in Table 7.1. This spectrum shows that the individual levels are very sharply defined in energy, even though the overall effect of the complexity, if observed at low resolution, is to broaden the spectrum into a band.

All of the transitions shown in Table 7.1 excite one unit of the asymmetric stretch vibration, confirming the idea of a band based on single excitation of the asymmetric stretch vibration. Clearly, the excitation energy of that vibration is near 2300/cm so that an improved value of the spring constant k can be obtained using an energy 2300/cm, that gives k = 1371 N/m.

Indexing of these transitions identifies six initial vibrational states that are (00^00), (01^10), (10^00), (02^20), (11^10) and (03^30). The superscript number is the angular momentum in the linear combination of deformation states. The highest initial J quantum number in the set is 44, that would correspond to an initial thermal excitation energy $2BJ(J+1) = 3960B = 189.9$ meV. The classical chance of this excitation is

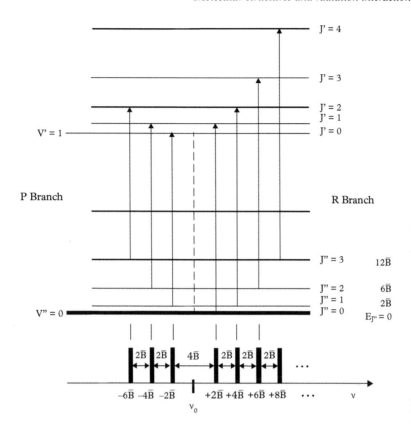

Fig. 7.5 *Vibration and rotation levels schematically, indicating transitions on the P and R branches. Transitions in Fig. 7.6 below and Table 7.1 are of this type, with one-unit increase in excitation of asymmetric stretch vibration $\Delta v = +1$. If rotation quantum number J increases transition energy is increased, on the R branch. If rotation quantum number decreases, on the P branch, the energy is smaller than the basic vibrational energy change. Temperature affects the population of initial $J > 0$ states (B with the super bar indicates cm^{-1} units)*

$$\exp(-\Delta E/kT) = \exp(-189.9/26) = 6.7 \times 10^{-4} \text{ at } 300 \text{ K.}$$

The absorptions shown in Fig. 7.3 and discussed are also evident in Fig. 3.2, where carbon dioxide absorptions near 15 μm (667 cm^{-1}) (bending deformation); 4.3 μm 2349 cm^{-1} (asymmetric stretch), 2.7 μm (3700 cm^{-1}) (linear combinations) and also absorption near 2.0 μm (about 5000 cm^{-1}) are seen. According to Dennison (1940) p. 180, combination vibrations are indexed at 4859, 4981 and 5108 cm^{-1}. It appears that the rotational states adequately account for the broadening into bands of the transitions seen in Figs. 3.2, 7.3 and 7.5. In conclusion, the properties of carbon dioxide are very well known, the system is complex even at three atoms, and the addition of linear combinations and overtones (multiples) of fundamental vibrations plus their rotational broadening make for many absorption wavelengths beyond the basic vibrations.

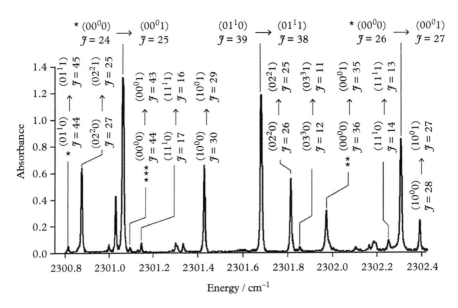

Fig. 7.6 *High resolution infrared spectrum for carbon dioxide, in a portion of the asymmetric stretch band usually quoted as 2349 /cm. The assignment (111 1), e.g., specifies an initial vibrational state with one unit each, respectively, of symmetric stretch, deformation, angular momentum, in the linear combination of deformation, and asymmetric stretch. The asterisks indicate isotopic variations: * has carbon 13, ** has O 18 and *** has O 17. J is the rotational quantum number (Castle, 2007) Fig. 1*

Because CO_2 has no electric dipole moment there are no pure rotation bands as there are for water, for example. This means that there is no absorption by carbon dioxide for wavelengths longer than 15 μm.

7.3.4   The water molecule H_2O

The next molecule to consider, because of its importance in atmospheric absorption, is water, H_2O, prominent in Fig. 3.2. We can expect that the deformation/bending mode will be at higher frequency than in carbon dioxide, because the hydrogen is much lighter than oxygen, and it seems in Fig. 3.2 that there is a broad band around 6 μm (1667 cm^{-1}). This bending mode is shown in Fig. 7.7, and indicated as corresponding to 6.25 μm = 1595 cm^{-1}.

Because of the electric dipole moment (that is oriented in the plane of the molecule bisecting the 105° angle between the two OH bonds) water has absorptions in the far infra-red extending from the lowest vibrational mode at 6.25 μm all the way to about 500 μm. As suggested in Fig. 3.2, the pure rotational absorption of water reaches a peak in the vicinity of 50 μm and decreases to shorter wavelengths until the onset of the bending/deformation band starting at 8.5 μm. The moments of inertia of water are all different, putting the molecule into the class of asymmetric rotators. The axis for the

Table 7.1 *Energies and assignments for CO_2 absorptions in the narrow range 2300/cm − 2302.4/cm shown in Fig. 7.6 In these assignments, e.g., $(11^1\ 1)$ specifies an initial vibrational state with one unit each, respectively, of symmetric stretch, deformation, angular momentum in the linear combination of deformation states, and asymmetric stretch. The asterisks indicate isotopic variations: * has carbon 13, ** has O 18 and *** has O 17. J is the rotational quantum number.*

Transition Energy/cm^{-1}	Assignment
2300.815	$*(01^10)\ J = 44 \rightarrow (01^11)\ J{=}45$
2300.874	$(02^20)\ J = 27 \rightarrow (02^21)\ J{=}26$
2300.999	$(03^30)\ J = 13 \rightarrow (03^31)\ J{=}12$
2301.031	$**(00^00)\ J = 37 \rightarrow (00^01)\ J{=}36$
2301.064	$*(00^00)\ J = 24 \rightarrow (00^01)\ J{=}25$
2301.096	$***(00^00)\ J = 44 \rightarrow (00^01)\ J{=}43$
2301.146	$(11^10)\ J = 17 \rightarrow (11^11)\ J{=}16$
2301.428	$(10^00)\ J = 30 \rightarrow (10^01)\ J{=}29$
2301.681	$(01^10)\ J = 39 \rightarrow (01^11)\ J{=}38$
2301.814	$(02^20)\ J = 26 \rightarrow (02^21)\ J{=}25$
2301.855	$(03^30)\ J = 12 \rightarrow (03^31)\ J{=}11$
2301.974	$**(00^00)\ J = 36 \rightarrow (00^01)\ J{=}35$
2302.107	$***(00^00)\ J = 43 \rightarrow (00^01)\ J{=}42$
2302.253	$(11^10)\ J{=}14 \rightarrow (11^11)\ J{=}13$
2302.309	$*(00^00)\ J = 26 \rightarrow (00^01)\ J{=}27$
2302.394	$(10^00)\ J = 28 \rightarrow (10^01)\ J{=}27$

intermediate moment of inertia is along the electric dipole. Two other rotation axes are in the plane perpendicular to the electric dipole moment (smallest) and out of the plane through the center of mass (largest).

The lowest purely rotational states of H_2O inferred from spectra are listed in Table 7.2. The selection rule $\Delta J = \pm 1$ or 0 applies, as well as symmetry related rules $(+\ +) \leftrightarrow (-\ -)$ and $(+\ -) \leftrightarrow (-\ +)$. Following these rules, it appears that the lowest-energy transition is $42.36{-}23.76\ \text{cm}^{-1} = 18.6\ \text{cm}^{-1} = 538\ \mu\text{m}$. This exact absorption was in fact measured by Hall and Dowling (1967), shown in their Fig. 1 and Table I as $18.57\ \text{cm}^{-1}$ with an indexing $1_{01} \rightarrow 1_{10}$.

We can connect these numbers to the geometry and masses in the water molecule by noting that the three principal moments of inertia, using Eq. (7.7a) are (by convention, the smallest moment is designated A):

Water vapor molecule

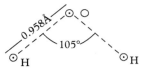

Vibratory states

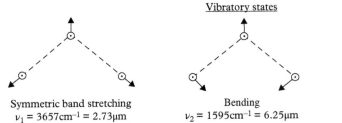

Symmetric band stretching	Bending	Asymmetric band stretching
$\nu_1 = 3657\text{cm}^{-1} = 2.73\mu\text{m}$	$\nu_2 = 1595\text{cm}^{-1} = 6.25\mu\text{m}$	$\nu_3 = 3756\text{cm}^{-1} = 2.66\mu\text{m}$

Fig. 7.7 *Structure and basic vibrations of the water molecule (Peixoto and Oort, 1992) Fig. 6.5, p. 107*

$A = 1.09 \times 10^{-47}$ kg m², $B = 1.91 \times 10^{-47}$ kg m² and $C = 3.0 \times 10^{-47}$ kg m². (These calculations have to be done with care[2], because the principal axes run through the center of mass, that is slightly displaced from the oxygen toward the hydrogens along the line bisecting the 105° angle.)

The three rotational energies for $J = 1$, following Dennison (1931) p. 321 are:

$$E_{1,1} = (\hbar^2/2)\ (1/A + 1/B) \tag{7.24a}$$

$$E_{1,0} = (\hbar^2/2)\ (1/A + 1/C) \tag{7.24b}$$

$$E_{1,-1} = (\hbar^2/2)\ (1/B + 1/C) \tag{7.25}$$

Putting the moments of inertia numbers A, B and C into Eqs. (7.23–7.25) we find:

$$E_{1,1} = (\hbar/2)\ (1/A + 1/B) = 5.517 \times 10^{-69}\ (9.17 \times 10^{46} + 5.24 \times 10^{46})$$
$$= 7.95 \times 10^{-22}\ \text{J} = 39.94\ \text{cm}^{-1}$$

$$E_{1,0} = (\hbar/2)\ (1/A + 1/C) = 5.517 \times 10^{-69}\ (9.17 \times 10^{46} + 3.33 \times 10^{46})$$
$$= 6.90 \times 10^{-22}\ \text{J} = 34.73\ \text{cm}^{-1}$$

$$E_{1,-1} = (\hbar/2)\ (1/B + 1/C) = 5.517 \times 10^{-69}\ (5.24 \times 10^{46} + 3.33 \times 10^{46})$$
$$= 4.72 \times 10^{-22}\ \text{J} = 23.76\ \text{cm}^{-1}.$$

[2] Carl W. David (2006). "The Tensor of the Moment of Inertia". http://digitalcommons.uconn.edu/chem_educ/21/, accessed Jan 27, 2018

Table 7.2 *Lowest-observed rotational states of water H_2O, in cm^{-1}. J is the total angular momentum and subscript τ orders the states in energy and appears to represent the projection of that angular momentum on a particular direction. The $+\,-$ symbols describe the symmetry of the state, with a selection rule for transitions that $+\,+$ can change only to $-\,-$ (and vice-versa) and $+\,-$ can change only to $-\,+$ (and vice-versa). After Dennison (1940), p. 189.*

	J_τ	W/hc
$++$	0_0	0
$+-$	1_1	42.36
$--$	1_0	37.14
$-+$	1_{-1}	23.76
$++$	2_2	136.15
$+-$	2_1	134.88
$--$	2_0	95.19
$-+$	2_{-1}	79.47
$++$	2_{-2}	70.08
$+-$	3_3	285.46
$--$	3_2	285.26
$-+$	3_1	212.12
$++$	3_0	206.35
$+-$	3_{-1}	173.38
$--$	3_{-2}	142.30
$-+$	3_{-3}	136.74

These numbers are quite close to those derived from the spectra, as collected in Table 7.2.

From these numbers, and selection rules, the lowest transition energy is 39.94 − 23.76 = 16.18 cm⁻¹, that is close to the observed value 18.57 cm⁻¹.

To understand the small differences, the paper of Hall and Dowling (1967) Table VI gives values of rotational constants for water, based on $E = \hbar^2/2I$, as 27.1107, 14.599 and 9.512 cm⁻¹. These correspond, respectively, to moments of inertia 1.0246×10^{-47} kg m², 1.902×10^{-47} kg m² and 2.92×10^{-47} kg m². It appears that our adopted value of A is larger than the accepted value 1.0246×10^{-47} kg m², accounting for most of the discrepancy.

In fact there are actually 3 other even lower energy, but very weak, microwave rotational transitions in water, at 6, 11 and 15 cm⁻¹, observed (Lichtenstein et al, 1966) and indexed, respectively, as J: 2→3, J: 4→5 and J: 3→4.

These transitions can be seen as allowed differences in energy levels in Table III of Dennison (1940), part of which is Table 7.2. For example, Lichtenstein et al (1966) found a transition at 183310 MHz that corresponds to 6.11 cm^{-1} that they index as J: $2_2 \rightarrow 3_{-2}$. Looking at Table 7.2, these two energies are, respectively, 136.15 and 142.30 cm^{-1}. So, the difference energy is 6.15 cm^{-1}, in excellent agreement. These new long wavelength absorptions come from small differences in rather larger energies corresponding to higher J values. The Formulas (7.23–7.25) apply only for J = 1, for larger J different rules apply as listed by Dennison (1931) Table II p. 321. The rules are very complicated and lead to widely varying transition energies.

A sample of the absorption spectrum from rotation levels of water is given in Fig. 7.8. These absorptions due to purely rotational transitions are possible only with an electric dipole moment, present in water but not in CO_2. Lichtenstein et al (1966) find that the dipole moment of water is 1.884 Debye units, each equal to 0.2082 e Angstrom.

The wavelength corresponding to f = 1000 GHz = 10^{12} Hz is $\lambda = c/f = 3 \times 10^8/10^{12} =$ 300 μm.

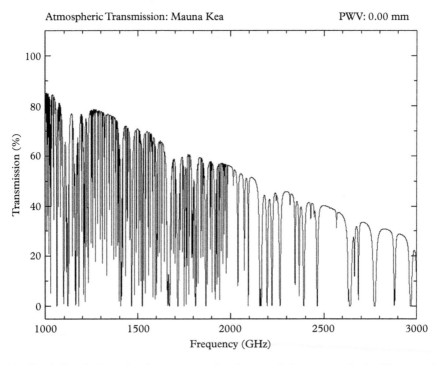

Fig. 7.8 *Far infrared absorption from pure rotational states of the water molecule. The range 1000–3000 GHz corresponds to 33.3 cm^{-1} to 100 cm^{-1} https://commons.wikimedia.org/wiki/File:Atmospheric_terahertz_transmittance_at_Mauna_Kea_(simulated).png*

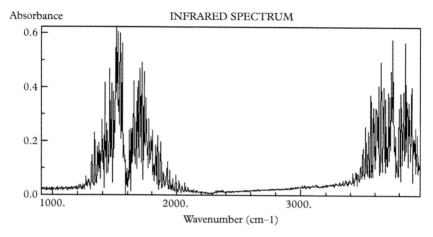

Fig. 7.9 *Infrared absorption spectrum of water vapor, based on vibrations. The bending absorption band is seen on the left and the symmetric and asymmetric stretching bands are seen on the right. Webbook. nist. gov*

Water is important in the absorption by the Earth's atmosphere, as is evident from Fig. 3.2 and Fig. 7.1. The absorptions for wavelengths shorter than 10 μm (energy greater than 1000 cm^{-1}) require a vibrational excitation, with the bandwidth arising from a distribution of thermally excited rotational states as suggested in Fig. 7.5. The main vibration-related absorption bands of water are shown in Fig. 7.9.

7.3.5   Ozone O$_3$

The structure and vibration states of ozone O$_3$ are shown in Fig. 7.10.

Ozone is similar to water, as a non-linear triatomic molecule, but has generally lower vibrational frequencies since the oxygen mass exceeds the hydrogen mass. The electric dipole moment of ozone is reported as 0.53 Debye units, so it has a pure rotational spectrum. The rotational constants A,B and C (e.g., A = $\hbar^2/2I_A$) for ozone are reported, respectively, by Gora 1959, as 3.55, 0.44, and 0.394 cm^{-1}, leading to a rotational absorption spectrum extending from 0 to about 90 cm^{-1}. The main location of ozone in the atmosphere is in the stratosphere where the temperature is near 220 K. (The origin of ozone is related to the optical excitation of molecular oxygen in the Schumann–Runge bands will soon be described.) The thermally excited J values range up to about 40 but values near J = 20 account for the strongest lines in the pure rotation spectrum of ozone. As in the cases of carbon dioxide and water, the thermally excited range of rotational states accounts for the width of the main absorption bands. These bands for ozone lie near 14.3 μm (701 cm^{-1}) for bending; 9.0 μm (1110 cm^{-1}) for symmetric stretch; and 9.6 μm (1045 cm^{-1}) for anti-symmetric stretch. The ozone absorptions at

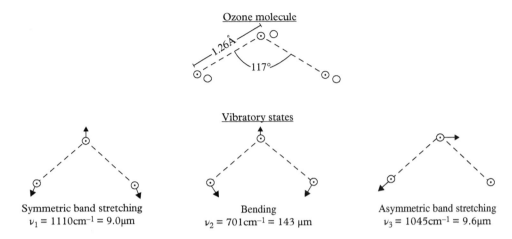

Fig. 7.10 *Structure and vibrational states of ozone (Peixoto and Oort, 1992) Fig. 6.5, p. 107*

9 and 9.6 μm are particularly noticeable in Fig. 7.1 because they lie in the "window" of low-atmospheric absorption centered around 10 μm.

Other absorptions for ozone in Fig. 7.1 appear at about 0.27 μm and about 0.65 μm, that can be regarded as overtones or linear combinations of the basic ozone vibrations. Other bands arising from ozone will be mentioned shortly.

7.3.6   Absorptions by molecular oxygen O_2

Molecular oxygen is found to strongly absorb in the ultraviolet, as seen in Fig. 7.12, and weakly absorb in the visible and near infrared, in spite of its symmetry, that implies lack of an electric dipole moment. Absorptions at 0.762, 0.688 and 0.629 μm are shown in Fig. 7.11.

All of these absorption lines represent transitions from the ground state to the first electronic excited state of the oxygen molecule, that lies higher in energy by 6.12 eV. An approximate band diagram for the oxygen molecule is shown in Fig. 7.13. The excited state is labeled $^3\Sigma_u^-$ (also referred to as B – $^3\Sigma_u^-$, or just B) while the ground state is referred to as X – $^3\Sigma_g^-$ or just X. The ground state is known to have 35 separate vibrational levels and the excited state at least 17 separate vibrational levels. A portion of the strong Schumann–Runge absorption system in shown in Fig. 7.14 that corresponds to transitions from the lowest vibrational level of the ground state to vibrational levels numbered 12 through 17 of the electronic excited state.

One can see from Fig. 7.14 that the spacing of the vibrational levels in the excited state B – $^3\Sigma_u^-$ is around 250 cm^{-1} near v' = 12. The spacing of vibrational levels in the ground state is on the (Bytautas et al 2010) order of 1100 cm^{-1}, on the order of the width of this plot.

The nature of these densely packed absorption bands is shown in Fig. 7.14. The transitions are from the lowest vibrational level of the ground state of the oxygen molecule to excited vibrational states labeled v'. Both states are further split into rotational

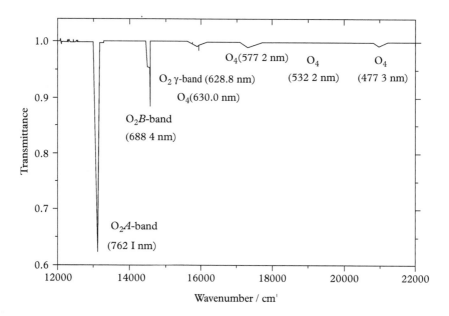

Fig. 7.11 *Survey spectrum of near-infrared and visible bands of molecular oxygen. Weak absorptions are attributed to O_4, interacting pairs of oxygen molecules (Newnham and Ballard 1998) Fig. 2*

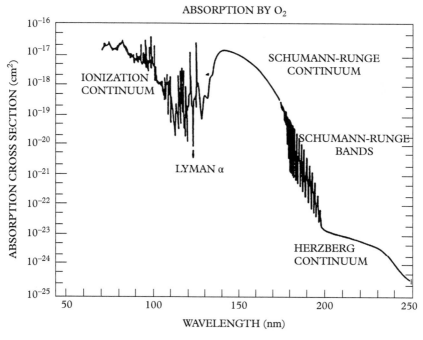

Fig. 7.12 *Overview of absorption bands due to molecular oxygen. The prominent Schumann-Runge system of molecular oxygen consists of 14 bands nearly equally separated in the range 202.6 nm (6.12 eV) to 175.0 nm (7.08 eV). At higher energies a strong continuum is seen, apparently arising from dissociated states including free oxygen atoms. At still higher energies direct ionization of the molecule at 102.7 nm leads to strong absorption*

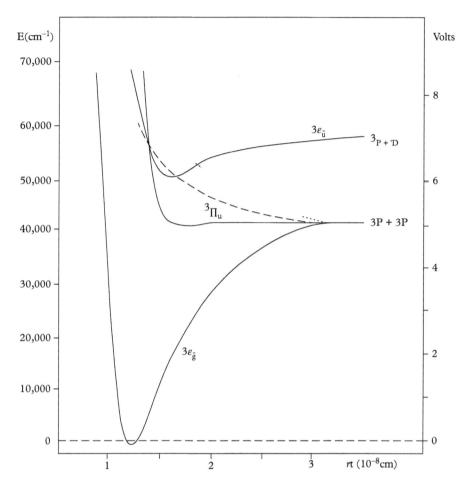

Fig. 7.13 *Suggested energy levels of oxygen molecule, in cm^{-1} on left and in electron volts on right. Vibrational levels, not shown, exist up to at least 35 in the ground state $X - {}^3\Sigma_g{}^-$ and to at least 17 in the highest state shown, $B - {}^3\Sigma_u{}^-$. The intermediate state suggests a decay of the upper electronic state by dissociation into two oxygen atoms. This is an important process leading to ozone in the upper atmosphere. The upper state is suggested to dissociate at large spacing into two oxygen atoms, one in an excited ${}^1D$ state (Wilkinson and Mulliken, 1957) Fig. 3*

substates labeled by J or N, suggested by Fig. 7.5. The letter N is used in molecular spectroscopy for angular momentum specifically excluding spin contributions, but for oxygen spin is zero and N means simply J. J values up to 23 are suggested in the indexing.

The indexing of these lines is complicated, but well understood. According to (Yoshino et al 1984) the Schumann-Runge system altogether has 14 branches, and the initial and final states are both regarded as triplets. All of the final states are in the

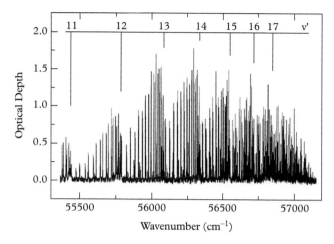

Fig. 7.14 *Measured absorption of molecular oxygen arising from transitions from the lowest vibrational level v = 0 of the ground state to vibrational states v' = 11–17 in the electronic excited state B – $^3\Sigma_u$ –. The high density of lines comes from the many rotational levels, labeled by N = J, that are possible for each vibrational state. This is about 25 percent of the Schumann–Runge absorption band system of oxygen extending from 48780 to 57143 cm^{-1} (Matsui et al, 2006) Fig. 1a*

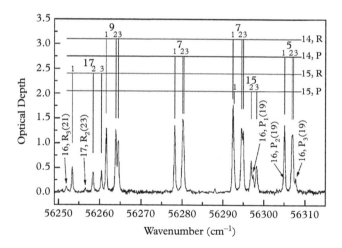

Fig. 7.15 *An expanded portion of the spectrum of the (14,0) band in the region 56250–56310 cm^{-1}. This plot includes some high J lines of the (15,0) and (16,0) bands (Matsui et al 2006) Fig. 1b*

uppermost band shown in Fig. 7.13. Most of the observed absorption occurs in the six principal branches. In terms of of rotational quantum numbers J (representing total angular momentum) and N (representing total angular momentum exclusive of spin) the fine structure levels have, if N > 1, J = N + 1, N, N − 1, and are designated F_1, F_2, F_3, respectively. The selection rules that apply to this case are $\Delta N = \Delta J = \pm 1$ for the six principal branches R_1, R_2, R_3, P_1, P_2, P_3. Different rules, $\Delta N = \pm 1$, $\Delta J = 0, \pm 1$; $\Delta N \neq \Delta J$ apply for the six satellite branches $^R Q_{21}$, $^R Q_{32}$, $^P P_{31}$, $^P Q_{12}$, $^P Q_{23}$, $^P R_{13}$. Further, completing the 14 branches, are "forbidden" $^T R_{31}$ and $^N P_{13}$ with $\Delta N = \pm 3$, $\Delta J = \pm 1$. The branch line symbols previously used summarize the quantum number changes shown in the symbol $^{\Delta N} \Delta J_{ij}$ where the subscripts i, j represent, respectively, the upper-F_i' and lower-F_j'' fine-structure levels defining the line transition. For principal branches, $i = j$ and $\Delta N = \Delta J$ so that a symbol with a single subscript (and no superscript) is adequate.

A list of all line assignment in the (12,0) band is shown in Table 7.3, from Matsui et al (2003).

The numbers are energies in wavenumbers of observed lines (due to oxygen molecules) that have been indexed as (12,0), indicating 12 as excitation number in the excited band and zero for vibrational level in the ground state. This energy is for the transition is labeled N = 1, R_1 at 55774.253 cm^{-1}. This corresponds to an energy 6.19195 eV, that is well above the known energy 5.115 eV for dissociation of the oxygen molecule into atoms.

The large number of entries in this table arise from the wide range of thermally allowed rotational energies in the ground state, subtracting from the initial energy, and rising to N = 21, corresponding to an energy increase in the initial state (reduction of line energy from 55784 cm^{-1}) of the order of B × 21(21+1). If B = 0.563 cm^{-1} this amounts to 260 cm^{-1}.

In summary, the oxygen Schumann–Runge bands lie in the range 202.6 nm to 175.046 nm (6.12 eV to 7.08 eV). There are 14 bands, separated by approximately

Table 7.3 *(Matsui et al, 2003) Table 1.*

N	R_1	R_2	R_3	P_1	P_2	P_3
(12,0) band						
1	55784.253	55785.083b		55780.703		
3	55777.781	55778.593b	55778.922b	55769.950	55770.707b	55771.260b
5	55764.225	55765.063b	55765.380b	55751.945	55751.719b	55751.975b
7	55743.605	55744.525b	55744.892b	55726.883	55727.694b	55727.973b
4	55715.943	55716.954b	55717.387b	55694.790	55695.668b	55696.007b
11	55681.214b	55682.348b	55682.846b	55655.627	55656.617b	55657.026b
13	55639.420b	55640.661b	55641.264b	55609.433	55610.540b	55611.017b
15	55590.502	55591.869b	55592.604b	55556.173b	55557.394b	55557.977b
17	55534.482	55536.000b	55536.804b	55495.809	55497.156b	55497.872b
19	55471.235	55471.921b	55473.839b	55428.351	55429.857	55430.643
21	55400.804b	55402.590b	55403.556b			

constant energy shifts. These bands represent transitions from vibrational levels v of the ground state to vibrational levels v' of the excited bound state of the oxygen molecule, the uppermost solid curve in Fig. 7.13, with a minimum energy around 6.2 eV. The intermediate solid curve represents two oxygen atoms interacting only weakly, as by van der Waals forces, lying at the dissociation energy 57128 cm^{-1} = 5.115 eV. According to Wilkinson and Mulliken (1957) the v' = 12 level (in the excited electronic state) lies 6.916 eV above the lowest J = 0, v = 0 level of the ground state.

The absorption in these molecular oxygen bands to some extent limits the Sun's spectrum seen at the Earth's surface, by absorbing light on the ultraviolet side of the solar spectrum. This absorption is seen in Fig. 3.2 and Fig. 7.1. A second importance of the oxygen molecule absorption is that it leads to generation of ozone O_3. Although the upper level in Fig. 7.13, B – $^3\Sigma_u$ is bound, it is also metastable, lying above the

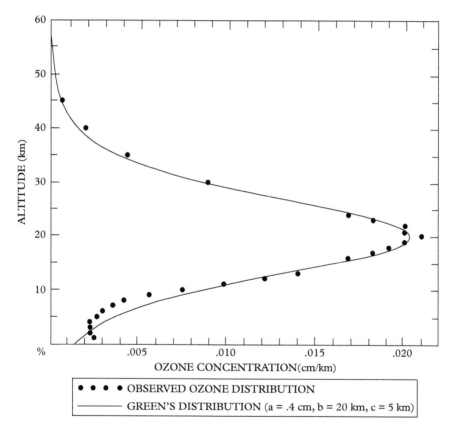

Fig. 7.16 *Measured (dots) and modeled distribution of ozone, seen to peak 20 km, in the stratosphere. This distribution is typical for mid-latitude winter. The model solid curve assumes parameters a, the total ozone amount in a vertical column and b is the altitude of maximum concentration (Lacis and Hansen, 1974) Fig. 7*

energy of two separate oxygen atoms, (middle curve in Fig. 7.13) and it is known to dissociate by by interaction with repulsive states to produce two oxygen atoms in the ground state (Matsui et al, 2003). In turn, the free oxygen atoms combine with molecular hydrogen to produce ozone. The splitting of the oxygen bond in O_2 by ultraviolet solar radiation is the primary step in the formation of O_3 and hence essential for the ozone layer (Bytautas et al, 2010). The dissociation of oxygen molecules by absorption into the excited state is the major source of oxygen atoms in the stratosphere (height 11–48 km) and especially in the mesosphere (height of 48–90 km). Atomic oxygen plays an important role in the photochemistry of these regions. The ozone produced plays a role in both the absorption of incoming sunlight and in absorbing and reradiating black-body radiation from the Earth's surface. A measurement of a typical stratospheric ozone distribution is shown in Fig. 7.16.

7.4   Molecular contributions to the greenhouse effect

We turn now to the overall effects of the impurity molecular absorptions on the amount of light reaching the ground and also on the emission of infrared light, black-body radiation, that is reduced by the greenhouse effect. Looking at the Earth from outer space, at its radiation spectrum, we recall that the effective emission temperature is reduced by absorption and reradiation from atmospheric impurity layers including water vapor, ozone, and carbon dioxide. As shown in Fig. 7.1, the expected spectrum emitted by the Earth (there assuming 250 K) extends from roughly 5 to 70 μm. The details of the actual spectrum are important, because the area under this curve represents the energy emitted by the Earth to balance its large input of energy from the Sun, and thus sets the Earth's temperature. This is complicated by the fact that the atmosphere has a temperature profile, and the outermost layer in the atmosphere that radiates directly to space is the most important. The atmosphere has roughly two parts, the lower part, the troposphere, up to about 17 km in mid-latitudes (but 20 km in the tropics and 7 km in polar regions), and the stratosphere. The troposphere establishes its temperature profile fixed at the bottom by the surface temperature, approximately set by incoming radiation from the Sun, and falling with altitude approximately at the rate $g/c_p \approx 10$ K/km, following ideal gas laws, as suggested by Eq. 3.33. The minimum temperature in fact is near 220 K, and this appears at the tropopause, the top of the troposphere. In this region the pressure is about 100 kPa, one-tenth of an atmosphere. In the region above the tropopause, the stratosphere, where the pressure falls another factor of ten, the temperature begins to rise with altitude. The temperature rises above the tropopause and is set more by radiation energy balance than by the properties of an equilibrium gas. In this region ozone plays an important role, since its absorption band as discussed earlier, lies in the "window" region of wavelengths where the atmosphere is otherwise basically transparent. Ozone absorption in the ultraviolet strongly heats this region in the stratosphere. At the same time, the ozone contributes to the greenhouse effect by absorbing and re-radiating black-body radiation coming up from the Earth's surface.

The radiation spectrum of the Earth seen from above (from outer space) depends on the temperature of the closest (highest) atmospheric layer in direct radiative equilibrium with outer space. Radiation from deeper layers has been absorbed, and what is seen from space is mostly re-radiated energy, except for the "window" region. For example, the top of a high thundercloud is composed of ice particles that form an opaque layer, and the low temperature of this layer, the effective black-body temperature, sets the spectrum and rate of outward black-body radiation from that region, following Eq. 3.19a. In the absence of clouds, at a wavelength in the "window" region, so that surface radiation goes directly to outer space, the effective black-body temperature may be 285 K at ground level. In the specific part of the window region where ozone absorption peaks, blocking radiation from below, but radiating upward, the temperature of the opaque ozone layer will be the effective radiating temperature. How opaque the ozone layer is will depend on the specific wavelength of radiation in question, since the absorption of ozone is sharply peaked, and in the wings of its absorption it will allow more infrared light to come through. All of this contributes to a greenhouse effect, a net outward radiation from the Earth smaller than if there were no atmosphere, when the outgoing radiation would all come directly from the surface at 285 K.

Satellites carrying infrared cameras looking down at the Earth have allowed direct measurement of the spectrum of outgoing radiation from the Earth. The results from one such measurement, the Atmospheric Infrared Sounder (AIRS) are shown in Fig. 7.17 (Pierrehumbert, 2011).

Fig. 7.17 shows the spectrum of outward of radiation from Earth in three ways. The solid curves are the expected black-body flux in Watts/m² steradian vs. photon energy in cm^{-1}, assuming unit emissivity and source temperatures from 220 K to 285 K. The data points from the left vertical arrow at 670 cm^{-1} to 1600 cm^{-1} are overlapping satellite measurements (in red) and modeling (in blue). For values less than 670 cm^{-1} only the blue modeled data are available. The region from about 800 to 1230 cm^{-1}, (about

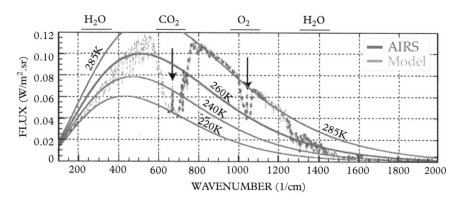

Fig. 7.17 *Infrared emission from the Earth, compared to blackbody spectra. Red points are measured by Atmospheric Infrared Sounder (AIRS) satellite and blue points are interpolated from model. The arrows point to interesting emission dips with centered emission spikes arising from carbon dioxide and ozone (Pierrehumbert, 2011) Fig. 3a*

8 μm to 12.5 μm) apart from the dip at ozone, shows emission close to that expected directly from the surface of the Earth at temperature 285 K. So, this is the "window" region mentioned earlier, and evident in Fig. 7.1. In this plot we see regions marked "H_2O" below 600 cm^{-1} and above 1300 cm^{-1} that are related, respectively, to its pure rotation spectrum (mentioned in Fig. 7.1), and the bending mode-absorption band shown on the left side of Fig. 7.9.

The dips in the observed emission spectrum of Fig. 7.17 below the black-body curve for 285 K are direct measurements of the greenhouse effect: reduced emission to outer space by absorption and re-radiation processes occurring in the atmosphere. The most prominent dip, marked by the arrow on the left side of the figure, near 667 cm^{-1}, corresponds to the deformation or bending mode of carbon dioxide (shown on the right in Fig. 7.4). The wide dip in the spectrum of Fig. 7.17 centered at 667 cm^{-1} indicates (Pierrehumbert, 2011) that a CO_2 layer as cold as 220 K, well below the surface temperature 285 K, is the effective black-body radiator in a spectral region about 100 cm^{-1} wide.

But the central spike observed indicates that the effective radiating temperature precisely on the peak absorption is higher in temperature, perhaps 240 K. This means that the layer is so heavily absorbing, at the center wavelength, that only the top of the layer (higher in altitude, and thus a bit warmer) is directly seen from outer space. As the wavelength detunes from the peak, emission radiation from deeper and cooler carbon dioxide can escape directly to space and the satellite camera. A similar effect is noted at the center of the ozone absorption around 1077 cm^{-1} (stretching vibrations shown in Fig. 7.10). This indicates that the bulk of the ozone is near 260 K while at the center of the absorption the local temperature, due to a higher altitude, is higher. At this wavelength the deepest that the satellite spectrometer can see down into the atmosphere is a location in the stratosphere because of the high-absorption of Sun-heated ozone. This still amounts to a greenhouse effect because the emission temperatures are all lower than the surface temperature 285 K.

7.5   Molecular absorptions and the ground-level solar spectrum

The upper solid curve of Fig. 7.18 is directly comparable to Fig. 1.4, and the peak value shown here, 2100 W/m^2 μm is very close to the level in that figure. The value for sea level depends on parameters that are shown, one is "air mass 1.5" that means 1.5 times the direct vertical path, that occurs when the Sun is 48° away from vertical. The assumption in the construction of this realistic modeled spectrum is that the water vapor was equivalent to 2 cm of liquid water in the total air column, and similarly 0.34 equivalent cm of ozone. The integrated power density from this constructed spectrum is 814.9 W/m^2. This number is close to 850 W/m^2 that was directly measured in the experiment described in Section 3.2.1. The parameters α and β represent very small amounts of particulate aerosol matter in the sky, corresponding to a very clear day in a

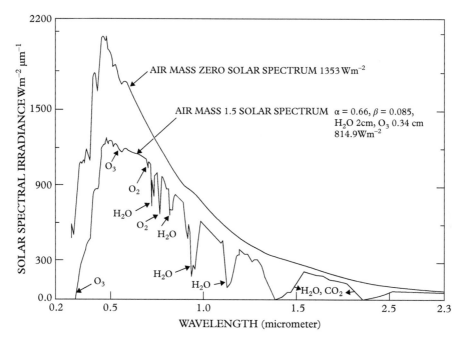

Fig. 7.18 *Modeled solar spectrum at Earth's surface (lower solid curve, with annotated molecular dips) with Sun at 48° corresponding to air mass 1.5, compared to a measured spectrum outside the atmosphere (Mecherikunnel and Richmond, 1980) Fig. 7*

non-polluted location. The molecular oxygen bands at 0.6 and 0.8 µm are shown, but not the Schumann-Runge bands, that appear only for wavelengths less than 0.2 µm. The modeling to produce Fig. 7.18 included careful consideration of Rayleigh scattering that was mentioned in Section 1.2, see Eq. 1.2. Fig. 9.1 shows similar data plotted vs. energy. The official ground-level solar spectrum agreed on by workers in the photovoltaic field is available here[3].

---

[3] http://rredc.nrel.gov/solar/spectra/am1.5/ (accessed May 24, 2017).

8

Wind, hydro and tides
Fully sustainable energy

An introduction to these topics was given in Sections 2.5 and 2.6. We now go into detail on some aspects of these topics. As these forms of energy come either from the Sun (in an indirect fashion) or from the motion of the Earth and Moon, they are available on an indefinite term into the future.

8.1   Wind-turbine technology

An introduction to the wind power resource, with power density $P = \frac{1}{2}\,\rho v^3$ W/m², with ρ the air density and v the air speed, was given in Section 2.5.1. This flow energy is created from uneven heating of the air by the Earth's varied surface, leading to air motions, and is a secondary effect of Sunlight. There is a general tendency for the flowing air mass to be slowed down very near the surface, and a crude estimate of the vertical gradient of horizontal airspeed v near the surface, is:

$$v = v\,(10)\,(z/10)^{1/7}. \tag{8.1}$$

This approximation normalizes windspeed to a reference height z of 10 m (Peterson and Hennessey Jr, 1977).

A straightforward analysis of an idealized wind turbine showed that extraction of this flow energy cannot be more efficient than 0.59 (see Eqs. 2.4, 2.5, the limit is known as Betz's Law). However, the analysis, at first sight, gives no advice on the design of the rotor to achieve the best energy extraction. It was an important contribution of Betz to extend the analysis to guide the design of efficient rotor blades.

8.1.1   Rotor design

Extraction of wind energy by rotating blades is now well advanced, as is illustrated in Fig. 2.3, a three-bladed wind turbine rated at 7.5 MW. The evolution of rotor design culminating in a device such as in Fig. 2.3, is worth some consideration.

Physics and Technology of Sustainable Energy. E. L. Wolf, Oxford University Press (2018).
© E. L. Wolf. DOI: 10.1093/oso/9780198769804.001.0001

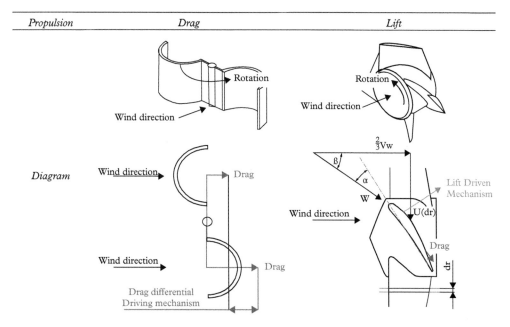

Propulsion	*Drag*	*Lift*

Fig. 8.1 *The two mechanisms of energy extraction from wind, based on drag, (left) and lift, (right). Incorporated in the lower right vector diagram is the deduction of Betz's analysis that the wind speed in the plane of the rotor is reduced to 2/3 of its initial value (Schubel and Crossley, 2012) Table 1., p. 3427*

The extraction of windpower, even apart from the use of sails for ship propulsion, is an ancient exercise. Windmills were used for irrigation in the era of King Hammurabi of Babylon, in 1792 BC (in Mesopotamia, that later became Persia). For an excellent review see Gasch and Twele (2012) p. 15.

These earliest windmills, referred to as Persian windmills, were based on the mechanism of drag, whereas modern windmills use the principle of lift. (The well-known Dutch windmills also used a lift principle.) These two mechanisms are illustrated in Fig. 8.1. An elementary observation related to the lower efficiency of the drag-based device, exemplified by the cup anemometer, is that no part of the device can move faster than the wind, while in modern lift-based machines a ratio of tip speed to windspeed in the vicinity of five is often observed while ten is considered high.

In the left portion of Fig. 8.1, the cup-shape of the rotor gives more drag when the open side faces the wind. In the Persian windmill versions, that had vertical axes, a wall closed off half of the rotor from the wind, so that drag was much reduced in the shielded upwind motion. It is evident that such a rotor cannot move faster than the windspeed. Turning to the lift mechanism on the upper-right side of Fig. 8.1, the blades are seen, by their lift force, to cause counter-clockwise rotation. In the lower-right panel, an edge-on view of a single blade is shown, with the hub and other blades shown behind. The attack-angle α between the local air flow W and the aerofoil plane is a crucial parameter. The large angular difference shown here between the horizontal wind direction and the local airflow vector **W**, called the incidence angle β, arises by the assumed motion of the aerofoil (portion of the blade) itself in the rotation. At the end of the rotor, one can realize that incident air velocity **W**, on the tip can be nearly at right angles to the horizontal wind direction, if the tip is moving fast enough. (The optimum attack angle, between local wind and aerofoil, for maximum lift is in the vicinity of 15° as shown in Fig. 8.2.) Evidently, the incidence-angle β increases with distance from the hub along the blade, recognizing the local rotational speed as v = ωr, with ω the angular velocity and r the radius along the blade from the hub. This variation can be accommodated in blade design by a built-in twist to the blade. However, it is only the component of the lift force along the local direction of rotation that adds power. The drag nearly opposes the desired vertical force in the limit of high tip speed. Thus, the important role of the aerofoil is to minimize drag, and the best shape depends on the anticipated windspeed.

In thinking about the shape and orientation of the blade to optimally get torque from the horizontal wind, it may be helpful to think of the airplane wing that gets lift largely from its angle of attack, producing an upward force on the bottom surface. The airplane wing is often also cambered in shape so that a longer air path, and thus a higher speed, occurs on the upper side. The Bernoulli principle's reduction in pressure with the higher windspeed on the upper side of the airplane wing is the source of some of the lift. But the larger utility of the smooth aerofoil is in its reduction of the drag force **D**. In the wind turbine case the force on the blades, corresponding to the lift on

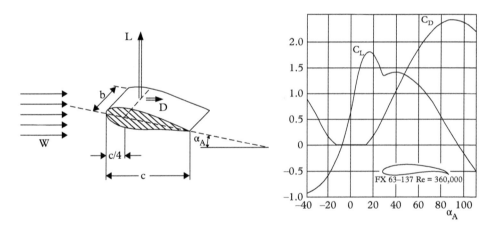

Fig. 8.2 *(Left) Lift and drag forces on an aerofoil as function of attack angle* α_A *of leading edge above local windflow* **W**. *These forces act perpendicular and parallel, respectively, to the local wind vector* **W**. *(The vector direction of the local windflow* **W** *is strongly dependent on the radius r along the blade that controls the transverse velocity ωr.) The lift force* **L** *acts at about ¼ of the chord length c back from the leading edge. It is the vertical component (tangential to the rotational motion) of the lift force* **L** *that contributes to the rotational torque and energy extracted. The role of the aerofoil shape is largely to reduce the drag* **D**. *(Right) numerical coefficients calculated for the indicated aerofoil for lift and drag forces as function of attack angle (Gasch and Twele, 2012). Fig. 2–24 p. 40*

the airplane wing, gives a rotational torque. The angle of the rotor blades (the pitch) is usually adjustable in the wind turbine installations, and the first precaution to slow the rotor down in an oversize wind speed is to reduce the angle of attack to zero or negative values.

The curves in Fig. 8.2 give numerical coefficients c_L and c_D respectively for lift and drag in the formulas for forces L and D:

$$L = c_L \, \rho/2 \; w^2 cb \tag{8.2}$$

$$D = c_D \, \rho/2 \; w^2 cb \tag{8.3}$$

acting on an area cb (chord c and length element b) of aerofoil with incident airspeed w. With ρ the density in kg/m³ these formulas give Newtons, taking dimensionless values of c_L and c_D. It is important to note in Fig. 8.2 that the operating range for the aerofoils is narrow, say α_A between 0 and 15 degrees, and in this range the drag force is nearly zero with a well-designed aerofoil as is sketched. In this range the lift increases approximately linearly with attack angle.

The design of the blades was proposed in detail by Betz (1926), and has been refined by later authors, as described by Gasch and Twele (2012). The design gives the chord

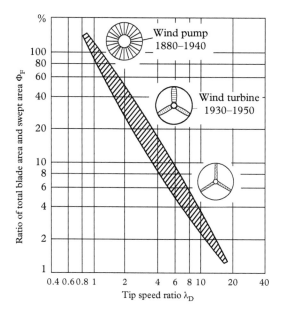

Fig. 8.3 *Evolution of blade design to smaller "solidity", area of blades vs swept area as function of tip speed ratio* $\lambda_D = \omega R/v$, *with v the windspeed and R the outer radius (Gasch and Twele 2012) Fig. 5.15 p. 184*

(blade width) and the twist angle as a function of radius r from the axis, maximum value R.

The design of Betz is closely based on the efficiency formula giving the best possible value for extraction of the flow power as 16/27 = 0.593. This formula recognizes that the optimum design will allow the air past the turbine to retain 1/3 of its initial speed. In the design of the blade dimensions, Betz therefore assumes that the wind speed in the actual plane of the rotor is 2/3 of the initial speed, as indicated in the lower-right panel of Fig. 8.1.

From Fig. 8.5 one sees that the angle φ can be expressed as:

$$\phi = \tan^{-1}(2/3 \ v_1/\omega r), \qquad (8.4)$$

depending, through the rotation at rate ω, upon position along the blade.

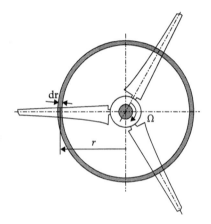

Fig. 8.4 *Element of area in Betz derivation of optimum blade shape (Gasch and Twele, 2012) Fig. 5.14 p. 182*

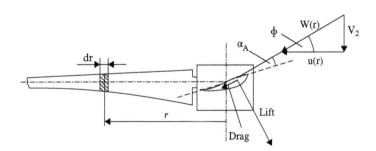

Fig. 8.5 *Aerodynamic forces on blade element **dr** at radius **r**, where the rotational velocity is u(r) = ωr and the effective wind velocity v_2 is taken as 2/3 of windspeed v_1, based on the Betz analysis. Angle of attack is designated α_A (Gasch and Twele, 2012) Fig. 5.13 p. 181*

The power that can be extracted from the radius range dr at r, following the Betz formula, (2.4) is:

$$dP = 16/27 \; \rho/2 v_1^3 (2\pi r dr). \tag{8.5}$$

The increment of power from the aerodynamic forces on radius element dr at radius r (assuming z blades) is:

$$dP = z \; dU \, \omega r, \tag{8.6}$$

where dU is the projection of the aerodynamic force along the direction of rotation. Neglecting the drag force (with a good aerofoil oriented with appropriately low attack angle), dU reduces to the projection of the lift force L:

$$dU = dL \; \sin\phi = \rho/2 \; c_L \; w^2 c(r) \; dr \; \sin\phi. \tag{8.7}$$

Here c(r) is the chord width to be determined at radius r.

The corresponding increment of power dP is then,

$$dP = z \; \omega r \, \rho/2 \; c_L w^2 c(r) \; dr \; \sin\phi. \tag{8.8}$$

Equating the two expressions for the increment of power, Betz found the chord-width relation c(r) (in meters), with z the number of blades:

$$c(r) = z^{-1} \; 16/27 \; 2\pi r/c_L \left(v_1^3 / w^2 \, \omega r \, \sin\phi \right). \tag{8.9}$$

Using the relations $v_1 = 3/2 \; w \sin\varphi$ and $u(r) = \omega r = w \cos\phi$, (see Fig. 8.5), this can be rewritten as:

$$c(r) = 2\pi R \; 1/z \; (8/9 c_L) \; \lambda_D^{-1} \left[\lambda_D^2 (r/R)^2 + 4/9 \right]^{-1/2} \tag{8.10}$$

where λ_D is the tip speed ratio, and z is the number of blades of radius R. Finally, a near working formula for large tip speed is obtained by dropping the 4/9 in the square root of Eq. (8.10) to get (Betz, 1926),

$$c(r) = 2\pi R \; 1/z \; (8/9 c_L) \; 1/\left[\lambda_D^2 (r/R) \right]. \tag{8.11}$$

One sees that the chord width drops off inversely with position r along the blade, and also as the inverse square of the tip speed. These effects are seen in Fig. 8.3, where the scatter is due to some variation in the choice of lift coefficients c_L near 1.0.

It is a working assumption that the blade is twisted along its length so as to ensure that each aerofoil section dr sees an appropriate small attack angle with its local airflow, to reduce drag. Under these conditions the drag force, as seen in Fig. 8.2 (right panel) is negligible, from the point of view of power generation.

The corresponding design twist angle, as illustrated in Fig. 8.7, is:

$$\beta(r) = \phi(r) - \alpha_A(r) = \tan^{-1}[2R/3r\lambda_D] - \alpha_A(r). \tag{8.12}$$

The rotor blades designed in this way in principle will deliver the Betz efficiency 0.593. In practice, largely following the design outlined, present turbines reach efficiency 0.5. One loss of efficiency arises from a simplification in the Betz analysis, that neglects the tendency of the air downwind from the rotor plane to carry away rotational kinetic energy. This must occur as rotational angular momentum is imparted to the rotor. This correction (Schmitz, 1956) is not large, especially for a large tip speed. A second loss arises from the tendency of the tips to induce local vortex motions in the air. A third loss is from aerodynamic drag from the blade surfaces. These effects, that can be made small in present practice, were analyzed by several authors, and are summarized by Gasch and Twele (2012).

A summary of performance, based on the extended analysis of Schmitz (1956) is given in Fig. 8.6. In Eq. 8.11, the blade number z has been carried along without being specified. It seems that the blade number z enters the efficiency correction only as tip losses, and the choice in practice is based on the cost of additional blades weighing against a tendency of a rotor with three or more blades to rotate more smoothly than the one- or two-bladed rotor, reducing time-dependent forces on the support structure. Two-bladed rotors are historically well-known, but more recent practice has nearly settled on three blades.

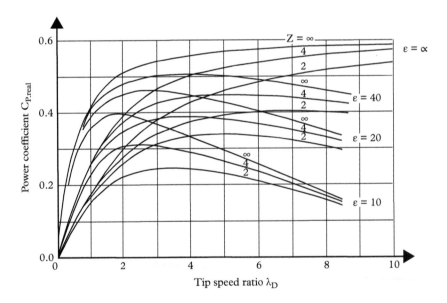

Fig. 8.6 *Summary of available wind turbine efficiency (power coefficient) based on the Schmitz (1956) extension of the Betz theory. Blade numbers z are shown as 2, 4 and ∞. The lift to drag ratio c_L/c_D is indicated as ε (Gasch and Twele, 2012) Fig. 5.25 p. 200*

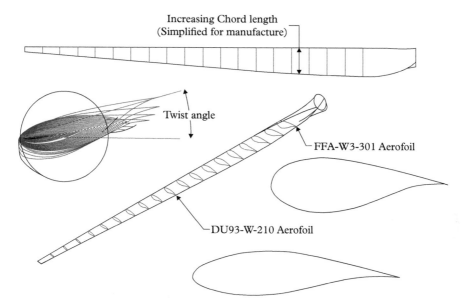

Fig. 8.7. *Sketch of typical modern horizontal axis wind turbine blade with multiple aerofoil shapes, twist, and linear chord length increase. Aerofoil shapes are chosen to minimize drag at the expected local windspeed (Schubel and Crossley, 2012) Fig. 8, p. 3440*

8.1.2 Blade construction

The performance of the modern turbine blade has benefited from development of strong lightweight materials, especially epoxy glass fiber and epoxy carbon fiber composites. These materials are also known as glass fiber reinforced plastic and carbon fiber reinforced plastic. Respectively, the Young's modulus of the two materials are in ranges near 25 GPa and 70 GPa, while the respective tensile strengths are in ranges near 440 MPa and 600 MPa. The carbon fiber material is more expensive. Both materials are commonly available as cloths, for example, in large rolls 50 inches wide, sold by the yard.

The blade as sketched in Fig. 8.8, is largely hollow, formed by bonding together two rigid shells: the "suction side" upper and the "pressure side", lower. Two separate molds are used to form the two shells. In one construction method, each blade shell is formed by laying into the correct mold successive layers of fabric made of glass fiber, carbon fiber, or a combination of these, creating a laminate. Once the design thickness of the shell wall is reached by layering the woven fiber fabrics, the mold is sealed and flooded with either polyester resin or epoxy resin. The assembly is then subjected to a heat cycle to harden the resin. The two rigid shells are then bonded together at the leading and trailing edges, as shown in Fig. 8.8.

Beyond the aero-dynamic efficiency questions related to the design of the rotor, and its construction, the conversion of rotational power into electrical power is in the realm of conventional electrical generator technology. Light weight is desirable, as well as

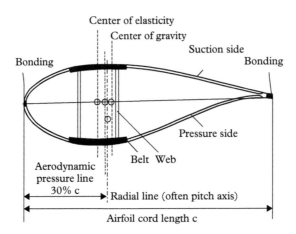

Fig. 8.8. *Construction of the blade by bonding its separate upper and lower molded shells, with inclusion of internal bracing elements ("Web") as needed. The circles in this diagram locate (clockwise from upper left) the center of elasticity, the aerodynamic pressure line, the center of gravity and the radial line. In some cases, balsa-wood is added to the interior as a spacer (Gasch and Twele, 2012) Fig. 3.9, p. 54*

long lifetime and freedom from maintenance, especially for installations offshore in shallow water, and a final arbiter is cost. The conventional generator rotates at a high speed, and the initial, and still predominant, approach is to use gear sets in the turbine nacelle to raise the rotational speed far above the slow rotational speed of the huge rotor as shown in Fig. 2.3.

8.1.3   The large-scale turbine

A sketch of a conventional drive train of a large scale horizontal axis turbine is shown in Fig. 8.9. (Actually, a direct drive train without gears is used in the large Enercon machine shown in Fig. 2.3.)

The large turbine conventionally has three blades upwind from the support pillar, and the yaw angle, rotating the huge nacelle atop the support pillar, is adjusted to keep the horizontal axis facing into the wind. Sensors are placed on the top rear of the nacelle to monitor the wind speed and direction. It has turned out that the gear train is a problem requiring oil and sometimes failing, and the corresponding innovation is to use a direct drive, requiring a larger annular (ring) generator, as used in the turbine of Fig. 2.3. Fig. 8.10, shows the direct drive nacelle of the smaller 1.8 MW Enercon E-66 turbine. This machine has a rotor diameter of 70 m and a tower height of 98 m.

This generator is classified as a synchronous direct-drive ring generator with electrical excitation. Here the large diameter stator coil is fed a current to generate the needed magnetic field. This machine has no gearbox, is designed for a windspeed of 12 m/s and will start operating at windspeed 2.5 m/s. The rotor speed is in the range 10 rpm–22 rpm. The nacelle weight of this 1.8 MW machine is 58.8 tons, while the rotor weight, total of three blades and hub, is 31.7 tons.

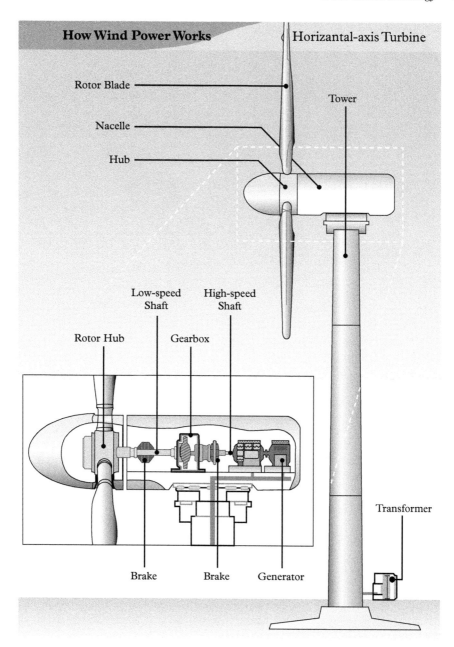

How Wind Power Works   Horizantal-axis Turbine

Rotor Blade

Tower

Nacelle

Hub

Low-speed
Shaft

High-speed
Shaft

Rotor Hub   Gearbox

Transformer

Brake   Brake   Generator

Fig. 8.9 *Sketch of drive train of conventional large horizontal-axis three-blade turbine, such as those made by Vestas, with gearbox speed reducer and conventional generator http://science.howstuffworks.com/ environmental/green-science/wind-power2.htm*

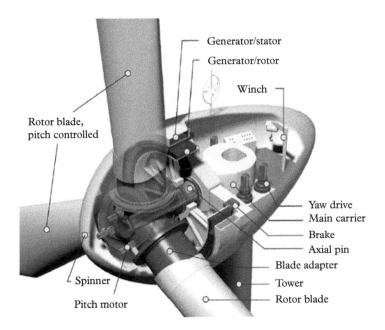

Generator/stator

Generator/rotor

Winch

Rotor blade,
pitch controlled

Yaw drive

Main carrier

Brake

Axial pin

Blade adapter

Spinner

Tower

Pitch motor

Rotor blade

Fig. 8.10 *Nacelle with direct drive turbine assembly for the Enercon E-66 1.8 MW model, showing the large-diameter annular generator. This generator, also called a ring generator, rotates only slowly requiring many copper wire coils and sliding electrical contacts between the generator rotor and generator stator, at the outer diameter, that provide the Faraday law emf $\varepsilon = - \, d\Phi_m/dt$. A variation on this design (by manufacturers other than Enercon) uses permanent magnets to generate magnetic fields, rather than current-fed copper coils, in the large-diameter annular generator (Gasch and Twele, 2012) Figs. 3–30, p. 72*

The magnetic field at the stator, see Fig. 8.10, can alternatively be generated by permanent magnets.

One major manufacturer, Goldwind, markets a line of turbines in the permanent magnet direct drive (PMDD) category.

Referring to Fig. 8.11, top, (the system indicated in Fig. 8.9), is a constant speed system with a three-stage gearbox and a squirrel-cage induction generator directly connected to the utility grid. This system is also referred to as "the Danish concept". An attraction of this system is that it consists of simple off-the-shelf components and, therefore, it is inexpensive. According to Polinder et al (2013), during the 1980–2000 period, the power rating of turbines of this type increased to about 1.5 MW.

After about 1996 many manufacturers changed to a variable speed system with a doubly fed induction generator (DFIG) for wind turbines with power levels above roughly 1.5 MW (second from the top in Fig. 8.11). This system consists of a multi-stage gearbox, a relatively low-cost standard DFIG, and a power converter. The power rating of the converter is about 0.25, enabling variable rotor speeds in the range 0.6 to 1.1 of the rated speed.

In the third panel of Fig. 8.11 we see the direct drive DD system, first introduced in 1992, (see also Figs. 2.3 and 8.10), using a synchronous generator with no gearbox.

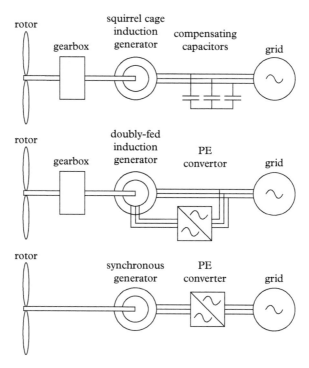

Fig. 8.11 *Three commonly used generator systems. Here PE stands for "power electronic", and "doubly-fed induction generator" may be referred to as DFIG. After (Polinder et al, 2013) Fig. 2*

In this case a fully rated power-electronic PE converter is needed for the grid connection. At first the DD systems, like those of Enercon, used electrical excitation in the generator. When permanent magnets (PM) of higher performance became available, some manufacturers, such as Goldwind, switched to PM generators. The main disadvantage of the DD generator is that the low-speed generator, shown as the large diameter ring in Fig. 8.10, is large, heavy and expensive, and perhaps less efficient than the high-speed generator. These factors are balanced by mechanical simplicity, reliability, and probably less need for maintenance.

Polinder et al (2013) also discuss a potential variant of the DD system using high temperature superconductor wire for the field windings in synchronous generators. The potential advantage is that large fields up to two to three Tesla can be achieved compared to less than 1 T for the permanent magnets, in principle allowing the volume of the superconducting machine to be reduced by a factor two to three compared to one with permanent magnets below 1 T. The disadvantage, that has so far prohibited any prototype device, is the need for refrigeration. The high temperature superconductors would allow the thermal insulation and cooling system to be relatively simple, and cryocoolers in this temperature range can be purchased off the shelf. A review of superconducting wind turbine generator development is given by Jensen et al (2013).

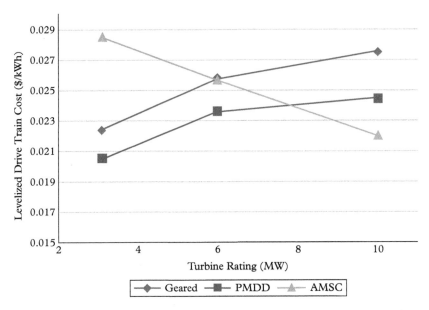

Fig. 8.12 *Levelized drive train costs in $/kWh for three types of turbine: conventional geared, permanent magnet direct drive and proposed superconducting. (Maples et al, 2010) Fig. 15*

A comparison study of these approaches, including the further proposed class of superconducting generators, has been provided by Maples et al (2010), see Fig. 8.12.

The superconducting generator, as proposed by the firm American Superconductor, would use a stator coil of a high-temperature superconductor that requires cooling to liquid nitrogen temperature. As mentioned previously, such coils can be light in weight, consisting of the wire only with no iron core, and are capable of very large magnetic fields, allowing the generator to be smaller in size. A new fixed cost and complexity is added to by the need to cool the assembly with a refrigerator to 77 K or so. Probably because of the refrigeration requirement, there is no working example of such a turbine even though it is favored by this cost analysis in the limit of very large turbine installations.

Another recent impetus, even though the idea is old, is a two-blade system,[1] especially for the large offshore turbine.

The firm 2-B Energy has designed a 6 MW Turbine of this type, an advantage being that access to the nacelle could be accessed by helicopter when the blades are stopped in the horizontal position. Sea installations are a challenge for the service technician, needing a boat and climbing many stairs, and this is one response to that problem. A 6 MW two-blade turbine with the blades down-wind from the tower and helicopter deck has been installed by Ming Yang Wind Power, a major Chinese manufacturer, following a design licensed from the German firm Aerodyn Energie system.[2]

---

[1] Fairley, P. (2014 MIT Technology Review) https://www.technologyreview.com/s/528581/two-bladed-wind-turbines-make-a-comeback/

[2] Weston, David (2015 Windpower Monthly) http://www.windpowermonthly.com/article/1337932/ewea-offshore-aerodyn-6mw-connected-grid?HAYILC=TOPIC

The advantages of the two-blade system are simplicity and lower weight. It is said that two-blade systems will rotate faster, possibly making more noise, but this is not a problem in an ocean installation.

8.1.4   Growth patterns for utility scale wind power adoption

Global capacity increased 17 percent in 2015 (GWEC Global Wind 2015 Report) p. 13, see Fig. 8.13.

If the 17 percent rate per year reported for global wind capacity by the GWEC Global Wind 2015 report were to continue, it would take 14.7 years, or until 2031, for the total capacity to increase by a factor 10. The capacity has been growing rapidly in China. A recent article (Hernandez, 2017) reports that China now has 92,000 wind turbines, with a capacity of 145 GW, and is adding turbines at a rate of more than one per hour.

Fig. 8.14 indicates that the installed electric generating capacity from wind in 2015 was 7 percent of global electric capacity. The fraction of electric power obtained from wind, called the penetration, varies among countries. At present the largest penetration, 42 percent is in Denmark, followed by Portugal 23 percent, Spain 15 percent, Ireland 16 percent, UK 11 percent, Germany 8 percent, United States, 4.5 percent, and China 3.3 percent. The fraction of the time a given turbine is operating at nameplate capacity is called the capacity factor, and is typically in the vicinity of 30 percent.

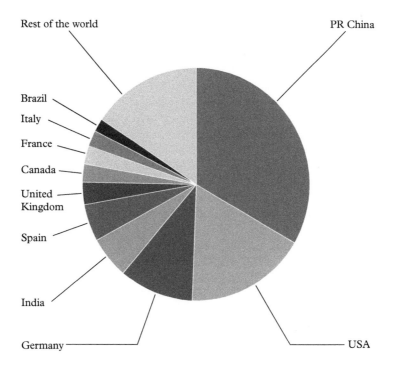

Fig. 8.13 *Total global wind capacity installed Dec. 2015 is 432.9 GW, distributed as shown*

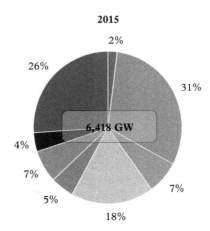

2015

Fig. 8.14 *Global installed electric capacity by technology 2015. Clockwise from top right: coal 31 percent, wind 7 percent , hydro 18 percent, nuclear 5 percent, other 7 percent, solar 4 percent, gas 26 percent and flexible capacity 2 percent. Total capacity 6.42 TW corresponds at 7 percent to 449 GW installed wind capacity, in good agreement with Fig. 8.3. The total electric capacity stated as 6.4 TW can also be compared to the 17.3 TW average total global power consumption derived from the BP Statistical Report. (Bloomberg New Energy Finance, "New Energy Outlook 2016") Fig. 1 https://www.bloomberg.com/ company/new-energy-outlook/*

8.1.5   The importance of siting

The windpower increases as the cube of the windspeed. This gives a large benefit to siting turbines where the windspeed is larger. A map of the typical windspeed in the US was given in Fig. 2.2. Not shown is a map of larger windspeeds generally available offshore. A daytime scenario for a larger wind turbine just offshore is suggested in Fig. 8.15. A discussion of the vertical wind updraft in the thundercloud was given in Section 3.3.1.1. The storm cloud is an engine for driving a vertical flow of warm moist upward, an element in the circulation indicated in Fig. 8.15.

Wind turbines offshore are common in Scandinavia and England and are beginning to be adopted in the United States. According to Werdigier (2010), the Thanet Windfarm, consisting of 100 turbines of height 377 feet off the coast of Kent, in southeast England, is rated at 300 MW. The 100 turbines are located on 13.5 square miles of shallow ocean. The offshore turbines are typically large. For example, a recent announcement of an 8 MW offshore direct drive turbine with 154 m rotor by Siemens mentions that this firm has already manufactured and placed 600 offshore wind turbines.[3]

A small start for offshore turbines in the US (Cardwell, 2017) includes five turbines now operating in the waters off Rhode Island and a plan of the Long Island Power Authority to build 15 turbines about 35 miles offshore from Montauk Point, at the eastern tip of Long Island.

---

[3] http://www.power-eng.com/articles/2017/01/siemens-installs-8-mw-wind-turbine-prototype.html

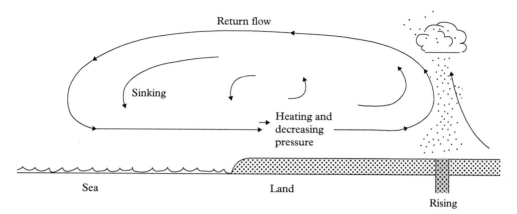

Fig. 8.15 *Sketch of the sea-land breeze as it exists during the day. This is one reason for siting wind turbines just offshore (Gasch and Twele, 2012) p. 117. Fig. 4.3*

It is common for the ocean depth to remain rather small for a shelf extending even a hundred miles beyond the actual shore line. By far the most common practice is to locate offshore turbines on the shallow ocean bottom in such regions.

Historically the cost of power from offshore turbines has been higher than from those on land. A recent report (Reed, 2017) gives $0.083/kWh as the recent cost of electricity from windfarms offshore from The Netherlands. This article also states that

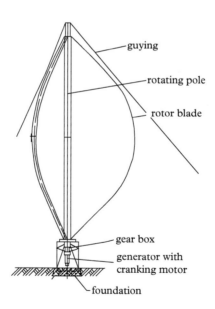

Fig. 8.16 *Vertical axis wind turbine of the Darrieux design, here with 3 blades (Gasch and Twele, 2012) Fig. 2.4b, p. 17*

offshore installations of wind turbines are less than 10 percent of overall wind turbine installations.

8.1.6   Vertical axis wind turbines

The earliest wind turbines in Mesopotamia rotated about a vertical axis, as already mentioned, and were based on the relatively inefficient drag principle. Modern vertical-axis aerofoil lift-based turbines, see Fig. 8.16, are well developed but have not found utility scale adoption. Such turbines are more often seen in urban locations, perhaps on the roof of a 590-foot apartment building as at 388 Bridge Street in Brooklyn, NY. These small Darrieux-style turbines are receptive to wind gusting from different directions, are not noisy, and do not involve heavy equipment suspended above the building, as a horizontal axis turbine would. The working parts that may require maintenance are at the floor level, rather than at the top of a tower. The most advanced vertical axis machine is

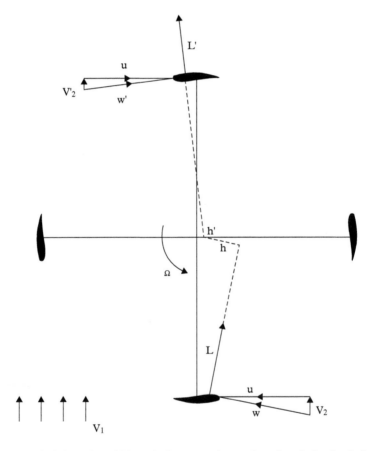

Fig. 8.17 *Wind and lift forces **L** and **L'** producing torque in top view of vertical axis wind turbine of the Darrieux design with 4 blades (Gasch and Twele, 2012) Fig. 2.4b, p. 17*

the Darrieux Turbine, as sketched in Fig. 8.16. The largest such turbine was a 4 MW two-blade unit 110 m tall that was operated for a period of time at the Eole Windfarm in Canada. This turbine was shut down after being damaged in a high wind. It appears that this design does not have a mechanism other than braking to reduce the wind torque in a storm, to compare to the blade-pitch control in the horizontal axis wind turbine.

8.2   Water turbines in tides or flowing streams

As mentioned briefly in Section 2.5.1.1 the same flow analysis of Betz applies to the kinetic energy of flowing water. Compared to flowing air, the nearly thousand-fold increase in density of water similarly raises the kinetic energy per unit volume. Water turbines that look like scaled versions of air turbines are available in small sizes. However, the sites for such turbines are quite restricted and hard to work with, and the technology has not found wide application.

Perhaps the most publicized installation (not yet complete) is in the East River in New York City, at Roosevelt Island, where free flow water turbines have been tested by Verdant Power. This is known as the Roosevelt Island Tidal Energy RITE project because the flow of the East River reverses with the change of tide. The first set of turbines installed in 2002–6 suffered blade degradation, and a new set of 30 turbines to produce 1.05 MW was proposed as indicated by news release[4] in 2012 from the US Department of Energy.

A summary[5] offered by Verdant Power described testing of its "Gen 4" water turbine that extracts power from both the outgoing and the incoming tide, called a passive yawing system. The device and power train are quite similar to the air turbine shown in Fig. 8.9, except that the rotor is downstream from the mounting pillar. A complete technical report of the testing in 2012 is also available.[6] The licensed commercial application, in progress, will involve "Gen 5" turbines, with blade diameter 16 feet. It appears that flowing debris in a river can be a significant hazard, more than any equivalent threat to an air turbine.

8.3   Hydropower and pumped storage

A brief summary of hydropower and pumped storage has been given in Section 2.5.2. We here give more details of this most important renewable electricity source, starting with a discussion of a leading form of water turbine.

---

[4] https://energy.gov/articles/turbines-nyc-east-river-will-provide-power-9500-residents
[5] https://web.archive.org/web/20140224004159/http://verdantpower.com/what-initiative/
[6] Corren, D. "Improved Structure and Fabrication of Large, High-Power KHPS Rotors", at: https://www.osti.gov/scitech/servlets/purl/1084212

8.3.1   The Francis turbine

A key component for extracting power ghdm/dt from water, with a height h and flow rate dm/dt, is a water turbine. The Francis turbine (dating from 1848, this is a very mature technology) is the most widely applied type. (A competing widely used design is the Kaplan turbine, to be discussed in Sect. 8.4.) An admirable history of the Francis turbine is given by (Lewis et al, 2014). It appears that the earliest turbine of Francis, ca 1848, had a measured efficiency of 71 percent and that present-day forms can have efficiency exceeding 95 percent.

In modern installations of Francis turbines, flow of pressurized water at depth h, (the "head") below the top of the dam, is released radially into the angled vanes of the turbine "runner", and causes the runner to rotate. The exhaust water exits axially below at reduced velocity from the rotating turbine blades of the runner. The radially injected water comes from a spiral-shaped pipe, the "spiral case" or "inlet scroll", that encircles the turbine's rotating runner. The water enters the runner radially, from the outside, while in most other designs, including the Kaplan, the water flows along the axis. Low-speed exhaust water exits the Francis Turbine vertically down from the runner, in the usual case of a vertical axis. The water enters the spiral casing at high pressure and low velocity. There are regularly spaced openings (with guide vanes in more modern Francis designs) on the inside of the spiral casing allowing the water to spurt at high velocity against the blades of the encircled turbine runner. The geometry is fairly clear from a photograph taken in 1941 (Fig. 8.18) of the inlet scroll (spiral casing) of one of the Francis Turbines at the Grand Coulee Dam, that was shown in Fig. 2.4. In Fig. 8.18, the inlet scroll is attached to the "penstock" (inlet pipe, diameter 18 feet) at the upper right and steadily decreases in diameter to the end of the snail-shaped "scroll". The central opening, where the runner turbine will be placed, called the turbine pit, can be seen to have equally spaced rectangular openings on its inner face, to provide equal water flow into all sides of the turbine runner. (A modern large-scale runner turbine is shown in Fig. 8.19.) Detailed managing of the flow rate is provided by "wickets" (not shown), adjustable to maintain the running turbine exactly at 120 rpm, the speed needed in the Grand Coulee installation to create alternating current at exactly 60 Hz.

Momentum transfer occurs from the moving water to the runner blades and causes its motion. The design of such devices is dependent on the specific situation, characterized by the head (water height, and thus inlet pressure) and available water flow. The possible head range is from 40 m to 600 m and the connected power generator may be as large as 800 MW.

As a specific recent example of a small hydroelectric installation, the City of New York is adding hydropower, using 4 Francis turbines, to the existing Cannonsville Reservoir Dam on the West Branch of the Delaware River, that is part of the city water supply[7].

The drainage area of the 12-mile long Cannonsville Reservoir on the west branch of the Delaware River is 454 square miles. The head measured from the top of the spillway

---

[7] "Cannonsville Hydroelectric Project, FERC Project No. 13287", Feb. 2012, pp. 142. www.nyc.gov/hml/dep/pdf.../cannonsville_1_exhibits_a-e.pdf.

Fig. 8.18 *Inlet Scroll for Francis Turbine at Grand Coulee Dam (see Fig. 2.4) photographed in 1941, before the Francis turbine runner was installed. The equally spaced rectangular openings around the (empty) "turbine pit" allow water to spurt radially onto the turbine blades equally from all sides http:// users.owt.com/chubbard/gcdam/html/photos/construction.html*

at the dam is 122 ft, 37.2 m. The nominal flow rate out of the reservoir guaranteed to parties downstream is 1750 cubic feet per second (cfs), and the full capacity of the proposed hydropower installation is 1500 cfs. The facility will have four horizontal Francis Turbines: two runners with diameter 1.76 m and two runners with diameter 0.89 m, with rated speeds, respectively, of 257.1 rpm and 450 rpm. The four turbines have individual inlet valves. The rated capacities of the connected generators are 1.185 MW and 5.855 MW, giving a total power capacity of 14.08 MW. A nominal efficiency 0.908 can be inferred from these numbers, power output 14.08 MW, flow rate 1500 cfs = 42.48 m³/s, water density 1000 kg/m³, g = 9.81 m/s² and head h 37.2 m, using Eq. 2.6.

The Cannonsville dam, located northwest of New York City, is an example of small recent hydropower installation. The world's largest installation, also quite recent, is that at the Three Gorges Dam in China.

According to a planning document of the Chinese National Committee on Large Dams, the power was to be generated by 26 Francis Turbine units of individual capacity

Fig. 8.19 *One of 32 Francis turbine runners for the Three Gorges Dam in China. Water is spurted into the blades radially from the outside, and leaves axially below the runner. Image courtesy of Voith-Siemens https://en.wikipedia.org/wiki/Three_Gorges_Dam#/media/File:Sanxia_Runner04_300.jpg "Three Gorges Project"*

700 MW, total to 18.2 GW. The runner of one such unit is shown in Fig. 8.19. The completed installation has capacity 22.5 GW, with 32 main turbines plus two 50 MW turbines to power the plant itself.

8.3.2   Pumped hydroelectric storage

The hydropower turbine-generator equipment is in principle reversible. The generator basically can be run backward as a motor, and the turbine can be reversed to pump water to some higher reservoir for storage. Only a small fraction of the installed hydropower is in the category of "pumped hydro". Francis turbine-generator combinations stated as reversible, specifically for pumped storage installations, are commercially available, up to capacities as large as 500 MW. The basic use of the pumped storage facility is to allow the utility to meet peak demand by storing energy during periods of low demand and releasing the energy when demand is high. The overall efficiency in the transformation is typically on the order of 80 percent.

An example of pumped storage is the additional installation at the Grand Coulee Dam on the Columbia River in the state of Washington. From the discussion near Fig. 2.4, extended in Fig. 8.18, the Grand Coulee Dam was built in 1942, a third power house was added in 1966,[8] and the capacity is rated as 6.8 GW, the largest in the US.

The added pumped storage capacity for 600 MW pumping and about 300 MW generation was completed in 1984. The pumping is from Lake Roosevelt, immediately behind Grand Coulee Dam, at maximum elevation 1290 feet, via a feeder canal to the higher artificial Banks Lake, at elevations varying from 1567 to 1570 feet above sea level, depending on the storage level.

An additional large stand-alone pumped storage facility is that at Ludington, MI, see Fig. 8.20, close to Lake Michigan in the US.

A refurbishment of the facility, that was originally built in 1973, is underway, replacing all motor/generators with somewhat larger units to bring the new capacity of the plant to 2.172 GW. Toshiba is providing the new Francis Turbines,[9,10] and a review[2] states that the redesigned Francis turbine runner is a one-piece fabrication of stainless steel with nine blades that are welded in place. The diameter of the new runner is 8.4 m and the weight of each unit is 270 tons. The new generating capacity per turbine is 360 MW vs 312 MW. The Ludington facility is jointly owned by Consumers Energy and Detroit Edison, utility companies in the state of Michigan.

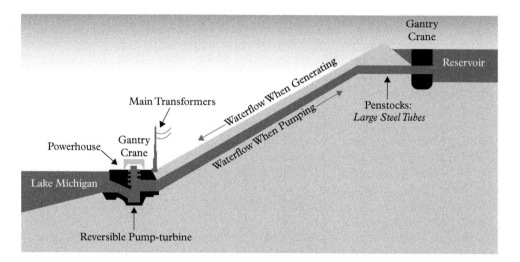

Fig. 8.20 *Schematic diagram of the Ludington pump storage installation. The elevation change is 363 feet and the artificial reservoir has an area of 842 acres. There are 6 individual penstocks of 28 feet maximum diameter feeding 6 reversible Francis pump-turbines, at initial capacity 312 MW per turbine. https://www.consumersenergy.com/content.aspx?id=6985*

[8] https://www.nwccouncil.org/media/10864/MarkJones.pdf
[9] https://www.toshiba.co.jp/about/press/2011_02/pr0801.htm
[10] https://www.toshiba.com/taes/cms_files/hydro_review.pdf

A somewhat larger pumped storage facility is located in Bath County, Virginia, with a total capacity 3.003 GW. The individual reversible Francis turbine-generator units in the recent upgrade of this facility are made by Voith and are rated at 500.5 MW as generators and 480 MW as pumps.

According to the Energy Storage Association[13] the pumped storage capacity in the US is 22.2 GW, that is only 2 percent of total electric generating capacity. In Europe the pumped storage capacity is given as 44 GW, representing 5 percent of capacity, while in Japan the pumped storage is 25 GW, or 10 percent of capacity. The global pumped storage capacity is given as 104 GW. According to the US Deptartment of Energy "Grid Energy Storage" document of December 2013, the US has 24.6 GW of storage, 2.3 percent of total electric capacity, and of that 24.6 GW, 95 percent, or 24.4 GW is pumped storage hydro. We will return to the subject of energy storage in Chapter 11.

8.4   Tidal energy installations

Tidal energy was first discussed previously in connection with Fig. 2.9. This energy comes from the orbital motions of the Moon around the Earth and the Earth around the Sun, that distort the distribution of water in the oceans in a rhythmic fashion. Tidal power is a form of hydropower which harnesses the sea level changes with respect to the land, due to the effect of the gravitational field of the Moon and the Sun. The combined result of the two fields causes the oceans to be constantly moving from high tide to low tide, and then back to high tide, producing a twice-a-day ebb and flow of the Earth's tides. The positions of the Earth, Moon and Sun in the strongest tides, Spring tides, and in the weakest tides, Neap tides are shown in Fig. 8.21.

There are two tidal peaks and two tidal troughs per day, each one separated by six-and-a-quarter hours, due to the fact that the Moon completes one orbit in our sky every 25 hours. The movement of the Moon around the Earth is also the reason why, each day, the times for high and low tides change by 50 minutes. Because of the relatively small distance between the Moon and the Earth, the Sun's gravitational force on Earth accounts only for 46 percent that of the Moon, even with the huge difference between the two masses.

The first form of energy available from tidal flows is potential energy from the rise and fall of the water level, that in favorable locations, as in the Bay of Fundy in Nova Scotia, may be as high as 17 m. The power of release of water with head h was given in Eq. 2.6. The second form of energy is kinetic as the tidal stream of water flows in and out of a narrow estuary. Here the kinetic energy is described by Eq. 2.4, with the understanding that the flow is of water rather than air.

The type of device for harvesting the tidal potential energy is a dam, across the tidal estuary, known as a "tidal barrage", with water turbines deployed similarly to the case of hydropower discussed previously. There are not many installations of this sort, the first is in La Rance, France in the province of Brittany. The devices exploiting kinetic

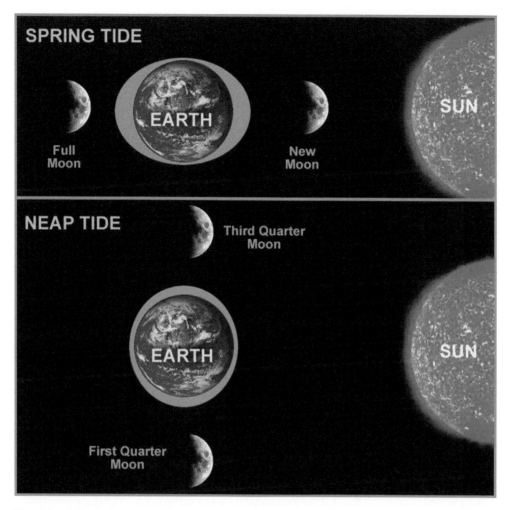

Fig. 8.21 *Schematic diagram shows deployment of Sun, Earth and moon in the strong "Spring Tide" configuration (top) and in weak "Neap Tide" configuration (below). The strongest tide is Spring Tide with the New Moon, when the moon is between the Earth and the Sun. The linear configuration with the moon on the opposite side, the full moon, is also an occasion of strong tides. The effect of the moon on the oceans is greater than that of the Sun. https://www.windows2universe.org/earth/Water/images/tides_lg_ gif_image.html*

energy, analogous to wind turbines, are called "tidal stream" devices, as described in Section 8.2. A summary of tidal installations is given in Fig. 8.22.

Fig. 8.22 lists also installations of "Ocean Wave" type, that include the Pelamis device discussed in connection with Eqs. 2.11–2.23, and "Thermal Gradient" as illustrated in Fig. 2.10, called also OTEC, ocean thermal energy conversion. It appears that

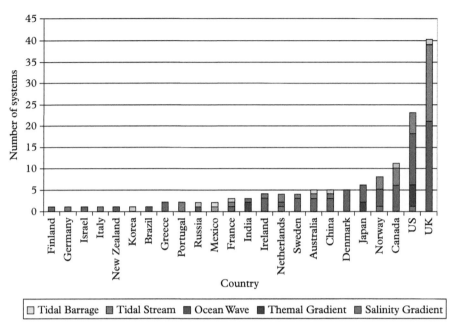

Fig. 8.22 *Summary of tidal energy installations as of 2009. Tidal barrage installations (light color at top of respective entry) are shown for Korea, Russia, Mexico, France, Australia, China, Canada and United Kingdom (Khan and Bhuyan, 2009) Fig. 1*

the most important large-scale device is the Tidal Barrage, the dam on the tidal estuary. In this case it is possible to run turbines either at low tide, on the outflow only, or on both the inflow at high tide and the outflow at low tide.

The tidal barrage hydropower station of La Rance (Brittany, France) was the first tidal station in the world and second largest at the present, after the Sihwa Lake station, South Korea. The station has delivered electricity since 1967, producing about 70 percent of Brittany's electric power at present. The plant is based on a dam 330 meters long and in a basin of 22 km² area. The vertical range of the basin is 8 m. A set of 24 Kaplan turbines, of 5.4 m diameter, is installed on the barrage, with a total power of 240 MW, and directly connected to France's national electricity grid. This plant produces power for four-hour periods twice daily.

The original La Rance installation uses Kaplan turbines, contained within the large pipe or penstock carrying water at release or at inflow at high tide, that turn under axial flow of water. A moderate size Kaplan turbine is shown in Fig. 8.23. Kaplan turbines are typically smaller than the Francis turbines described previously, and have a feature that the pitch of the blades can be adjusted while the machine is in operation. These turbines are commonly used with water heights of 3 m but can still be operated with head as low as 0.3 m.

Fig. 8.23 *Photograph of a Kaplan turbine runner. The strong water flow enters axially from below in its containing pipe or penstock (not shown). The pitch angle of the blades is adjustable while the device is in motion. Turbines of this type are capable of 90 percent efficiency. Viktor Kaplan Turbine Technisches Museum Wien. https://commons.wikimedia.org/wiki/File:Kaplan_Turbine.JPG#/media/File:Kaplan_Turbine.JPG*

In summary,[14] as of 2016 the leading installations of tidal power are South Korea, 511 MW; France, 246 MW; United Kingdom, 139 MW; Canada, 40 MW; Belgium, 20 MW; China, 12 MW; and Sweden, 11 MW. This totals to world installed capacity 0.979 GW. Looking to the future, the Minas Passage in the Bay of Fundy, Nova Scotia, is viewed as a potential source of as much as 2.5 GW tidal power.

9

Solar Thermal Energy
Limits and Technology

Having learned quite a bit about how the Sun's energy is created, and how that process might be reproduced on Earth, we turn now to methods for harvesting the energy from the Sun as a sustainable replacement for fossil fuel energy. The Sun's energy most obviously heats what it falls upon, reducing, for example, the need to run an oil or gas furnace in a house in winter. Heat energy is the easiest form to store and locally distribute and is useful for heating water and home heating. (The advantage of solar water heating is that the vexing storage and distribution issues are absent in this use of sunlight. The saving in electricity from the grid is equally as valid as adding power from solar farms.)

The more technical challenge is to turn the Sun's energy into electricity, for consumption, or perhaps convert its energy into hydrogen as a portable fuel. Two main tracks using the Sun's radiation to produce electricity are through a heat engine running a generator; and direct photovoltaic conversion. An additional track is to use the Sun's energy to directly create a fuel, hydrogen, through a photo-catalytic water-splitting process, that bears some similarity to photosynthesis. In this chapter we first summarize the Sun's spectrum on Earth and how it can be focused for optimal use. Next, we consider the heat engine that extracts mechanical energy from a hot reservoir, that can be heated by the Sun, rejecting excess heat at a lower temperature reservoir. The limiting efficiency of such energy conversion is given by the Carnot efficiency formula, mentioned briefly in Section 2.6.3.

9.1   Sun as an energy source, spectrum on Earth

The main properties of the Sun as an energy source are its energy density, $1366\,\mathrm{Wm^{-2}}$ at the top of the Earth's atmosphere; the highly directional nature of its radiation, and its spectrum. For simplest estimates we will take the Sun as a black body at 6000 K, but for most purposes we must consider the ground level spectrum as it emerges below the Earth's atmosphere. Locally the spectrum and power density are modulated by cloud cover, for example.

Physics and Technology of Sustainable Energy. E. L. Wolf, Oxford University Press (2018).
© E. L. Wolf. DOI: 10.1093/oso/9780198769804.001.0001

Starting from the energy density of the ideal black-body radiation given in Chapter 3,

$$u(v)dv = [8\pi\, hv^3/c^3][\exp(hv/k_B T) - 1]^{-1} dv, \tag{3.18}$$

we multiply by $c/4$ to get a power density, in $\mathrm{Wm^{-2}}$, and then convert to wavelength using $c = \lambda v$, to get the power density per unit wavelength (as plotted in Fig. 1.4):

$$P(\lambda)\, d\lambda = \left[2\pi\, hc^2/\lambda^5\right]\left[\exp(hc/\lambda k_B T) - 1\right]^{-1} d\lambda. \tag{3.19a}$$

As mentioned earlier, the peak in this function is at λ_m, such that $\lambda_m T = \text{constant} = 2.9$ mm-K. The peak values $\lambda_m = 486$ nm and most probable photon energy 2.55 eV for the above-Earth solar spectrum lie in the visible corresponding to $T \approx 5973$ K.

The sunlight reaching Earth is diminished by specific absorption from water and other minority molecules in the atmosphere, and by Rayleigh scattering of the sort that makes the sky blue, as discussed in Section 1.2, following Fig. 1.4. 1000 W/m² is the approximate peak illumination at the Earth surface. The details of the spectrum are more important for semiconductor devices, and for this purpose the spectral power expressed per unit energy is preferable. Fig. 9.1 shows the clear-day noontime spectrum for the northern hemisphere, a standard spectrum called "AM-1.5", meaning 1.5 air mass traversal, at 48° from the vertical. Most of this intensity is direct light, with a small diffuse contribution from the Rayleigh scattering. Most of this intensity can be

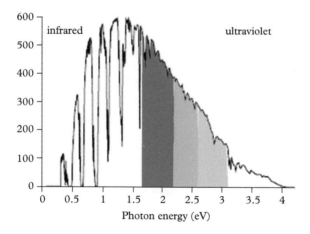

Fig. 9.1 *Earth level peak Sun's power spectrum, units W/(m² eV). This is the "AM 1.5" spectrum characteristic appears at noon on a clear day for moderate climates, and 48° from normal incidence. The area under this curve represents about 1 kW/m². The shaded areas in range 1.7 eV to 3.1 eV represent red, green and blue parts of the visible spectrum About half the energy is in wavelengths longer than visible, i.e., energies less than 1.7 eV. A small portion of this light intensity comes indirectly by Rayleigh scattering (MacKay, 2009) p. 57*

concentrated as will be described below. On a cloudy day the light intensity may not be too different, but the light is diffuse and cannot be focused. Focusing systems are almost useless on a cloudy day, while flat solar panels may still operate at 30 percent to 40 percent of their normal capacity.

9.2   Heat engines, Carnot efficiency

As a preface to this discussion, we repeat that direct heat absorption to heat water and for home heating is extremely important in economic terms, perhaps larger than the electric conversion systems that are our primary interest. These direct solar heating systems reduce the use of fossil fuels and also the load on the electric power grid, and are correctly classified as important examples of renewable energy. Building homes in several countries, notably Israel, by law requires solar heating to be used for domestic hot water and to partially heat the home.

In utilizing the Sun's energy, our first main topic is the heat engine, typically a steam turbine, or possibly a Stirling engine, running an electric generator. A heat engine contains a fluid, in the simplest case a gas, that works between a hot reservoir, T_s and an ambient reservoir T_a, and has a limiting efficiency,

$$\eta_{Carnot} = 1 - T_a/T_s \tag{9.1}$$

that applies to the Carnot cycle. In the Carnot cycle, as represented on a PV diagram, an enclosed area representing the work done in the cycle, is enclosed between two isothermal curves, where $P = RT/V$ and two adiabatic curves represented by $P = $ constant/V^γ. The latter, adiabatic, constant entropy, curves fall more steeply with increasing V on the PV plot because the exponent $\gamma = C_p/C_V$ exceeds unity. The exponent value is 5/3 for a monatomic gas like He and 7/3 for diatomic gases like hydrogen, nitrogen and oxygen. On the adiabatic curves the temperature T varies as $1/V^{\gamma-1}$. The efficiency results from straightforward evaluation of the area $\int PdV$ enclosed between the four curves, divided by the heat energy taken in at the hot reservoir.

If we evaluate the Carnot efficiency $1-T_a/T_s$ at $T_s = 6000$ K for the Sun and $T_a = 300$ K for ambient, we find 0.95. As a second example, the inlet temperature of a steam turbine, as is used in concentrating solar plants, is commonly 540°C, or 813 K. In this case, with 300 K ambient, the ideal Carnot efficiency is 0.631. In practice such steam turbines, operating on the Rankine cycle to be described subsequently, give efficiency on the order of 0.4.

Practical heat engines deviate from the Carnot cycle, but its efficiency formula can still be used as a guide. The Stirling cycle can be roughly approximated by two constant volume lines connected by the isothermal curves at the higher and lower temperatures. The steam turbine is more complicated because steam and water are two different phases, gas and liquid. In all of these cases, however, the efficiency is raised by reducing the ratio of the two temperatures, T_a/T_s.

9.2.1 Focusing sunlight: tracking

The practical application of a heat engine to capture solar power uses an optical focusing system to increase the power density on the hot reservoir of the heat engine. The ability to focus direct sunlight is large because the Sun is nearly a point source.

In more detail, the Sun's radius $R_s = 0.696 \times 10^6$ km, and its distance D_{es} about 93 million miles (1.496×10^8 km) from Earth make its angular radius seen from the Earth very small, about 0.266°. We can infer the total power density at the Sun's surface as:

$$P = 1366 \text{ W/m}^2 \times \left(D_{es}/R_s\right)^2 = 6.312 \times 10^7 \text{ W/m}^2, \tag{9.2}$$

where the geometric ratio of concentration is 46,200. It is this power density that corresponds to the black-body power density $\sigma_{SB} T^4$, with $T = 5973$ K, where $\sigma_{SB} = 2\pi^5 k^4/(15h^3 c^2) = 5.67 \times 10^{-8}$ W/m^{2}K^4.

The concept of the black-body radiator is a surface that emits the power density equally into all directions, as discussed in Section 3.2. A hemisphere centered above a black-body emitter will receive equal power on each point of its surface. The hemisphere represents solid angle 2π. (In reverse, a black surface will absorb equal power from directed beams originating at each point on the hemispherical surface.)

The model for a black-body radiator is a small opening into an enclosed cavity at temperature T, suggested by a pizza oven glowing red on the inside. The opening into the cavity acts as a perfectly black absorber because a light ray entering the small opening has no chance of coming out the small opening, independent of the direction it took entering the cavity. Inside the equilibrium cavity photons are propagating in all directions, and any one of these can come out through the hole. The light emitted from the hole is directed equally into each angular range defined by increments of polar and azimuthal angles $d\phi$ at ϕ and $d\theta$ at θ. The Sun acts as a black-body radiator, and many types of surface act to a good approximation as black-body absorbers and radiators.

Let us imagine a small area of black-body surface, with angle from the surface normal taken as θ and azimuthal angle ϕ. The Sun is at a particular angular location and represents an angular diameter of 0.53°. How can we design an optical system that will maximally concentrate the Sun's rays onto this small area of black-body surface? The problem is equivalent to designing an optical system that will take all of the light emitted by the black-body surface (into all 2π radians) and focus it into the angular range defined by the Sun. An example of an optical system that can approach this is illustrated in Fig. 9.2.

A concentrator system is perhaps more commonly based on a parabolic mirror as the primary element, and a system of this type has demonstrated a concentration ratio 84,000. Literally, this focused light is more intense than the surface of the Sun, and would have intensity $1366 \times 84000 = 114.7$ MW/m^2 if operated above the atmosphere.

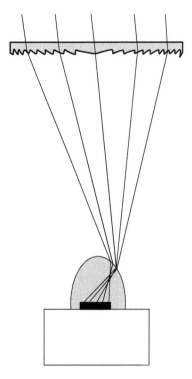

Fig. 9.2 *Two-stage optical system taking wide angular range emission from hot black surface (bottom of figure) and focusing it into a small angular range. In reverse, this will focus Sun's rays onto black surface making use of a large angular range. This system has a primary Fresnel lens and a secondary imaging element made of glass of high index of refraction n. The concentration ratio in such a system can in principle reach $n^2 (D_{es}/R_s)^2$, with concentration ratio 84,000 experimentally attained. (Marti and Luque, 2004) Fig. 13.7, p. 307*

9.2.2   Available efficiency in solar concentrator heat engine

This analysis can be extended to simulate a solar thermal power plant. In such a case, a focused light system puts sunlight onto an intermediate collector heat reservoir, that rises to an intermediate temperature T_c. The collector reservoir then is the hot reservoir for the heat engine. In this case radiation back from the intermediate collector to the Sun is an important part of the analysis. In Fig. 9.2, one can imagine the intermediate reservoir being the black surface at the bottom of the field, which we now say will come to temperature T_c and radiate a power density $\sigma_{SB} T_c^4$ back to the Sun through the optical system, that is shown as a Fresnel lens in Fig. 9.2, but in many cases, will be a mirror system. In this circumstance, that was investigated by Landsberg and Tonge (1980), the best possible efficiency is expressed as:

$$\eta_L = (1 - T_c^4 / T_s^4)(1 - T_a / T_c) - \delta. \tag{9.3}$$

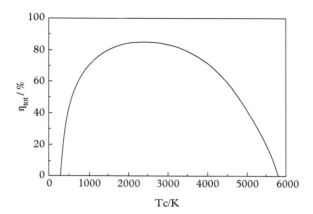

Fig. 9.3 *Plot of system efficiency η_L (mechanical power output divided by Sun power input) vs collector temperature (referred to as T_c in text), assuming Sun at 6000 K and ambient exhaust at 300 K. The peak efficiency is 0.854 at collector temperature 2544 K. (Marti and Luque, 2004) Fig. 3.1, p. 52*

The maximum value of this efficiency, $\eta_L = 0.854$, setting the correction δ to its minimum, zero, with $T_s = 6000$ K and $T_a = 300$ K, occurs at collector temperature $T_c = 2544$ K. (We should note that this is 2271 C, unattainable in conventional technology.) The terms in this expression clearly relate to radiation equilibrium between the Sun and the collector, and to Carnot efficiency of the heat engine working between the collector at T_c and the ambient exhaust temperature T_a. The correction δ is related to entropy generation, $\delta = T_a \, (dS/dt)/P_s$ where P is the power from the sunlight. Entropy $dS = dQ/T$ is accepted from the Sun by black-body absorption, some is returned to the Sun by the black-body emission from the collector, and a contribution $dS_c/dt = P_c/T_c$ is created by black-body radiation from the collector to the Carnot engine.

A plot of the Landsberg limiting efficiency as a function of the collector temperature T_c is shown in Fig. 9.3.

This type of system is known as concentrated solar power, (CSP), with two main forms, the large central receiver and the parabolic trough system. Large central receiver solar thermal electric power plants are shown in Fig. 9.4. The design of the mirrors and their tracking is a feat of engineering. Only in a large system does it seem likely that the theoretical Landsberg efficiency 84.5 percent could possibly be approached, because of the difficulty in maintaining components and heat-carrying fluids at temperatures approaching 2544 K. In such high temperatures it seems that molten salts, or a molten metal such as lithium, with boiling point 1615 K, are possible candidates, certainly oils will decompose. From Fig. 9.3, and Eq. 9.3, the efficiency at 1615 K is 0.81, much higher than found in photovoltaic systems, as we will see. In fact, the PS-10 system, left in Fig. 9.4, has 11 MW capacity, with 624 metal-glass mirrors of 120 m² area, and uses pressurized steam at 250°C and 40 Bar to run a conventional steam turbine.

Fig. 9.4 *Early central receiver CSP thermal solar power installations PS–10 (left) and PS–20, near Seville, Spain. These systems of accurately tracking mirrors (functionally similar to Fig. 9.2) image the Sun onto the central receivers, that are described as the collector at temperature Tc in Eq. 9.3. Following the ideal Landsberg analysis, maximum efficiency η_L (mechanical power output divided by Sun power input) will occur at $T_c = 2544$ K, a higher temperature than is currently possible. Achieving such a high receiver temperature is an as-yet unmet challenge in materials science and engineering (http://www.solarpaces.org/Tasks/Task1/ps10.htm) https://en.wikipedia.org/wiki/File:PS20andPS10.jpg*

9.2.3 The role of the steam turbine

The PS-10 system, uses a steam turbine working at a relatively low inlet temperature, 250° C, where the Landsberg theoretical system efficiency is in the vicinity of 43 percent, by looking at Fig. 9.3 and using Eq. 9.3.

An empirical efficiency estimate for PS-10 can be made easily: the receiving area is 624×120 m², assuming arbitrary input 1000 Watts/m², leading to $P_s = 74.9$ MW. At 11 MW rated output, the overall efficiency is 14.7 percent, much smaller than the ideal limit near 43 percent for 250°C inlet temperature or 81 percent for the theoretically optimum inlet temperature of 2544 K. This seems to be a situation, unlike that in silicon solar cells that we discuss later, where a large increase in efficiency is potentially available. In practical terms, the operating efficiency of this plant has also been given as 18 percent, that may partially reflect that the plant has been designed to provide nameplate output, 11 MW, even when the Sun is lower in the sky.

To go on with an optimistic "blue-sky" estimate of potential added value that would come from the hypothetically added 49.7 MW potential capacity, let's assume all the power is put into the electric grid at 0.1 $/kWh. On a yearly basis the added revenue from the upgrade, assuming the Sun power is available 0.33 of the time, would be 0.33 × 49.7 MW × 3.15 × 10^7 × 0.1$/3.6 × 10^6 = $14.4 Million/year. This seems a large payoff for upgrading a portion of the system. (The land area and the mirrors would be unchanged in this upgrade.) The reason that this was not done may be that the technology for the (tungsten receiver/molten lithium heat fluid) system is not available, but one can see potential for such an advance. The design of a turbine that will convert the higher temperature into mechanical energy would also be needed.

A European Community Report (of 2006) suggested that the total investment in PS-10 was 35 million Euros, about $52.5 M. Roughly, the cost per Watt is $52.5 M/11 MW = $4.77/W, for the original plant in Spain. This is a favorable value. It is clear that this site was well chosen from points of view of sunlight and cost of land.

The theoretical Landsberg efficiency of a system as we have described is dependent on attaining a high collector temperature. However, the main practical route to electricity generation is through the steam turbine running a generator. At present, steam or other turbines are not available that can utilize the high temperature potentially present from concentrated sunlight.

The steam turbine, rather than the Stirling engine, is the practical choice for existing large-scale installations of concentrated solar power. The steam turbines are typically based on the Rankine cycle that is outlined in Fig. 9.5.

The steam Rankine reheat cycle is outlined in the lower portion of Fig. 9.5, in a T-S diagram. Note that the area of a closed loop in this diagram, namely $\int T dS$, represents work done and is the basis for an efficiency calculation. The dark solid curve in Fig. 9.5 is the T-S characteristic of water, liquid on the left and vapor on the right. The water enters the boiler (steam generator) at point 6 from the high-pressure pump following the condenser and is heated until the horizontal upper line is reached, representing evaporation at constant temperature and pressure. The later rising section ending at point 1 (inlet to the turbine) represents superheating of the steam to the typical inlet values 540°C and 115 Bar. The Rankine reheat cycle, shown in an idealized form, is represented by the lighter closed solid line connecting points 1 through 6. The calculation of efficiency is accomplished by summing the power output of the two turbines minus the work needed for the pump, divided by the total heat input. In general terms, the efficiency is less than that of the Carnot cycle, but can be written in a similar way.

$$\eta_R = 1 - T_{m1}/T_{m2},$$

with two mean temperatures.

In the Rankine cycle, the mean temperature at which heat is supplied is less than the maximum temperature T_C of the Carnot cycle, so the efficiency is less than the Carnot efficiency. In this regard, the reheat section of the augmented cycle is a benefit in lowering the mean temperature at which heat is rejected, namely in the condensation process following the low-pressure turbine, in path 4 to 5 of the diagram. In the Rankine cycle,

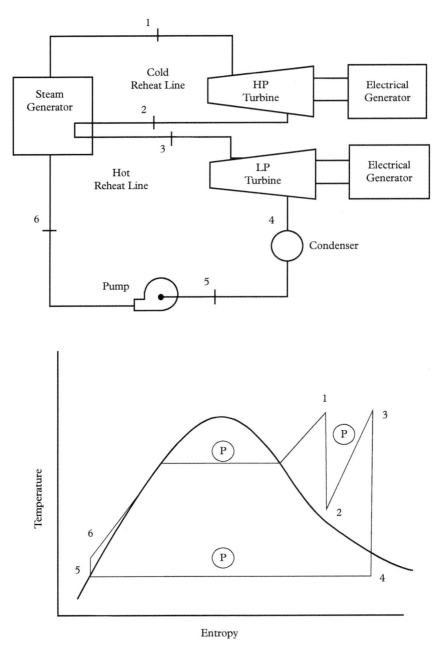

Fig. 9.5 *Outline of steam Rankine cycle, including re-heat turbine, similar to that used in leading CSP plants. Upper panel shows high pressure pump on bottom, that pressurizes boiler (steam generator) on left, and two-section steam turbine on the right. The exhaust steam from the high-pressure turbine, at an intermediate pressure, is reheated by a second pass through the boiler section, where its temperature again comes to the high-pressure inlet value, for example 540°C. Lower panel shows T-S diagram for the reheat Rankine cycle. Point 1 locates the inlet to the high-pressure turbine, with steam parameters typically 540°C and 115 atmospheres. The rising line leading to point 1 represents the superheating of the steam leaving the boiler. The temperature drops within the turbine, points 1 to 2, followed by steam reheat at a reduced pressure between points 2 and 3, to reach the inlet of the low-pressure turbine. The steam at the exit of the low-pressure turbine is condensed back to water from points 4 to 5 on the lower diagram. The high-pressure pump between points 5 and 6 also raises the temperature (K. Weston, 1992) Fig. 2.8 on p. 55*

increasing the pressure in the boiler automatically raises the temperature for boiling. This raises the average temperature of the heated the steam and thus raises the efficiency.

Carefully considered efficiency values for Rankine cycle steam turbines with parameters close to those obtained in concentrated solar power tower plants are quoted by Sargent and Lundy (2003), namely 37.5 percent at inlet temperature 390°C and 40.5 percent at inlet temperature 500°C. It is stated that there are numerous steam turbines in operation in conventional power plants with inlet pressures over 250 Bar and inlet temperatures over 590°C, where the turbine efficiency is over 44 percent. Steam turbine efficiency is well understood but cannot be substantially raised.

In contrast to the usual steam turbine design, a leading example of a smaller high-performance system is shown in Fig. 9.6, known as a Dish–Stirling system. A higher operating temperature is available in a closed-cycle Stirling engine using helium or hydrogen gas, with no need for a heat fluid transport. In a Stirling engine the same limited volume of gas cycles back and forth in a repetitive fashion.

A report[1] of $\eta_L = 0.31 = P_{electric}/P_s$ efficiency in a separate Stirling engine system suggests that a substantially higher collector temperature is available in Stirling engines.

Fig. 9.6 *Small array of Stirling engine solar concentration electrical generators. These engines have been demonstrated by Sandia National Laboratories to provide 31 percent efficiency defined as electrical power output divided by solar power input, under high illumination, considerably higher than the power tower systems such as those in Fig 9.4 that utilize steam turbines https://share.sandia.gov/news/resources/releases/2008/solargrid.html*

[1] https://share.sandia.gov/news/resources/releases/2008/solargrid.html

The reported dishes (see Fig. 9.6) each used 82 mirrors focused on 7-inch apertures of the hydrogen-based Stirling engines. The efficiency 31 percent was demonstrated as electric power output divided by solar energy input, and the record was set on a cold day when the cold port of the Stirling engine could be maintained at 23°C = 296 K. Over a period of several hours during the day the dish system maintained an electrical output 26.5 kW, or about 36 HP. A substantial exhaust heat flow is also generated at 23°C = 73°F, so evidently $P_{exhaust}$ = 0.69 × 26.5 kW/0.31 = 59 kW. If water cooling was employed, a substantial supply of water at 73°F on a winter day might be useful for home heating, but typically such exhaust energy is wasted. A system like the Dish–Stirling shown, with a highly concentrating parabolic mirror, requires accurate two-angle tracking to achieve its potential performance.

All thermal solar power systems, unlike non-concentrating photovoltaics, require cooling. In practice for large plants this often means that water is used. It is noted that in California the use of drinking water for this purpose is illegal, suggesting that common practice is to dump the cooling water.

Smaller installations based on Stirling engines heated by plastic Fresnel lenses are possible, e.g., for a remote household power supply, typically to charge batteries. An acrylic Fresnel lens of 0.5 m diameter and 2 mm thickness is inexpensive, and it is possible to set up a Stirling engine so that fossil fuel may alternatively heat the Stirling intake when the Sun is not shining. Based on the performance noted previously, peak Sun power would, at 30 percent efficiency, generate at best 60 W. While this might be more expensive than a larger area of photovoltaic cells, the latter would not allow fossil fuel backup.

The Swedish company Ripasso Energy has tested Dish–Stirling systems in the Kalahari Desert of South Africa. An efficiency of 34 percent is reported for the Ripasso system[2] using an advanced Stirling engine that was earlier developed by the Swedish firm Kokums, and using a 12-meter diameter mirror.

9.3   Solar thermal electric power

The high efficiency Dish–Stirling engine system described has not been widely deployed. The reasons for this failure are probably rooted in a higher cost for the installation and the recently falling price of solar power panels. More common Concentrated Solar Power CSP systems are built either with parabolic troughs or with heliostats focused on a power tower. Early installations generally did not allow any heat storage, including notably the Ivanpah, California, 392 MW power-tower installation with 173,500 heliostats that became operational in February 2014.

As an introduction to heat storage with molten salt, we consider the Andasol-1 parabolic-trough solar thermal electric power system. A good description is available from Dunn et al (2012), of this system that incorporates heat storage in molten

---

[2] (Barbee, 2015). https://www.theguardian.com/environment/2015/may/13/could-this-be-the-worlds-most-efficient-solar-electricity-system

salts based on commonly available fertilizers, to continue electric power generation overnight.

A system diagram for this form of concentrated solar electric power (CSP) installation, based on parabolic reflecting troughs, and including molten salt energy storage, is shown in Fig. 9.6. The trough geometry provides a concentration ratio limited to about 100, while the power-tower geometry, shown for PS-10, duplicated in the Ivanpah plant and also the Tonopah plant to be described, provides concentration in excess of 1000. The Andasol-1 50 MW installation of 2008, in the province of Granada in Spain, indicated in Fig. 9.7, notably allows for energy storage. The diagram shows that oil is fed through pipes at the focus of the parabolic reflectors, where it reaches an operating temperature 393°C. Some of the oil goes directly to a heat exchanger to provide steam at 100 Bar and 377°C for the steam turbine. However, some of the hot oil is diverted to an oil-to-molten-salt heat exchanger, filling a tank of molten hot salt at 386°C. For overnight electric operation, some of this molten hot salt is re-routed, via a heat exchanger, to heat oil that then goes to an oil-to-steam heat exchanger, again providing steam for the turbine. The lower temperature salt storage tank is maintained at 292°C to avoid any chance of its crystallization that would damage the piping system.

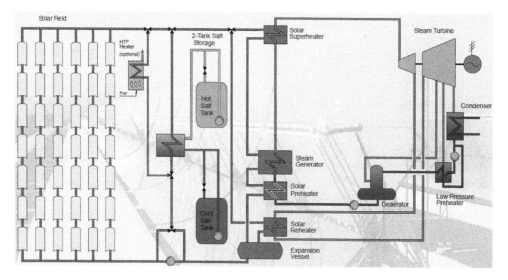

Fig. 9.7 *System outline for Andasol-1 50 MW solar electric power installation in Granada, Spain based on 6 linear arrays of parabolic reflecting troughs. Pipes carrying oil run along the focal point of the parabolic troughs. The hot oil at 393°C is partially fed to create hot molten salt at storage locations and partly fed to heat exchangers to provide high temperature steam for the steam turbine, using the Rankine re-heat cycle, at right side of diagram. Note the heat exchangers to provide the superheating and re-heating present in the Rankine re-heat cycle shown in Fig. 9.5. The heat stored in the hot molten salt can be recovered at night to extend duration of electricity generation (Dunn et al, 2012) Fig. 7*

Molten "solar salt", that is a eutectic mixture of 60:40 Na:K nitrates, has a melting point 220°C and decomposes at 600°C. These polycrystalline sodium and potassium nitrates are widely available at a low price as fertilizers sold by the 100-pound bag. The solar salt is typically operated in a temperature range 292°C to 565°C and provides a high heat capacity (at 300°C) of 1495 J/kg-K, according to Dunn et al (2012). The operating temperature range is limited at the low end by the freezing point and at the high end ultimately by the decomposition temperature, 600°C, but, in practice, to limit corrosion of the high-temperature pipes that are made of Types 304 and 316 stainless steel.

Several installations of the parabolic trough plants using hot oil and solar salt are in service, principally in Spain and in the southwest US. Their performance is limited by the relatively low maximum operating temperature of the oil, 393°C. A higher temperature (more efficient) and simpler system, with less extensive piping and fewer heat exchangers, uses only molten salt and a single tower rather than a field of parabolic troughs. This choice requires care in design and operation to maintain all of the piping and storage tanks safely above the freezing point of the solar salt, 220°C: in practice the entire molten-salt system is maintained above 292°C. Special care in using the salt as heat fluid is needed mostly at the initial commissioning, requiring melting of all the salt, and in the start-up each morning of the system. The related difficulties have been addressed and overcome in large part, with several systems in commercial operation. The limitation of the deployment of such plants is now more in the economic realm, depending on the prices of solar panels, oil, natural gas, and coal plants in the regulatory environment.

http://cleantechnica.com/2016/02/22/crescent-dunes-24-hour-solar-tower-online/.

The state of the art in CSP concentrated solar power facilities in 2015 is exemplified by the 110 MW power tower facility (with 10,347 heliostats, each with two-axis tracking) at Tonopah, Nevada, known as the Crescent Dunes Solar Energy Project. The facility has a contract with Nevada Energy to provide power for 25 years at a cost 13.5 c/kWh.

http://www.nrel.gov/csp/solarpaces/project_detail.cfm/projectID=60.

This facility uses only molten salt as its heat fluid, and is designed for 10 hours of storage. The molten salt in the plant totals to 32 million kg, that took 2 months to be fully melted, and is expected to remain molten over the lifetime of the facility, that is 25 years. The molten salt flows directly to the top of the 640-foot solar tower, where it is heated to 565°C. The heated salt fluid is partly contained in a hot salt tank that provides 1.1 GWh of energy storage. This plant, then, apart from a run of cloudy days that is very rare in its location, provides 24 hours per day of uniform power output, similar to a coal, gas or nuclear power plant. A heat exchanger from molten salt to steam is provided to run the Alstom 125 MW steam turbine with inlet temperature 540°C that turns the electric generator.

http://prod.sandia.gov/techlib/access-control.cgi/2013/131960.pdf.

An extended analysis of the performance molten salt power tower systems is given by Pacheco et al (2013).

According to these authors, the salt temperature at the tower and hot tank is at most 565°C. After the heat exchanger to steam, the turbine inlet conditions are 120 Bar at 540°C. The steam turbine efficiency in such a plant is estimated as 43 percent with water cooling and 41.3 percent with air condenser cooling. The solar-to-electric efficiency, however, defined by Sargent and Lundy (2003) as the product of the turbine efficiency with the solar field optical efficiency times the receiver thermal efficiency, is much smaller. Values for the solar field optical efficiency and the receiver thermal efficiency are given as 73 percent and 85 percent, respectively, that gives an estimate for solar-to-electric efficiency (with the air-cooled turbine) of 25.6 percent. This number should be comparable with the 31 percent that was observed for the Dish–Stirling system mentioned previously. A rough estimate for the solar-to-electric efficiency of the Ivanpah plant can be made using the full output power 377 MW, the mirror area 14.1 m^2 × 173,500 = 2.45 × 10^6 m^2 and the 850 W/m^2 direct sunlight in California measured by Raman et al (2014), an estimate of peak illumination. This efficiency is then 377 × 10^6/[2.45 × 10^6 × 850] = 0.18, for the Ivanpah plant under strong illumination. This is comparable to the 14.7 percent already estimated in a similar fashion for the PS-10 plant in Spain.

A similar estimate for the Tonopah plant, that has 10,347 heliostats of 116 m^2 area, is: 110 × 10^6/[1.2 × 10^6 × 850] = 0.108, again assuming full illumination. This is much less than the theoretical estimate previously of 25.6 percent for such a plant, and also less than the measured 31 percent for the Dish–Stirling system. This difference may indicate design of an over-sized heliostat system so that the full power of the turbine can be maintained over the whole day, including times near sunrise and sunset when the light intensity is reduced.

An additional efficiency measure that is used is the "annual solar to electric efficiency", calculated as yearly electrical output divided by the yearly solar input, both quantities integrated over the whole year. The design of the plant calls for full output over the whole period, while the solar input will vary over the day and over the winter as is characteristic for the plant location. Such an average solar input will be less than the peak value previously used, that increases the corresponding efficiency. The actual annual solar to electric efficiency of the Ivanpah plant is stated as 28.72 percent, well above the value we got, 0.18, based on estimated peak illumination.

In summary of the CSP technology, see Fig. 9.8. the power tower with salt energy storage as exemplified by the Crescent Dunes plant at Tonopah NV is important in effectively overcoming the intermittency problem, credibly competing in a sunny climate with a conventional coal or gas fired power plant. In a large electricity market, for example California, the plants such as Ivanpah without storage are viable for producing the extra power that is needed during the noon and afternoon hours of large demand. But the plants such as Tonopah are viable for producing the entire supply of electricity, leaving no need at all for fossil fuel electric power. These are huge installations compatible with the conventional power grid. For smaller off-grid locations the Dish–Stirling CSP systems may be viable, although likely more expensive than photovoltaics, and, like solar cells, will need a separate means of energy storage for a self-sufficient installation. These CSP plants however do not make sense except where cloudy days are very rare.

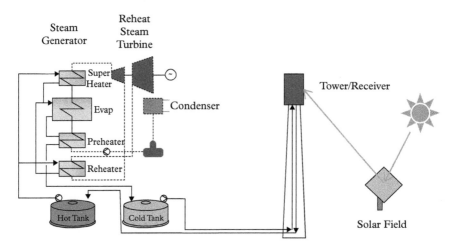

Fig. 9.8 *System outline for Molten Salt Power Tower system such as Crescent Dunes facility in Tonopah, Nevada. Four molten- salt/water heat exchangers are shown on the left feeding the re-heat Rankine-cycle steam turbine, such as the Alstom 125 MW reheat turbine used at Tonopah (Pacheco et al, 2013, Fig. 2)*

It appears that in the CSP categories the Dish–Stirling systems are more expensive than the trough and power tower systems. Beyond this it is clear that the solar thermal systems are more expensive than large scale photovoltaic and wind turbine systems, although neither of those can conveniently offer storage as can the trough and power tower systems. The important Lazard (2015) report[3] gives costs ranging from 11.9 c/kWh to 18.1 c/kWh for "Solar Thermal Tower with Storage", and gives $6.5/W for the capital cost of such facilities. (The cost of the Ivanpah facility is reported as $5.56/W). In contrast, the Lazard (2015) report predicts an electricity cost in 2017 from utility-scale photovoltaic solar at 4.5c/kWh, based on a mean of the costs of thin film and crystalline solar cell systems with single-axis tracking, and assumes a solar cell cost of $1.35/W. It appears that the recent drop in solar cell prices led the bankruptcy of firms that earlier manufactured the Dish–Stirling CSP systems.

A comparison of electricity prices resulting from solar thermal versus wind turbines, shown in Fig. 9.9, was recently offered in the Wall Street Journal (Spindle and Smith, 2016), where the information was extracted from Lazard (2015). The figure shows bands of costs, lying in the overall range 3 c/kWh to about 38 c/kWh, with separation of costs with and without government subsidies. The figure makes clear that wind power generally offers a lower electricity cost, as low as 3 c/kWh in some locations, with subsidy, and that solar thermal power costs can be as low as 9 c/kWh with subsidy in some locations.

---

[3] Lazard's Levelized Cost of Energy Analysis—version 9.0, November, 2015. https://www.lazard.com/media/2390/lazards-levelized-cost-of-energy-analysis-90.pdf

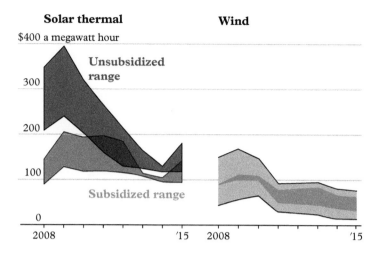

Fig. 9.9 *Summary from the Wall Street Journal of alternative energy costs from Solar thermal and Wind sources, over period 2008 to 2015, drawn from the report (Lazard, 2015). The bands of costs show separate estimates for subsidized power as available in the United States over this period. The article states, for example, that electricity in Houston TX costs 5.9 c/kWh from the vendor NRG Energy, Inc., the low price influenced by the large supply of wind power in the vicinity. In comparison, (Lazard, 2015) states electricity costs of 11.6 c/kWh and 15 c/kWh in New York City and Boston, respectively (Spindle and Smith, 2016)*

9.4 Solar tower for dissociating water to hydrogen

It is clear that very high temperatures can be obtained at the focus of a heliostat system. The world's largest solar furnace[4] is in France in the Pyrenees mountains having 10,000 mirrors and reported to reach 3000°C at its focus.

More recently the Hydrosol II test power tower facility[5] was built in Almeria, Spain demonstrating hydrogen production from water using high temperature from focused sunlight.

The chemical reaction used in the Hydrosol II project is described by Roeb et al (2006), in the proceedings of the World Hydrogen Energy Conference of 2006.

The thermo-chemical reaction to release hydrogen from water in the Hydrosol II project was a two-step high temperature water-splitting process of the following type (MO denotes a metal oxide):

$$MO_{oxidized} \rightarrow MO_{reduced} + O_2 \tag{9.4}$$

---

[4] http://www.atlasobscura.com/places/worlds-largest-solar-furnace
[5] http://www.iphe.net/docs/Renew_H2_Hydrosol-II.pdf

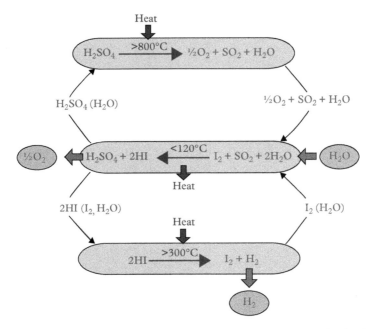

Fig. 9.10 *Iodine-sulfuric acid cycle proposed for thermo-chemical hydrogen production in power tower where temperatures in excess of 800°C can be obtained (Perret, 2011)*

$$MO_{reduced} + H_2O \rightarrow MO_{oxidized} + H_2 \qquad (9.5)$$

In the Hydrosol work described by Roeb et al (2006) the metal was iron.

Such solar thermo-chemical hydrogen (STCH) processes use high temperature and various catalysts to split water into hydrogen and oxygen. A report (Perret, 2011) gives several candidate reactions that could be used for this approach to creating free hydrogen. An additional example of such a reaction is shown in Fig. 9.10.

10

Solar Cell Physics and Technologies

Solar cells directly convert light into electrical power. They are based on the semiconductor PN junction and include several types that we will cover in this chapter. The simplest solar cell, with one PN junction, has a theoretical limiting efficiency of about 30 percent, with 24 percent the highest achieved, while series combinations of cells known as tandem cells can have efficiency above 50 percent, with 41.1–41.6 percent the highest achieved efficiencies at present.

10.1 Single junction cells, principles

The semiconductor properties described in Chapter 5 underlie the behavior of solar cells. Most relevant aspects are the formation and properties of PN junctions. In discussing leading types of single junction solar cells start with Section 5.7 "The PN junction, diode I-V characteristic, photovoltaic cell", Eqs. 5.79–5.89, and Figs. 5.17–5.18. The latter figure shows the I-V curve of a silicon solar cell under illumination. We begin with Silicon solar cells, that have evolved over time toward goals of improved efficiency and toward lower fabrication cost.

Fig. 10.1 sketches the bands of a Si N$^+$P solar cell, built to be illuminated from the left, a cell that might be similar in its characteristic to that shown in Fig. 5.18. The nomenclature and basic processes that go on in such a conventional single crystal Si device were shown in Fig. 5.17c. The device shown in Fig. 10.1 (upper panel) in the dark, but the lower panel shows the expected pattern of decaying intensity.

10.1.1 Light absorption

It is important to understand that the absorption process is strongly dependent on the wavelength of the light, that is related to the photon energy through the relation $E = hc/\lambda$. The hatched regions on right and left indicate metallic contacts to connect to a load, and details are omitted, such as how the left contact is made optimally transparent to incoming light.

In practice, the front contact is applied as narrow fingers of metal, to allow light to enter. The front surface may be textured to reduce reflection and promote internal

Physics and Technology of Sustainable Energy. E. L. Wolf, Oxford University Press (2018).
© E. L. Wolf. DOI: 10.1093/oso/9780198769804.001.0001

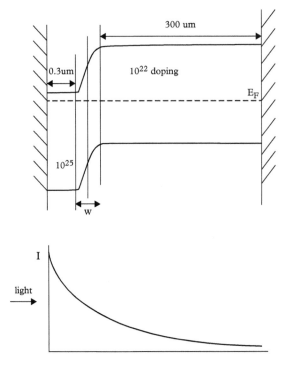

Fig. 10.1 *(Upper) Sketch of N⁺P junction solar cell, designed for illumination from the left side. The doping and dimensions are typical for a Si cell, see Fig. 5.18. (Lower) Light absorption into the cell follows law $I = I_o \exp(-\alpha x)$ where decay constant α strongly depends on photon energy. At stated N⁺ doping 10^{25} m⁻³ the Fermi level in the N⁺ region actually will lie in the conduction band, making the overall band bending larger*

trapping of light as discussed presently. In addition, the front surface may be coated with an anti-reflection coating (AR or ARC) typically with refractive index near two. Materials used on silicon include MgF_2, ZnS, Si_3N_4, Ta_2O_5 and TiO_2. The rear contact, here shown as a uniform metal layer, is sometimes replaced by an array of low-resistance contacts, with locally diffused P⁺ regions. In this case the larger portion of the rear surface will be oxidized to form SiO_2, quartz. The Si/SiO_2 interface reflects light back into the silicon and also, especially if passivated with hydrogen, reduces undesired recombination to a level less than would occur at a metal contact. The desired outcome for photogenerated electrons and holes is to be collected in the opposite terminals of the device and to drive an external current. Recombination internal to the device defeats this desired outcome. Recombination of photogenerated minority carriers is undesirable, and *surface recombination* can be inhibited by provision of a back-surface field (BSF). In the case of the rear contact to the P-region shown in Fig. 10.1, this could be provided by doping of acceptor impurities to produce a P⁺ P junction at the rear. This would have the effect of raising the conduction band edge at the rear of the cell and thus reflect minority electrons back into the bulk, reducing the chance for surface recombination.

It is important to understand that the active region of the classic PN junction device extends well beyond the depletion layer W, to include the hole diffusion length L_p on the left, and the (minority carrier) electron diffusion length L_n on the right. The action of light absorption is to greatly increase the minority carrier concentrations (p_n on the left, n_p on the right) leading to an enhanced reverse current density (Eqs. 5.88–5.89). The photoelectric effect is essentially that a photon of light disappears and gives its energy entirely to an electron, and in the semiconductor context to create an electron-hole pair.

The radiation-induced current density of a solar cell is $J_L = -Ge(W + L_n + L_p)$, where L is the diffusion length for the minority carrier and W is the junction width. Looking at the upper panel of Fig. 6.1, when under illumination, G will represent the rate at which electron-hole pairs are created by light photons of energy exceeding the local band gap energy. To be useful, the carriers have to be generated within a diffusion length L of the junction, where $L = (D\tau_r)^{1/2}$. Here τ_r is explicitly the minority carrier lifetime against recombination. L can be rewritten as, $L = (\tau_r \tau_s k_B T/m)^{1/2}$ making use of the relations $\mu = e\tau_s/m = e\,D/k_B T$. (The scattering lifetime can also be written as $\tau_s = \lambda/<v>$, using the mean free path λ and thermal velocity <v>.) A long minority carrier lifetime is achieved by minimizing defects, which would facilitate recombination of electrons and holes, with particular attention to recombination at surfaces, as mentioned previously.

10.1.2   Photovoltaic efficiency

The total current density is the sum of the light-induced density and the thermionic current:

$$J = J_o\left[\exp(eV/k_B T) - 1\right] - Ge\left(W + L_n + L_p\right). \tag{10.1}$$

The reverse current density $J_o = J_{rev}$, as discussed earlier, is strongly temperature dependent.

Setting V = 0 (short circuit condition), the square-bracket term is zero, so we recognize:

$$J_{sc} = -Ge\left(W + L_n + L_p\right), \tag{10.2}$$

where W is the depletion region width.

The open circuit voltage V_{oc} is obtained by setting J = 0 in Eq. 10.1, so that exp $(eV_{oc}/k_B T) = 1 - J_{sc}/J_o$, and,

$$V_{oc} = (k_B T/e)\ln\left(1 + J_{sc}/J_o\right). \tag{10.3}$$

From a basic point of view, this value cannot exceed E_g/e.

The output power per unit area is JV, and this is maximized at maximum power voltage V_{mp} (adjusted by the load), satisfying, $\partial(JV)/\partial V = 0$, that gives:

$$V_{mp} = V_{oc} - \left(k_B T/e\right) \ln\left(1 + eV_{oc}/k_B T\right), \tag{10.4}$$

and a corresponding current density J_{mp}, slightly reduced from the short circuit value.

The ratio between the available power at optimum load and the limiting power is defined as the "Filling Factor" (FF),

FF $= J_{mp} V_{mp}/J_{sc} V_{oc}$. In Fig. 5.18, the FF is simply the ratio of the areas of the inner hatched region defining the optimum power point and the larger rectangle, defined by the short circuit current density and the open circuit voltage. The basic efficiency is

$$\eta = J_{sc} V_{oc} \, FF/P_{inc}, \tag{10.5}$$

where P is the incident light power density, and,

$$FF = \left[eV_{oc}/k_B T - \ln\left(1 + eV_{oc}/k_B T\right)\right]/\left(1 + eV_{oc}/k_B T\right). \tag{10.6}$$

Looking at these expressions, one can make a few observations on the efficiency. One must maximize the short circuit current, that requires optimal absorption of photons and minimal recombination of the minority carriers before exiting the junction. One seeks a large open circuit voltage $V_{oc} = (k_B T/e) \ln (1+ J_{sc}/J_o)$, which requires, in addition to maximizing J_{sc}, *minimizing the reverse current,*

$$J_{rev} = J_o = e\left[n_p \left(D_n/\tau_n\right)^{1/2} + p_n \left(D_p/\tau_p\right)^{1/2}\right]. \tag{5.86}$$

Small reverse current density is promoted by making the thermal *minority* carrier concentrations small, favored by a low temperature (this is aided by making the *majority* dopings, $N_{D,A}$, large) and also by a long recombination life time. In practice, the open circuit voltage is limited by the built-in voltage V_B, because beyond that voltage the device no longer provides an exponential I(V) relation in the forward bias regime. The upper limit of V_B is E_G/e, although, if both sides of the junction are doped into the metallic regime, the value could be slightly larger than this, given by Eq. 5.98. To make an estimate of the optimal *filling factor* behavior, for 300 K and $E_G = 1.1$ eV for Si, we can take $eV_{oc}/k_B T = 42$. This gives FF $\approx [42 - \ln(43)]/43 = [42 - 3.76]/43 = 0.89$. One can see from this that large illumination, i.e., concentration of the light using mirrors or lenses, is advantageous, to increase the open circuit voltage and, hence, the conversion efficiency of the cell.

For the single gap junction, the efficiency η approaches 30 percent for favorable assumptions, a result that dates to Shockley and Quiesser 1961.

These authors also gave an "ultimate efficiency" of a more general nature, that is quoted as 44 percent, but is not as directly applicable to the single junction solar cell as is the basic efficiency η.

10.1.3 Quasi-Fermi levels

The absorption of light creates non-equilibrium concentrations of carriers, that can conveniently be described by "quasi-Fermi" levels, as sketched in Fig. 10.2.

These devices work by the photoelectric effect, the release of charge by annihilation of a quantum of light. Details of the absorption of light for important semiconductors are shown in Fig. 10.3. Since light photons have nearly zero momentum, the absorption process favors semiconductors with a direct bandgap, where the electron and hole have zero momentum. Fig. 10.3 shows the absorption coefficients for Ge, Si and GaAs as a function of photon energy. GaAs has the typical direct bandgap behavior, desirable for a solar cell, while indirect bandgaps for Si and Ge lead to low optical absorption, especially at energies close to the bandgap value, 1.12 eV for Si. This means that the thickness of a silicon layer to completely absorb light is larger than that in direct bandgap semiconductors by the relatively low absorption constant. Faceting the surface to

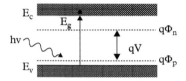

Fig. 10.2 *The increased densities of electrons and holes under illumination can be described by separate "quasi-Fermi levels" ϕ_n and ϕ_p for electrons and holes, respectively. It is the shift of these separate quasi-Fermi levels that creates the output voltage of the cell, as indicated in Fig. 10.2. The separation of the quasi-Fermi levels arises as a competition between photo-generation and recombination, which may occur in the bulk or at surfaces. Recombination is to be avoided. Internal trapping and absorption of the light photons to create electron-hole pairs is to be maximized*

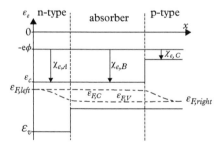

Fig. 10.2a *A generalized solar cell device, in which an interposed absorber layer creates electrons and holes from incoming light photons. Energy runs vertically, with zero at energy of free electron outside the device. This could be called a PIN device, with an I insulator layer. Work function ϕ shows highest energy of an electron in the device. Symbols χ are electron affinities in the different regions of the device, while E_C and E_V, respectively, are the lowest electron energy and the highest filled valence band energy. The effect of light is to separate the quasi-Fermi levels for electrons and holes, describing non-thermal light driven concentrations of electrons and holes. The resulting shift of quasi-fermi energies gives the output voltage of the illuminated device, Marti, A. and Luque, A. (2004) p. 58, Fig. 3.5*

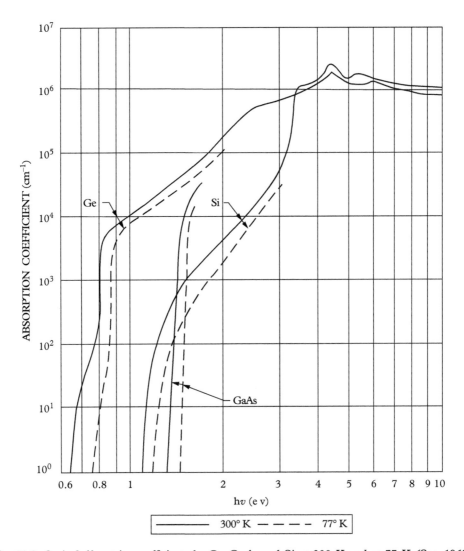

Fig. 10.3 *Optical Absorption coefficients for Ge, GaAs and Si at 300 K and at 77 K (Sze, 1969) Fig. 27, p. 54*

reduce reflection and to lengthen the path of the light within the cell is a means of overcoming this additional cost.

10.1.4 Silicon crystalline cells

The idea of surface texturing and its utility are illustrated in Fig. 10.4. It is seen that the "facet" makes two reflections necessary for backscattering of light, so that if

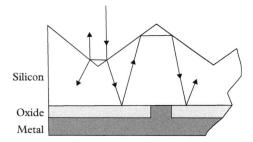

Fig. 10.4 *Surface-texturing strategem for enhancing optical absorption and reducing external reflection of vertically arriving photons by faceting of light-admitting surface. This has been applied to (100)-oriented single crystal Si cells, and a similar method to polycrystalline Si cells, by M. Green and coworkers (Wurfel, 2009) Fig. 7.6, p. 174*

the single reflection occurs with 0.1 probability, the backscattering probability is reduced to 1 percent, a large improvement. The photon that enters the silicon now has a larger angle to the normal, making its path in the silicon longer before encountering the back surface, and also increasing its chance for internal reflection at that surface.

The reflection coefficient R is calculated at normal incidence by the formula $R = (n_1 - n_2)^2/(n_1 + n_2)^2$. Taking values 3.5 and 1 for the index of refraction for silicon and air, respectively, $R = 0.31$, which is substantial, and an antireflection coating is usually applied as mentioned previously. The large index of refraction also implies that light inside the silicon is easily internally reflected. In fact, since the critical angle θ_c is given by $\sin\theta_c = 1/n_{Si}$ is small, $\theta_c = 16.6°$, all light except that within 16.6° of the surface normal is internally reflected. The corresponding angle at the Si/SiO_2 interface is 24.7°. Growing thermal oxide (quartz) on the back surface of the cell as indicated in Fig. 10.4 reflects light to increase the light path in the silicon and reduces the recombination rate of electrons and holes at the surface. Heating in a hydrogen atmosphere ("passivating") further reduces recombination at Si/SiO_2 interfaces, by filling "dangling bonds". Any step to increase the lifetime of minority carriers is beneficial to the operation of the solar cell, where the open circuit voltage (the separation of electron and hole quasi-Fermi levels as in Fig. 10.2) arises as a competition between photogeneration and recombination of carriers. (Fig. 10.4 is simplified, neither the PN junction nor the front anti-reflection layer are shown, but these are clearly shown in Fig. 10.5.) In Fig. 10.4 one of an array of small contacts to the rear is seen.

The most efficient single-crystalline Silicon solar cell, evolving from Figs. 5.17, 10.1–10.3 is shown in Fig. 10.5 Zhao et al (1998).

The faceting indicated in Fig. 10.3 has been implemented, showing the N^+P junction diffused into the faceted surface. The metal conductive fingers easily form Ohmic contacts with the N^+ layer, and then the upper surface is coated with a MgF_2/ZnS antireflection double layer applied above a thin and passivated native oxide (quartz) layer.

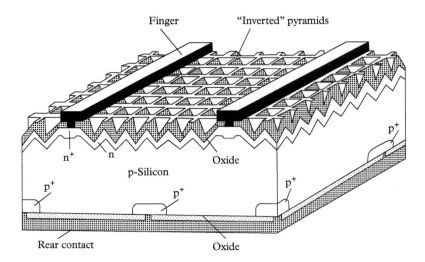

Fig. 10.5 *Structure of the best single-crystal Si solar cell, efficiency of 24.4 percent, developed by M. A. Green and coworkers. Note that entrance surface "oxide" is covered by an anti-reflection coating, and that the larger part of the rear surface is oxidized to reduce recombination and also to reflect light back into the Silicon wafer. The front top N-surface has locally diffused N^+ regions to facilitate forming a low resistance Ohmic contact to metal fingers and also to collect current from the N-layer with minimal voltage drop and reduced recombination. On the top surface the (thin-oxide plus antireflection layer) is shown as dark, while on the back surface thin thermal oxide (quartz) is indicated white and the rear metal contact is shown black. Locally diffused P contacts connect the bulk P-region to the metallic back contact (Zhao et al, 1995) Fig. 1*

This article also details a polycrystalline Si solar cell of 19.8 percent efficiency with a "honeycomb" texturing of the upper surface.

These authors found that completely enclosing the Si surface with thin thermal oxide to reduce recombination improved cell efficiency. However, the thermal oxide must be thin, on the order of 20 nm, so that the anti-reflection double layer applied above the oxide will still operate correctly. The polycrystalline version of this cell was grown on 1.5 Ω m, large-grained directionally-solidified P-type silicon 260 μm in thickness. The use of diffused highly doped material just below metallic contacts, on front and back, suppresses recombination by repelling the minority carriers.

It was also reported that the Si/SiO_2 interfaces could be improved, passivated to reduce recombination of electrons and holes, by exposure to atomic hydrogen. Thus, the record-efficiency cell is denoted "PERL" (passivated emitter, rear locally-diffused cell).

A large installation of single crystal solar cells, located at Nellis Air Force Base in the US, is shown in Fig. 10.6. This array provides 15 MW of power and is shown to track the Sun's motion in one direction.

Fig. 10.6 *Nellis Air Force Base solar panels track the Sun in one axis. Large conventional single crystal silicon installation http://en.wikipedia.org/wiki/File:Nellis_AFB_Solar_panels.jpg*

10.1.5   Gallium arsenide (GaAs) epitaxially grown solar cells

GaAs, as shown in Fig. 10.3, has a suitable bandgap and direct absorption structure for solar cells. GaAlAs layers can be grown epitaxially by liquid and vapor phase methods.

Features of this single junction multilayer AlGaAs/GaAs solar cell include a thin p-AlGaAs window, grown by low-temperature liquid phase epitaxy "LPE", and a prismatic cover made of silicone, (above the anti-reflection coating "ARC") which minimizes the optical losses caused by contact grid shadowing and reflection from the semiconductor surface. An efficiency of 24.6 percent was recorded with such cells, in a 100 times concentrated air mass zero AM-0 spectrum. Further details of the band structure in the cell are shown in Fig. 10.7a.

10.1.6   Single junction limiting efficiency, analysis of Shockley and Quiesser

Some crude estimates may be useful. For example, the open circuit voltage has an upper limit E_g/e, where E_g is the semiconductor bandgap. The analysis assumes a single junction and a single uniform bandgap energy. It is assumed that photons whose energy

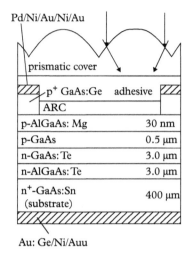

Fig. 10.7 *Structure of epitaxially grown GaAs solar cell as utilized in space-craft. (Alferov and Rumyantsev, 2003) Fig. 2.5 on p. 27*

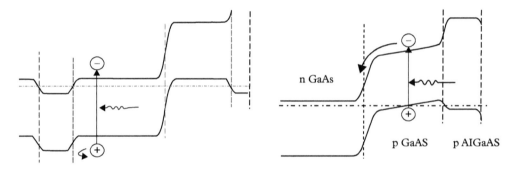

Fig. 10.7a *Band diagram of epitaxially grown GaAs solar cell, efficiency of 24.6 percent (left). The sloping bands (built-in quasi-electric field) in the p-GaAs front layer, where photo-absorption occurs, was formed by Zn diffusion during liquid-phase epitaxial LPE growth of the wide-bandgap AlGaAs window (right). Elaboration of cell with a back-surface field (BSF) provided by adding a highly doped N^+-GaAs layer. Light absorption in the rear N-layer is enhanced by the BSF, that inhibits recombination of minority holes with majority electrons at the rear surface. These diagrams make clear that light absorption primarily occurs in the field-free regions of dimensions L_p and L_n beyond the junction-forming depletion layer (Alferov and Rumyantsev, 2003) Fig. 2.4 on p. 25*

is less than the bandgap energy E_g do not contribute to the photocurrent. It is also assumed that the excess energy $hc/\lambda - E_g$ is lost to heat in the semiconductor. In practical terms the junction structure should be thick enough that all of the light of energy $hc/\lambda - E_g > 0$ is absorbed.

A disadvantage of the single-junction cell at bandgap E_g, is that all photons of energy less than E_g are lost, are not absorbed but pass through the cell. The energy output of the cell is the same for N photons of 2 E_g or 3 E_g as for N photons of energy E_g. For these reasons the efficiency of the single-junction cell is inherently limited and depends on the spectrum of photon energies that are incident. Roughly, the bandgap for best efficiency, within the single-junction assumption, should be close to the peak of the incident spectrum.

From a basic and simplified point of view, the light spectrum from the Sun is approximately that of a black-body of temperature 5973 K, that peaks at $\lambda_m = (2.9 \times 10^6/5973)$ nm = 486 nm, which corresponds to an energy 1240/486 = 2.55 eV. (The spectrum at ground level ("Air mass 1") is substantially altered, by strong absorptions from several minor constituents of the atmosphere, including ozone, water vapor, and compounds of nitrogen and carbon. It is also weakened and red-shifted by scattering in the atmosphere.) Adopting the hypothetical black-body spectrum, for simplicity, if the bandgap of Silicon is 1.12 eV, then the many photons in the range below 1.12 eV are completely lost, and an increasing fraction of the energy of the more energetic photons is also lost. The most probable photon at energy 2.55 eV, will contribute no more than will a 1.12 eV photon, the difference energy 1.43 eV only contributing to heat.

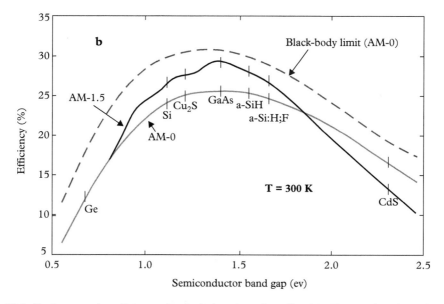

Fig. 10.8 *Basic conversion efficiency of a single-junction solar cell at 300 K vs. semiconductor bandgap energy E_G, originating in Shockley and Queisser 1961. [Note that this plot is not the incident light spectrum! (See Fig. 9.2)] The dashed curve shows the basic efficiency curve if the input spectrum is taken as an ideal black body spectrum for T = 6000 K, a convenient theoretical assumption. AM-0 means Sun spectrum in space, while "AM-1.5" is agreed test spectrum representing noon on a sunny day at a medium latitude, so the Sun is not directly overhead at noon and the air traversal path length is about 150 percent of the vertical height of the air mass. Dashed line represents best efficiency that could be obtained with fictitious black body Sun spectrum 6000 K, while AM-0 is measured spectrum above Earth's atmosphere (Lewis and Crabtree, 2005)*

The specialists in photovoltaic conversion have adopted an effective spectrum "Air Mass 1.5 G", correcting for an average daytime light path length 1.5 larger than when the Sun is directly overhead. This spectrum is reduced from the top-of-the-atmosphere spectrum, Air Mass 0 (AM 0), by about 28 percent, of which 18 percent is from absorption and 10 percent from scattering. Scattering is well known to follow a λ^{-4} law, which removes blue light from the direct path but adds in some blue light scattered from the blue sky. The resulting AM 1.5 G spectrum (which also includes a diffuse light contribution) is used to calculate the ideal conversion efficiency of a single-junction solar cell operating at 300 K, as a function of the semiconductor bandgap energy. The results for the AM- 1.5 G spectrum (labeled, in Fig. 10.7) peaks near 29 percent efficiency at a bandgap in the vicinity of 1.4 eV. This is close to the bandgap energy for GaAs, as is shown in Fig. 10.8.

The efficiency of the single junction cell, around 30 percent for one Sun intensity, was addressed by Shockley and Quiesser (1961).

The starting point for their analysis is the diagram of the bands in the PN junction, shown in Fig. 5.17c. If we imagine the sketched open circuit PN junction cell in thermal equilibrium at a temperature T, we realize that dynamical processes of thermal generation of electron hole pairs occur throughout the structure to maintain the equilibrium distributions of electrons and holes in the various regions of the junction structure. The reverse current density J_{rev} comes from diffusion of minority carriers located within the diffusion lengths L, to the junction, where carriers fall down the potential gradient. At open circuit, a positive bias appears under illumination, to exactly cancel the reverse current. Again, electrons in a p-type semiconductor are of limited lifetime, since they can fall into an acceptor site, give off light, and disappear. The statistics require a corresponding generation process to maintain the equilibrium concentration. The distance the minority electron in the p-region can diffuse before recombination is the minority carrier diffusion length L. Only carriers within such a distance of the actual junction can be usefully driven around the external circuit. To repeat, at open circuit a forward bias appears, such that the net current across the junction is zero. The generation processes in equilibrium are balanced by recombination of electrons and holes to create photons, that must generate a Planck spectrum. The recombination events can create photons of energy equal to or larger than $E_g = hc/\lambda$, where λ is the wavelength of the photon.

The insight of Shockley and Quiesser was to recognize that these internal processes must support emission of a black-body spectrum, at least for wavelengths shorter than hc/E_G from this open circuited thermal equilibrium structure. *The amount of light emitted, fixed by Planck's radiation law, is closely related to the reverse current density,* J_{rev}, and the spectrum of the light, required by Planck's radiation law will be related to the energy distributions of excited electrons and holes in the thermal equilibrium structure. If we recall Fig. 9.2 and imagine the small black area A at the bottom of the figure to be the open-circuit solar cell at temperature T in the dark. It will act as a black body and radiate power outward, and the two cases of interest are with and without the concentrating optical system. In the absence of the optical system, a tiny fraction of the black-body radiation will fall into an angular range comparable to that of the Sun seen from Earth,

$f_s = 2.16 \times 10^{-5}$. With the optical system in place, all of the black-body radiation will be focused into a solid angle comparable to that of the Sun as seen from Earth. According to Planck's law, the total power emitted by the junction of area A into the full hemispherical range of angles, solid angle $= 2\pi$, in the photon energy range E_1 to E_2, is, writing power $P = dE/dt = E'$,

$$P(E_1, E_2) = E'(E_1, E_2, T) = A(2\pi/h^3c^2)\int_{E1}^{E2} dE\ E^3/\left[e^{E/kT} - 1\right] \text{Watts}, \quad (10.7)$$

where $E_1 = E_G$. The upper limit on the energy may be approximated as large, $E_2 = \infty$, or may be estimated from the band structure of the semiconductor. The point is that this completely determined number of Watts comes from band-to band recombination in the solar cell, and gives a number that can be closely related to J_{rev} Eq. (5.86):

$$AJ_{rev} = I_0 = eA(2\pi kT/h^3c^2)\left[E_G^2 + 2\ kT\ E_G + 2\ (kT)^2\right]\exp(-E_G/kT). \quad (10.8)$$

It is also true that if the junction at temperature T acquires a voltage V between its terminals, then Eq. (10.8) is simply multiplied by $\exp(eV/kT)$. We now write the current I in terms of the rate of change:

$$dN/dt = N' = N'(E_1, E_2, T) = A(2\pi/h^3c^2)\int_{E1}^{E2} dE\ E^2/\left[e^{E/kT} - 1\right] s^{-1} \quad (10.9)$$

of photons exchanged between the area A junction at temperature T_c and the Sun at temperature T_s, as in Fig. 9.2.

$$I = Ae\ f_s\ N'(E_G, E_2 = \infty,\ T_s) - Ae\ f_c\ N'(E_G, E_2 = \infty,\ T_c)$$
$$\left[\exp(eV/kT) - 1\right] \text{ Amperes} \quad (10.10)$$

Set $f_s = 2.16 \times 10^{-5}$ to describe the absence of concentrating optics, and $f_c = 1$ to describe the area A radiating into all hemispherical directions 2π. The first term is the black-body spectrum of the Sun at T_s in the energy range above E_G (see Eq. 3.18 and Eq. 3.19a and related discussion). The second term is the returning black-body radiation of the cell at $T = T_c$ at its chosen open circuit voltage V. We are describing an open circuit cell, so $I = 0$. Taking the Sun at 6000 K and the cell at 300 K, Shockley and Quiesser (1961) thus found the best efficiency is 31 percent for $E_G = 1.3$ eV. The uppermost dashed curve in Fig. 10.7 describes this result, using the vacuum Planck's law solar spectrum.

Changing to the concentrating situation, where both f factors are 1.0, (input and output radiation both use a full hemispherical range of angles) the new equation is:

$$I = Ae\ N'(E_G, E_2 = \infty,\ T_s) - Ae\ N'(E_G, E_2 = \infty,\ T_c)$$
$$\left[\exp(eV/kT) - 1\right] \text{ Amperes} \quad (10.10a)$$

It is now seen that a larger open circuit voltage V is needed to balance the larger solar input from the concentrating optics. This means a larger efficiency, since the open circuit voltage times the current is the power. In this case Shockley and Quiesser thus find efficiency 40.8 percent at 1.1 eV for a fully concentrated single junction solar cell, see Fig. 9.2 for an example of such optics, where the imagined cell of area A is at the bottom of the field.

This analysis is based on the direct sunlight and does not include correction for the diffuse scattered light. This analysis is invalid for a cloudy sky.

In this way the complex situation was analyzed to provide the maximum efficiency of the single junction device as a function of the single bandgap energy, under assumed illumination conditions. The solid line plots of Fig. 10.8 are numerically obtained from the Shockley–Quiesser analysis using the below-atmosphere spectra, as illustrated in Fig. 9.1.

10.1.7   Advantage of concentration of sunlight

The equation for power per unit area, $P = JV$ is found to increase faster than linearly with incident intensity, because the open circuit voltage increases with incident light $J_{sc} : V_{oc} = (k_B T/e) \ln (1 + J_{sc}/J_o)$. This remains true even though the maximum power voltage is slightly smaller than V_{oc}.

As we will see later, the most straightforward means of improving the solar cell efficiency (at fixed intensity) is to put two or more single gap cells in *tandem* (series connection), so that the highest-energy photons are processed with the largest bandgap junction, and later cells process those photons whose energy was insufficient to generate electron hole pairs in the prior junctions. A cascade of tandem cells can approach the efficiency of the Carnot machine, in principle. In practice tandem cells of at least 40 percent efficiency under concentration have been demonstrated. The material science and engineering of these tandem structures makes them more expensive, but this can be counterbalanced by use of concentrating light systems. The cells are more efficient at higher light intensity, because the open circuit voltage increases with light intensity.

10.2   Single junction cells, thin film technology

An important class of solar cells is those made of thin films of semiconductors. These are cheaper because they use less of the primary material: a few micrometers rather than a fraction of a millimeter but are generally of lower efficiency because recombination occurs at the grain boundaries.

10.2.1   Single crystal vs. thin film cells

The grain size in polycrystalline films varies from millimeters to around a micrometer, but nanocrystalline films are also used with grains as small as 10 nm. Grain boundaries foster recombination of photogenerated electrons and holes, that occurs at defects such as dangling bonds at the surfaces. The dangling bonds can be mitigated by adding

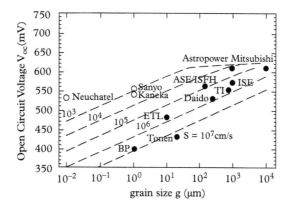

Fig. 10.9 *Empirical relation between open circuit voltage and grain size for polycrystalline (closed circles) and nano-crystalline (open circles) silicon solar cells. Note that the grain size is as large as 10 mm for the Mitsubishi polycrystalline cells. Manufacturers and originating laboratories are indicated in the Figure (Kondo and Matsuda, 2010) Fig. 8.1 p. 140*

hydrogen to fill empty bonds, that allows the nanocrystalline films to be viable. Nanocrystalline films have the advantage that lower processing temperatures may suffice, making fabrication easier.

Recombination reduces output voltage attainable from a solar cell. Recombination is characterized by a velocity that may vary from 10^7 cm/s to 1000 cm/s, as shown in Fig. 10.9.

Thin-film solar cells may also be made of organic compounds and polymers, as suggested in the right panel of Fig. 10.10. Such cells are more easily made flexible and are potentially much cheaper. They have historically had low efficiency and, further, usually have short lifetimes against degradation by the Sun's rays. As we will see, several types of organic solar cells, including tandem cells, are well advanced in technology and have efficiencies approaching 10 percent. A basic problem with organics is that the electron and hole mobilities are low, so the materials are of low electrical conductivity. Sometimes the polymers are loaded with conductive molecules such as C_{60} to improve the electrical conductivity. Such additions add to the cost.

Major opportunity for supplying competitive electric power from sunlight seems offered by thin film devices, that are appropriate to reach a lower cost. (This conventional assumption may be challenged by the combination of large concentrating mirrors with high efficiency multiple junction tandem cells, as will be discussed later.) Thin films use much less material, and polycrystalline silicon cells are well known. (Polymer and dye-sensitized cells are inexpensive but seem to be limited in efficiency, as are amorphous silicon cells, at about 10 percent.) We here focus on perceived opportunities to reach a large market. We describe next a class of device that is capable of mass production.

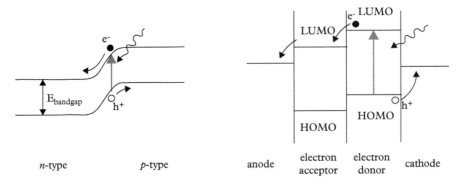

Fig. 10.10 *More general single bandgap devices. Schematic p-n junction (left) and an organic bilayer structure (right) under illumination. Illumination leads to output voltage and current in external circuit. In left panel, light can be usefully absorbed in wider region than the depletion layer, whose width includes the diffusion lengths for minority carriers in the semiconductor regions on both sides of the actual junction. The left panel applies to Si, GaAs and CuIn$_{1-x}$Ga$_x$Se$_2$ "CIGS" cells, and to individual elements of Tandem cells, that are generally of higher efficiency. The right panel applies to "organic polymer" cells, generally of low efficiency and low cost. A third type of cell, the "Dye-sensitized solar cell" is not shown. Multiple-junction (Tandem), "intermediate band" and quantum-dot-assisted cells are additional categories, see text (Global Climate and Energy Project, 2006)*

10.2.2   CuIn$_{1-x}$ Ga$_x$ Se$_2$ (CIGS) thin film solar cells

Low cost semiconductor cells with efficiency approaching 20 percent have evolved making use of alloys of selenides of copper, indium, and gallium (CuIn$_{1-x}$Ga$_x$Se$_2$, known as "CIGS"). These crystalline alloys are typically P-type semiconductors with appropriate band gaps, high absorption coefficients, and quite large diffusion lengths. The cell is typically in the form of a thin N-layer placed near the front of an "absorber" of CIGS with thickness about one micrometer. The PN junction, as in Fig. 10.9, left panel, sends electrons and holes in opposite directions. The efficiency limit in these cells is now 21 percent, with 17.5 percent available in laboratory modules. This is in the same range as polycrystalline Si cells, but it appears that the CIGS cells are easier fabricate on large scale and can be much cheaper to produce.

The historical origin of this line of "CIGS" cells, may have been the CdS/Cu$_2$S cell formed by chemical surface treatment of CdS. This cell fell out of favor partly because of the toxic nature of Cd.

10.2.2.1 *Printing cells onto large area flexible substrates*

The reasons for ascendency of CIGS include efficiency and durability, but the main reason is that these cells can be manufactured, without vacuum equipment, on a roll-to-roll mass basis similar to a modern printing press. Fig. 10.11 shows one possible step by step prescription for CuIn$_{1-x}$Ga$_x$Se$_2$ CIGS cell manufacture, which does not

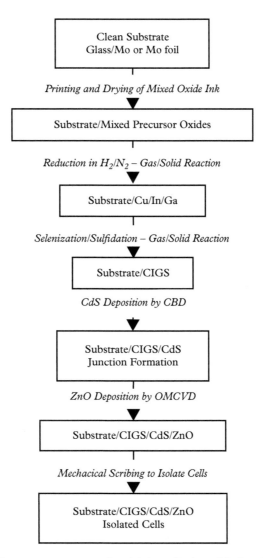

Fig. 10.11 *Ink-printed non-vacuum approach to fabricate GaAs and CuIn$_{1-x}$Ga$_x$Se$_2$ solar cells on flexible substrates. (Kapur, et al, 2003) Vol. 1, p. 465*

require vacuum equipment. (Many variations are possible, including those that avoid CdS and also those in which the substrate is simply aluminum foil.)

Fig. 10.11, top to bottom, shows steps in one version of an ink-printing approach to CIGS solar cell formation. Substrates are either glass coated with Mo, or a Mo foil, that can be thin and flexible. The "CIGS absorber layer" is formed in steps 2–5. The absorber is chemically Cu In$_{1-x}$ Ga$_x$ Se$_2$. The Cu, In, Ga metal components are printed onto the cell in the form of oxide nanoparticles suspended in liquid, i.e., a water-based ink. This is inherently efficient with regard to use of the expensive elements, and the formulation of the ink allows close control of the composition. The layer is reduced (the oxygen is removed) by heating in a reducing atmosphere of nitrogen and hydrogen, so that oxygen leaves the film in the form of water vapor. The resulting metallic alloy layer is transformed to the semiconductor alloy by gas reaction at modest temperature with selenium using H_2Se gas, producing the "CIGS Absorber". Speaking approximately, this is a P-type semiconductor of controlled bandgap and of thickness in the 1000 nm range. The thickness must be enough so that most of the light is absorbed. Again, speaking approximately, the device is a PN junction, and the next step, "Junction formation", accomplished by deposition of CdS (using CBD, or Chemical Bath Deposition). The transparent electrode to the N-side of the junction is ZnO (by OMCVD, OrganoMetallic Chemical Vapor Deposition). The highest efficiency these workers achieved was 13.6 percent using a Mo coated glass substrate. A variation of this process that does not use Cd is described forthwith.

This basic process is important because it uses the minimum amount of the expensive metals, because it avoids expensive high-vacuum equipment, and because it can be scaled up to large areas. Think of printing a newspaper, how many square meters of paper is printed by a major newspaper each day. The maximum efficiency reported for this type of cell is nearly 20 percent. This value is close to the value 20.3 percent mentioned for polycrystalline silicon cells, that are not amenable to a similar large scale non-vacuum fabrication.

The cell shown in cross section in Fig. 10.12 was formed on 12.5 µm thick commercial polyimide. The Mo back contact of 1 µm was deposited by dc sputtering. The CIGS layer was established following sequential evaporation of the metals Cu, In and Ga, followed by evaporation of Se, with a controlled temperature anneal. The CdS layer was then deposited in a chemical bath process. RF sputtering was used to apply

Fig. 10.12 *Scanning Electron Microscope cross-section image of CIGS cell grown on polyimide flexible substrate. Efficiency of 14.1 percent was achieved in cells of this type. Layers, bottom to top: polyimide (not shown), Mo back contact, CIGS absorber layer, CdS junction-forming layer, ZnO insulator layer, ZnO: Al conductive window layer (Bremaud et al, 2005)*

the upper window layer described as I-ZnO/ZnO:Al of 300 nm thickness. Finally, Ni-Al contact grids for better current collection were applied by electron beam evaporation. This is a complicated and expensive process but led to the most efficient CIGS cells, 14.1 percent, ever prepared on a flexible polyimide substrate.

The traditional process for forming the PN junction, use of a chemical bath to deposit a layer of CdS, is objectionable because Cd is toxic. An alternative means of making the CIGS PN junction without Cd has also been demonstrated by Negami et al (2002) achieving efficiency 16 percent, see Fig. 10.13. In this method the CIGS absorber layer is contacted by a CIGS:Zn junction-forming layer, $Zn_{0.9}Mg_{0.1}O$ "insulator" layer, ITO (indium tin oxide) conductive window layer, and current collecting grid.

The CIGS absorber was deposited by Negami et al (2002) on Mo coated glass by physical vapor deposition. The PN junction was formed by evaporating Zn onto the exposed surface of the CIGS, held at 300°C. The zinc is believed to diffuse to a depth of about 50 nm in an annealing time about 5 minutes. This forms an internal PN homo-junction within the CIGS. The $Zn_{0.9}Mg_{0.1}O$ "insulator" layer was grown on the CIGS:Zn surface by co-sputtering the oxides from separate sources with adjustable rf power to establish the desired ratio Zn/Mg = 9. This ratio was found to adjust the conduction band position in the new layer to match that in the CIGS:Zn layer, to allow photoelectrons to pass out to the ITO electrode without scattering. The resulting physical interface was studied by transmission electron microscopy, (TEM), and found to be epitaxial. The ITO transparent conductor about 100 nm thick was prepared by sputtering, followed by metal grid electrodes applied by evaporation.

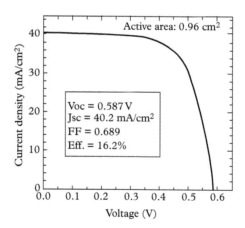

Fig. 10.13 *Current-voltage curve of CIGS cell grown using a dry process and avoiding use of Cadmium, on Mo coated glass substrate. Efficiency of 16.2 percent was achieved in cells of this type after applying MgF_2 anti-reflection coating. Layers, bottom to top: glass, Mo back contact, CIGS absorber layer, CIGS:Zn junction-forming layer, $Zn_{0.9} Mg_{0.1}O$ insulator layer, ITO (indium tin oxide) conductive window layer (Negami et al, 2002)*

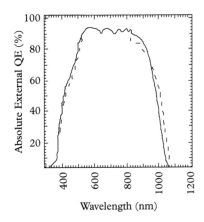

Fig. 10.14 *Quantum efficiency of CIGS solar cells: Solid curve, CIGS thickness 1 μm (17.2 percent efficiency), dashed curve CIGS thickness 2.5 μm (18.7 percent efficiency). These cells benefit from a final anti-reflection coating of 100 nm MgF₂ after depositing the 200 nm thick window layer of Al-doped ZnO, which exhibited a sheet resistance 65–70 Ω/square, and a Ni/Al grid to collect current (Ramanathan et al, 2006) p. 380*

High-efficiency CIGS cells are compared in Fig. 10.14. (The same group later reported a record efficiency for this type of cell as 19.9 percent.) These cells use a co-evaporation method for the CIGS layer. The traditional wet process is used to form the PN junction. This is described as growing 50–60 nm CdS films on the CIGS layer by immersion for 13 min. in a 60°C bath composed of 1.5 mM $CdSO_4$, 1.5 M NH_4OH and 75 mM thiourea.

The authors report nuances in the method of co-evaporation of the absorber to accomplish bandgap grading, and also suggest that for CIGS layer thickness below 1 μm the deep level density increases. For this reason, mobility and lifetime are reduced when the thickness is reduced to 0.5 μm. This specific process is not scalable, but demonstrates a high efficiency, 19.9 percent, in single junction CIGS solar cells.

The European firm Avancis produced 15.8 percent efficient CIGS modules using a more conventional linear process. However, this firm was sold to Chinese manufacturer CNBM in 2014. It appears that the total production of the CIGS modules is rising and is now about half that of the CdTe modules that presently dominate the thin film solar cell market.

The Japanese firm Solar Frontier[1,2] in 2015 announced a 22.3 percent efficient CIGS cell. This company is presently producing modules of CIGS cells at costs below $0.50/W.

---

[1] http://www.solar-frontier.com/eng/news/2015/C051171.html. (Accessed 5 April 2017).
[2] https://www.bloomberg.com/news/articles/2015-10-19/solar-frontier-eyes-lower-panel-costs-with-production-overhaul-ifxiq6ti. (Accessed 5 April 2017).

10.2.3   CdTe thin film cells

The largest thin film solar cell supplier is First Solar, manufacturing thin film cells of CdTe. According to the New York Times (Woody, 2009) First Solar, an American firm based in Tempe Arizona, signed an agreement with the Chinese government for a 2 GW photovoltaic farm to be built in the Mongolian desert.

The photovoltaic farm of area 25 square miles is part of a 11.9 GW renewable energy park at Ordos City in Inner Mongolia. The overall project is to include 6.95 GW of windpower, 3.9 GW of photovoltaic power, and 0.72 GW of solar thermal farms. Further to the plan for Ordos City are biomass operations, fueled by organic materials like wood chips and straw for 0.31 GW, and 70 MW from hydro storage, a load-balancing technology that uses off-peak power to pump water to a high reservoir from which it can be released to turn turbines at peak demand periods.

First Solar was expected to build a plant in China to make thin-film solar panels. According to the article, the 2 GW solar farm as built in China would be likely to cost significantly less than the $5 B to $6 B if it were built in the US.

It is commented that the CdTe solar cells of First Solar are less efficient than the standard crystalline silicon solar cells made by companies like Suntech, Trina or Yingli, but they are significantly less expensive to build. It is commented in this article that the Chinese project is atypical of large solar projects, that have generally been awarded to solar thermal technology, that deploys mirrors to heat a liquid to create steam that drives an electricity-generating turbine, rather than to straight photovoltaic projects. It is pointed out in this comparison that the straight photovoltaic projects generally have fewer environmental impacts, such as requiring cooling water, and can be brought online faster than solar thermal plants.

In fact, First Solar later canceled its contract in China, but it has completed two similar projects in Califonia, "Topaz" and "Desert Sunlight", with total 1.1 GW capacity to two California utilities. An aerial view of the Topaz installation, containing 9 million CdTe modules, is shown in Fig. 10.15. These are simple fixed module installations of 0.55 GW in each solar farm. The firm has also announced[3] an efficiency 22.1 percent achieved in a variation of its CdTe process, see Fig. 10.15.

Basic forms of CdTe cell construction are shown in Fig. 10.16. The junction is between n-type CdS and p-type CdTe that appears to be a non-lattice matched heterojunction.

CdTe is a direct bandgap material with high absorption of light of energy beyond the bandgap about 1.45 eV. Recent work (Burst et al, 2016) has produced CdTe cells with an open circuit voltage just above 1 V.

This article explains the complicated materials science of the lattice mis-matched CdTe/CdS heterojunctions, but does not describe the cells as are deployed by First Solar.

Although Cd is toxic, the devices contain only a few micrometers thickness of CdTe, and the toxic material is encapsulated, as not to present an external hazard. The material is favorable from the point of view of cost of fabrication. It appears that CdTe cells are leaders in the competition for large and cost-competitive photovoltaic installations.

---

[3] Richard Martin (2016) https://www.technologyreview.com/s/600922/first-solars-cells-break-efficiency-record/. (Accessed 18 March 2017).

Fig. 10.15 *View of CdTe Topaz solar farm in San Luis Obispo County, California, completed in Nov. 2014. There are 9 million fixed modules with capacity 0.55 GW, provided by firm First Solar. Note the 1-km scale bar https://commons.wikimedia.org/wiki/File:Topaz_Solar_Farm,_California_Valley.jpg#/ media/File:Topaz_Solar_Farm,_California_Valley.jpg*

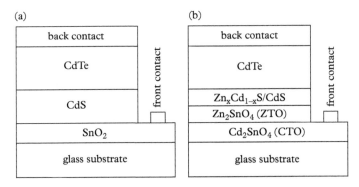

Fig. 10.16 *Two types of CdTe solar cell. In both cases the structure is built up from the glass substrate, and methods including Closed-Space Sublimation (CSS) and Chemical Bath Deposition (CBD) are used to deposit succeeding layers. In these devices the CdTe film is 1.5 to 3 μm in thickness, deposited using CSS especially for large area devices, and is subject to an anneal with $CdCl_2$ or other Cl^- containing compound to promote grain growth. (Deb, 2010) Fig. 2.5 on p. 251*

10.2.4   Dye-sensitized solar cells

Another inexpensive form of solar cell is the dye-sensitized solar cell. This approach is based on an absorber that is partly titanium oxide, TiO_2 in its anatase tetragonal form, and partly dye molecules. Anatase can be prepared in a nanoporous form (sometimes described as mesoporous, although microscopy shows particles of 10 nm–80 nm diameter) by various procedures, one described as hydrothermal processing of a TiO_2 colloid.

This oxide has a bandgap 3.2 eV, so that the maximum light wavelength that is absorbed is (1240 eV-nm)/(3.2 eV) = 388 nm. Thus, titania absorbs only a small part at the UV end of the solar spectrum. Spectacular extension of the light absorption range to at least 700 nm has been demonstrated by coating (sensitizing) the nanoporous anatase titania with dyes. The dyes have a relatively low specific absorption, so that a thick layer of dye is needed, each dye being in intimate contact with the anatase surface so that a photoelectron can reliably be transferred to the conducting anatase.

A conceptual schematic of a dye sensitized solar cell based on nanoporous anatase is shown in Fig. 10.17.

10.2.4.1   *Principle of dye sensitization to extend spectral range to the red*

The operation of this device follows the steps indicated in Fig. 10.17: Step 1 is absorption of a photon by a dye molecule. Typically, the dye is a metal-organic Ruthenium complex with response in the wavelength range 700–900 nm, light that is not absorbed by the titania. This is the idea of dye sensitization, to extend the absorption range of the photocell to utilize more of the solar spectrum. In Step 2, the excited-state electron jumps from the dye molecule into the conduction band of the TiO_2. The transfer is rapid from the dye only if the dye level lies ~ 0.2 eV above the conduction band, and reverse transfers from conduction band to dye are found to be slow. In Step 3 the electron diffuses through the porous nanoparticle layer, that may extend 10 μm, and enters the electrode, and in Step 4 the electron flows through the external load resistor R. In Step 5, the electron is transferred into the electrolyte, that is typically aided by a very thin platinum catalyst layer. In Step 6 the dye is regenerated into its ground state.

From the diagram, it appears that the open circuit voltage of the cell is approximately the photon energy divided by the electron charge, minus a voltage drop that occurs in the last step (Step 6) when the redox couple brings charge back to the dye molecule. This may be over-simplified. Other factors including a potential difference across the back contact and a Fermi level shift of the anatase nanoparticles, are discussed in an excellent review (Graetzel, 2001).

It is clear that these devices cannot be modeled in the conventional treatment outlined for crystalline semiconductor PN junction cells, where the diffusion lengths can be many μm, as outlined at the end of Chapter 5. It seems that low mobility and short diffusion lengths must necessarily limit the efficiency of dye sensitized and other organic types of solar cells.

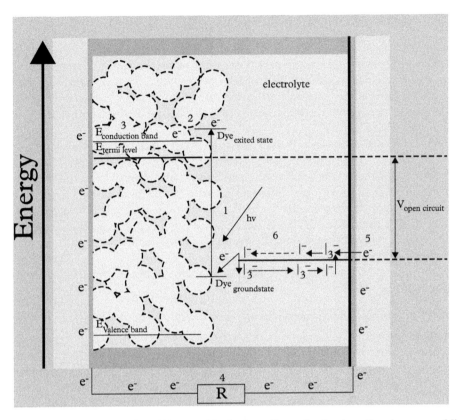

Fig. 10.17 *Principle of operation of Dye-sensitized solar cell, note load resistor R at bottom, and light excitation of dye by photon hv (Step 1, center of figure). Left, schematic of nano-porous deposit of TiO₂ (titania, anatase form) coated with dye molecules (dots) and immersed in electrolyte containing Iodine ions. Charge is carried to the dye by the iodide-triiodide reduction-oxidation couple in solution. Typically, the left side of the cell is commercial SnO₂ layered conductive glass, that has been coated with a Ti-nanoxide paste and fired (Junghanel, 2007)*

10.2.4.2   *Questions of efficiency*

State of the art performance of such a dye sensitized cell is shown in Fig. 10.18.

The dye sensitized solar cell represented in Fig. 10.18, under simulated AM 1.5 G irradiation exhibited short circuit photocurrent 20.5 mA/cm², open circuit voltage 720 mV and efficiency 10.4 percent. This efficiency is close to the best value ever obtained in this kind of cell. The dyes represented in Fig. 10.18 are termed "black dyes" because of their wide absorption range, and are complexes based on ruthenium. The structure of a slightly different dye, N945, is shown in Fig. 10.19.

Cells using this dye have been characterized by short circuit current 18.8 mA/cm², open circuit voltage 783.2 mV and efficiency 10.8 percent.

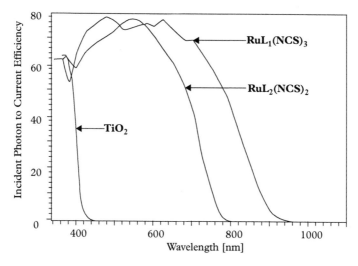

Fig. 10.18 *Extended spectral range by dye sensitization. Plots of incident photon to current efficiency, for pure TiO$_2$ (left trace) and titania (anatase) coated with two different dyes (right two traces). It is evident that photoelectrons generated in the dyes are efficiently transferred to the titania, to flow in the external circuit. Dyes that are optimum have high absorption that extends to long wavelength, and many of the best performing dyes are based on the metal Ruthenium. (Although the conversion of photons to electrons is efficient, these devices as solar cells exhibit relatively low efficiencies.) (Nazeeruddin, et al, 2001)*

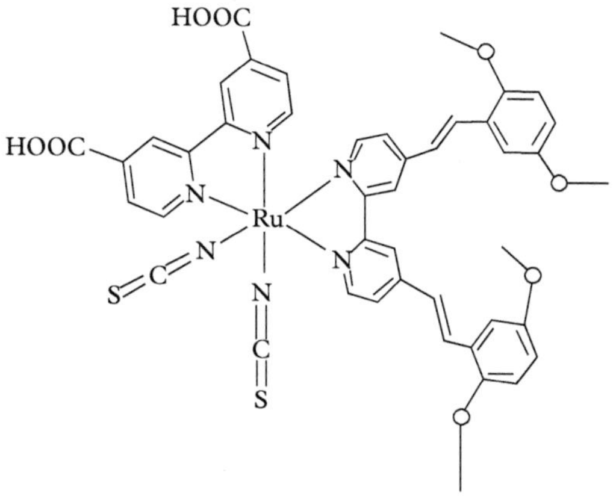

Fig. 10.19 *Ruthenium-based dye N945, that is one of the best sensitizing dyes for TiO$_2$. This dye has a strong absorption (a high extinction coefficient) allowing a more compact cell. (Nazeeruddin, et al, 2007)*

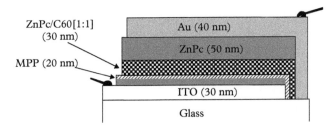

Fig. 10.20 *Schematic of organic solar cell structure. This is a cell of the type shown in Fig. 10.10, right panel. Here ZnP$_c$ refers to Zinc Phthalocyanine, and MPP refers to a methyl-substituted perylene pigment. Both are available from major suppliers, and were purified by sublimation. The ZnP$_c$/C$_{60}$ [1:1] layer was deposited by co-evaporation of ZnP$_c$ and C$_{60}$. The indium-tin oxide (ITO) coated glass substrate was obtained from a major commercial supplier. As described, this is not a low-cost process, because of the vacuum equipment needed (Rostalski and Meissner, 2000)*

10.2.5   Polymer organic solar cells

Probably the easiest type of cell to produce is the polymer organic cell, that involves coating a conductive substrate with two polymers and covering this with a transparent electrode.

10.2.5.1   *A basic semiconducting polymer solar cell*

The "ZnPc/MPP" organic solar cell depicted in Fig. 10.20 exhibited an efficiency of 1.05 percent when illuminated with AM 1.5 radiation at an intensity of 860 W/m². The measured short circuit current density was 5.2 mA/cm². This is an early example of a polymer-organic cell. This type of cell was shown in Fig. 10.10, right panel.

10.3   Tandem multi-junction cells

The single junction solar cell wastes a majority of the energy in the incoming light, as we have seen in detail in Section 10.1. Photons of energy greater than the bandgap lose the excess energy to heat, and photons of energy less than the band gap go through the cell without absorption.

An intuitive solution to the problem is the tandem array of junctions of varying bandgap, with the largest bandgap junction facing the Sun. In the limit of a *series of cells with graded bandgaps*, there is in the stack somewhere a semiconductor that will optimally convert the energy of each photon. By extension of the Shockley–Quiesser analysis, this picture can be verified. In the limit of an infinitely-subdivided tandem cell, or rather a series connection of an infinite array of single junction cells, the efficiency approaches, but does not reach, that of the Carnot cycle $\eta = 1 - T_c/T_s = 1 - 300/6000 = 0.95$. A concentrated infinite tandem cell is predicted in Fig. 10.21 to reach 86.5 percent,

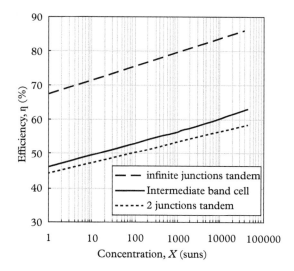

Fig. 10.21 *Efficiencies calculated for tandem cell combinations and the proposed "intermediate band cell". Note the common role of concentrated light intensity in improving the efficiency by raising the cell open circuit voltage (Algora, 2004) Fig. 6.7 p. 121*

close to the efficiency 85.4 percent called the Landsberg Limit in connection with thermal solar power installations in Section 9.1.

The curves in Fig. 10.21 are numerically based on an AM-1.5 spectrum as shown in Fig. 9.1. In the case of tandem cells, the calculation assumes that the power from each cell is extracted independently, while in practice the series-connected stack of cells is a two-terminal device and a constraint of equal current through each cell limits the power extracted. A small error may come from this difference, but in fact series-connected three-gap tandem cells have been demonstrated to operate at efficiency as high as 41.6 percent (King et al, 2009).

The record setting cell is a 3-gap GaInP/GaInAs/Ge lattice-matched cell produced by Spectrolab, a subsidiary of Boeing Corporation, operated at 364-Sun concentration. This is a structure grown epitaxially on a single crystal of Ge. We will discuss such cells presently. The junction technology is called liquid-phase organo-metallic epitaxy, and is an offspring of the ultra-high-vacuum method of molecular beam epitaxy (MBE), used with success for compounds based on GaAs. It is credible that the efficiency in concentrating tandem cells with up to five different bandgaps may reach 50 percent. What is open to question is the market price that would apply to large scale, complex, systems of this type. It may also be mentioned that apart from the practical tandem arrangement of multi-junctions, light-splitting optics can be used to illuminate a parallel array of junction devices. An efficiency record has been established by such an arrangement, although the cost is likely to be high.

10.3.1   GaAs-based epitaxial cells of high efficiency

The most successful tandem cells have been those in the GaAs system, similar to the one mentioned as having achieved 41.6 percent efficiency at a concentration of 364 Suns. Such a device would be used in a parabolic dish mirror concentrator, with an appearance similar to that of Fig. 9.6. Tracking of the Sun and water cooling are both implied.

The concept of the tandem cell is clearly shown in the spectral response curves published by the manufacturer Spectrolab, of the 3-junction cell, shown in Fig. 10.22.

The complex multi-junction GaAs-based cells are at a high state of development, building on experience in the Soviet space program (see Fig. 10.7) and in the wide application of 3–5 semiconductor(GaAs) materials in commercial electronics. An example close to the record 41.6 percent efficient cell is shown in Fig. 10.22. The three main layers are evident from top to bottom, whose absorption spectra were shown in Fig. 10.21. Complexities in this mature technology include use of low resistance PN junctions to series connect adjacent cells, and provision for back-surface-fields to inhibit recombination. The problem of maintaining the three cells at optimum operating points consistent with identical current through each junction has been apparently addressed successfully.

Cells of the type shown in Fig. 10.22 are termed "monolithic" and are two terminal devices. It is evident that these complex devices are at a mature stage of development. What is not clear is what cost levels would be attainable with such devices in mass production. The complexity of the junction device is compounded, in an actual system, with the need for high-accuracy optics, suggested by Fig. 9.2, and with a large and accurate tracking mirror or lens system suggested by Fig. 9.6. Nevertheless, it is argued that with high concentration the actual cell area needed decreases to the point where the large installation could be cost-competitive with larger area simple flat plate cells such

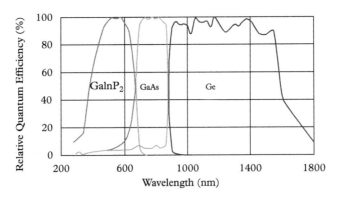

Fig. 10.22 *Relative quantum efficiencies (left to right) of the upper, middle and lower light-collecting layers of the GaAs-based 3-junction tandem cell. This cell is epitaxially constructed such that a common atomic crystal lattice runs through the whole multilayer system, accomplished by organo-metallic liquid-phase epitaxy. The sum of the three spectra is seen to include most of the Sun's spectrum (Spectrolab press release)*

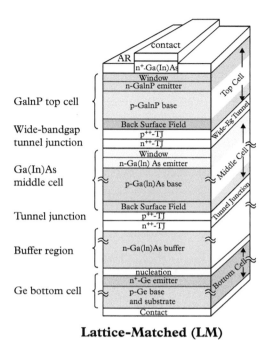

Lattice-Matched (LM)

Fig. 10.22a *Structure of single-crystal 3-gap GaInP/GaInAs/Ge Lattice-matched cell. Cells close to this configuration have operated at 41.6 percent efficiency at 364 Sun illumination. The manufacturer Spectrolab is able to project 4- and 5-junction versions of such cells approaching 50 percent efficiency under concentrated illumination. (King et al, 2002)*

as those made by First Solar. The manufacturer Spectrolab says the tandem cells are useful for point-focus concentrators as well as for dense arrays and linear concentrators.

It appears that the basic theory flowing from the work of Shockley and Quiesser has largely been validated by the results obtained from the GaAs-based tandem junction cells, as described.

10.3.2   Perovskite tandem solar cells

Perovskites of the form ABX_3, where A is typically Cs, methylammonium (MA: CH_3NH_3), or formamidinium (FA: R_2N-CH=NR_2) with R a substituent; B is Pb or Sn; and X is iodine, bromine or chlorine, have recently been explored (Eperon, McGehee et al, 2016) as potentially inexpensively fabricated, yet high-performance, light-absorbing layers in tandem solar cells. Perovskites are essentially ionic, with A and B cations, and can be grown from solution leading to high-quality crystals with high carrier mobility. An indication of the advanced state of perovskite solar cells, made by chemical deposition when possible, is suggested by the structure described in Fig. 10.26a,b, measured at 17 percent power conversion efficiency. In the cells shown in Fig. 10.26a, the 1.2 eV and 1.8 eV (bandgap) perovskites are, respectively, $FA_{0.78}Cs_{0.25}Sn_{0.5}Pb_{0.5}I_3$ and $FA_{0.83}Cs_{0.17}Pb(I_{0.5}Br_{0.5})_3$.

350 Suns, AM1.5D(35 W/cm²) 25°C

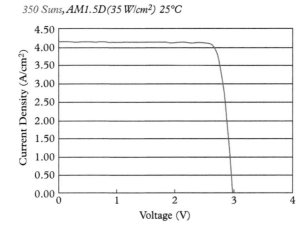

Fig. 10.23 *Current voltage curve of 3-junction GaAs cell under 350 terrestrial AM 1.5 Sun illumination at 25 C. The open circuit voltage in principle is limited by the sum of the bandgaps of the 3 junctions. (Spectrolab press release)*

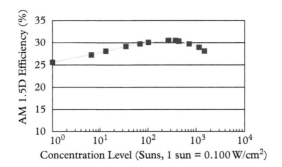

Fig. 10.24 *Measured efficiency as a function of illumination strength, for cell similar to that shown in previous figure. A slightly different cell has achieved 41.6 percent efficiency as opposed to the 30 percent shown here. Note that the peak efficiency is achieved at illumination on the order of 300 suns. (Spectrolab press release)*

The authors have succeeded to obtain efficient and stable perovskites with appropriate wide band gaps for front cells in tandem architectures by using a mixture of FA and Cs cations. They can control the band gap by tuning the Br:I ratio; thus $FA_{0.83}Cs_{0.17}Pb$ $(I_{0.5}Br_{0.5})_3$ has a 1.8 eV band gap that is well suited for the front junction of the two-terminal tandem cells.

The authors fabricated a series of planar heterjunction devices in an "inverted" p-i-n form, made of indium-tin-oxide (ITO)/poly(3,4-ethylenedioxythiophene)-poly (styrenesulfonate)(PEDOT:PSS)/$FASn_xPb_{1-x}I_3$/C_{60}/bathocuproine(BCP), capped with an Ag or Au electrode, as shown in Fig. 10.26a.

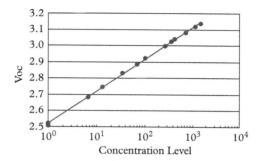

Fig. 10.25 *Measured open circuit voltage V_{oc} as a function of illumination strength, for cell similar to that shown in previous Figure. Note that these data support the enhanced power and efficiency in tandem cells predicted for concentrated illumination. (Spectrolab press release)*

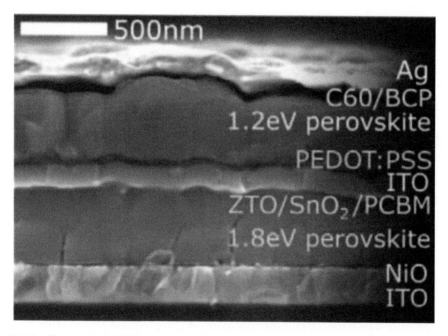

Fig. 10.26a *Two-terminal (2T) tandem double perovskite structure, shown in SEM image, based on perovskites of 1.8 eV bandgap (lower) and 1.2 eV (upper) see text. Here PEDOT:PSS is a highly conductive hole-transport layer and ITO is conductive Indium Tin Oxide. Light enters at the bottom through indium tin oxide/NiO transparent layer. (Eperon, McGehee et al, 2016) Fig. 3c*

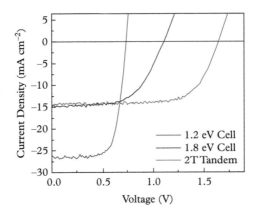

Fig. 10.26b *Measured current vs voltage at standard AM 1.5G illumination for the 3 perovskite cells implied in previous figure. The curves for the 1.2 eV, 1.8 eV, and 2-terminal tandem perovskite cells appear respectively in that order of open circuit voltage, in upper right of figure. (Eperon, McGehee et al 2016) Fig. 3c*

For the recombination layer in the two terminal (2T) tandem cell, the authors used layers of tin oxide and zinc-tin-oxide (ZTO) coated with sputter-coated ITO. This ITO layer protects the underlying perovskite solar cell from any solvent damage. It was found for perovskites containing pure Pb that heating to 170°C was needed to convert the film from the yellow room-temperature phase to the black phase.

The performance of the constituent cells and the tandem series connection are shown in Fig. 10.26b. The open circuit voltage of the 2T cell exceeds 1.7V. The authors state that they expect a main application of the perovskite cells to be as a front layer on a perovskite/Si cell, to inexpensively increase the efficiency of polycrystalline Si cells. It might be noted that the 2T cell at efficiency 17 percent falls far short of the attained efficiency 24 percent of the single crystal Si PERL cell shown in Fig. 10.5.

10.4   Spectral splitting cells

An elaborate approach to splitting the spectrum, then sent to multiple junctions, has been described by Barnett, Honsberg et al (2006) (see Fig. 10.27).

The method involves separating, by dispersive optics, the incoming light into "bins" of high energy, mid energy, and low energy. (In principle, the triangular glass prism of Isaac Newton would serve this purpose, and it appears that the efficiency of such optical devices can exceed 90 percent. In this case, a focusing action is also incorporated, to provide concentration of the light.) Light in each energy range is directed by such a "lateral optical system" to an appropriate, possibly multi-junction, solar cell. In the schematic of Fig. 10.27 two of the photon energy ranges are converted with tandem junction devices.

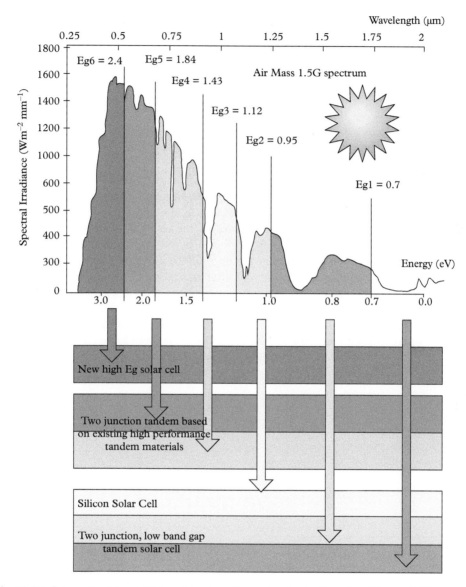

Fig. 10.27 *Schematic of the multi-junction solar cell grouping for the "lateral architecture". Upper panel shows the authors' version of the "AM 1.5 G" spectrum (See also Fig. 9.1) with the authors' division of the spectrum into 6 ranges, indicating an optimum single junction bandgap for each range. (Barnett, Honsberg, et al 2006)*

The same group (Honsberg et al, 2006) reported a record 42.8 percent efficiency for revised version of this "lateral optical system" approach, apparently involving five or six junctions in three separate laterally displaced devices, to which light is addressed by "spectral splitting optics".

Not shown in Fig. 10.27, is a semi-cylindrical dichroic lens system giving laterally displaced focal points for different wavelengths. The authors also mention arrays of these junction devices.

A more practical, inexpensive and manufacturable approach to spectrum splitting is based on a "broadband diffractive-optical element" (BDOE) as presented by Kim et al (2013).

As shown in Fig. 10.28, pairs of cells of differing bandgap receive different portions of the Sun's spectrum as split by the miniature BDOE element, one above each pair of junctions. The BDOE element, or "polychromat", as shown in the upper-right panel of Fig. 10.28 is simply an optical medium (glass) of refractive index larger than one, cut into different heights. The height pattern, found by a binary search algorithm, optimally directs light of different wavelengths into differing angles, reaching the different junctions described presently. The polychromat was fabricated using photoresist on a glass substrate using gray-scale lithography on the photoresist. The polychromat element requires spacing above the pair of junctions to allow the angular change to provide lateral displacement of the light. In the device shown in this paper the vertical distance was 14 cm. The authors say the vertical spacing could be reduced to 3 cm by reducing the pixel width (see upper-right portion of figure) from 10 μm to 1 μm.

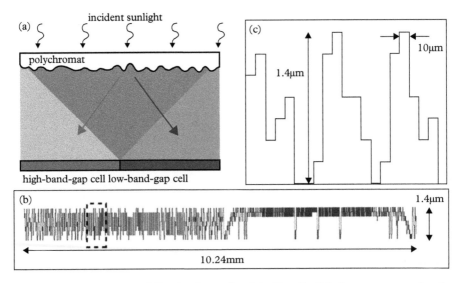

Fig. 10.28 *(a) The polychromat diffractive element bends incident Sunlight into two separate bands, each of which illuminate neighboring solar cells with matched bandgaps. (b) An example design of a polychromat represented by the height distribution of its pixels as etched into glass. (c) Magnified view of several pixels of the dashed region of the polychromat in (b). Note the eight height levels. (Kim et al 2013) (Fig. 1)*

10.5   Organic molecules as solar concentrators

Here, in the work of Currie et al (2008), solar cells are mounted vertically on the edges of the light collecting plates, that contain dissolved dyes. The upper plate collects blue photons, the dye re-emits (fluoresces) and that light is sent to the cells on the edge. The fluorescence quantum efficiency of isolated dye molecules approaches unity. (Total internal reflection helps keep the re-emitted light inside the plate, until absorbed in a solar cell.)

The redder light passes through and is caught in the second plate that has a red dye. The effective collecting area is multiplied by this scheme, and the scheme also effectively involves two separate bandgaps, leading to higher efficiency in principle.

In the collection of the re-emitted light from the dye molecules, two methods: Fluorescence (prompt response) and Phosphorescence (delayed response) are used.

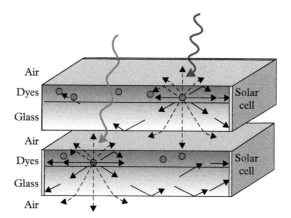

Fig. 10.29 *Light-concentration based on organic dyes that re-emit light to solar cells at edges of glass light-collecting layers. (Currie et al, 2008)*

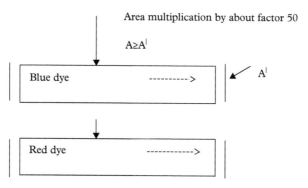

Fig. 10.30 *Area multiplication on the order of 50 by emission of fluorescent light and its capture in solar cells at the periphery. (After Currie, 2008)*

A dye molecule's characteristic color comes from the wavelength of its emission band. In the system shown in Figs. 10.29 and 10.30, dyes absorbing blue and red, respectively, are located in upper and lower plates. The emission bands of these dyes match the absorption properties of the solar cells mounted around the edges of the light collecting plates. A larger index of refraction in the glass promotes high efficiency in collection of the fluorescent and phosphorescent light by the solar cells, relative to its escape out of the glass into the vacuum. An effective area multiplication of around 50 is attained.

10.6   Concentrating optics: large vs. small scale

A small-scale conventional concentrator technology built on multijunction cells is marketed by SolFocus. The individual parabolic glass mirrors are smaller and are packaged in arrays provided with two-axis tracking of the Sun. A conceptual sketch for the SolFocus devices is shown in Fig. 10.31. The literature suggests concentration in the range of 500–650. The dashed curve at the bottom of Fig. 10.21 shows that a

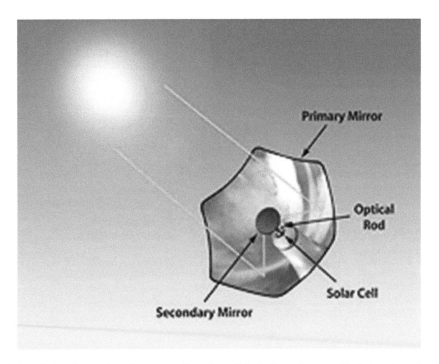

Fig. 10.31 *Artist's impression of concentrating solar cell including advanced GaAs based tandem junctions. SolFocus markets arrays of these devices provided with two-axis tracking of the Sun. There is a large change of scale between this approach and the approach represented by Fig. 9.6. SolFocus, Inc., 510 Logue Avenue, Mountain View, California, 94043 (2008). http://www.solfocus.com/en/technology/*

two-junction tandem cell at 500 Suns has the capability of about 55 percent efficiency, a large improvement over single cells. SolFocus states the efficiency of their multijunction solar cells as 38 percent.

The SolFocus concentrating cell design addresses the problem of heating of the cell's PN junctions, by insertion of a *non-focusing optical rod* to transmit the multiplied active spectrum, but not the un-utilized and deleterious infrared heat spectrum, to the tandem junctions. Concentration of light (by about a factor 500) with individual parabolic and accompanying secondary mirrors addresses the issue of cost and availability of elements, e.g., Ga, used in the cells, and of the multi-junction fabrication, since the concentration reduces the area of solar cell required. It appears that water cooling is not required to operate the tracking panels provided by SolFocus, because of the smaller scale and the use of the glass rod delivering the active spectrum from the mirrors to the cell.

A niche application of small-size concentrating solar cells on pontoons has been described (Sierra, 2011)[4] that exploits strengths and overcomes weaknesses of these devices. Solar cell efficiency decreases with temperature, that occurs in locations where the solar energy is strongest. Multi-junction tandem devices definitely need cooling, because of the concentration of energy, a factor of several hundred.

Concentrating devices need tracking which can be expensive. The pontoon-mounted concentrating solar cell arrangement shown in Fig. 10.32 addresses and exploits these aspects. There are many locations where standing water is left open to the Sun, for example at water reservoirs, above hydroelectric dams, in water treatment plants of various sorts, and in water transport systems. These areas, a bit like the top of a landfill, are often left open to the Sun and can be exploited without undue complications. The water surface is reliably horizontal and tracking of the Sun is accomplished by rotating the circular array of 24 floating pontoons around the vertical axis, which is easy to do. The device shown places the expensive solar cells in a bridge at the focus of parabolic mirrors, as shown, and water cooling with heat exchange coils under the pontoon is a central feature. So, this device will work in hot regions where the Sun is strongest, but only on a water surface. This is a niche application, a small market that may be relatively easy to open to concentrating solar cells. Worldwide there are many such locations, often favorably controlled with respect to gaining access.

For example, the Colorado River Aqueduct, part of the 400-mile California Aqueduct, is a 242-mile water conveyance system that provides a large portion of the drinking water in southern California, diverting water from the Colorado River. About 63 miles of the Colorado River aqueduct is an open canal, on the order of 100 feet wide, too shallow for boating, and open to the sky. Solar power, at 200 watts/m², falling on this area of water is then on the order of 724 MW. This large area of water is under the control of a single authority that might be persuaded to make use of its potential as photovoltaic power source. An account of planning toward exploiting this water-way for solar power appeared in the New York Times (Woody, 2011).

---

[4] Sierra Magazine, Sept./Oct. 2011. Innovate/solar flotilla.

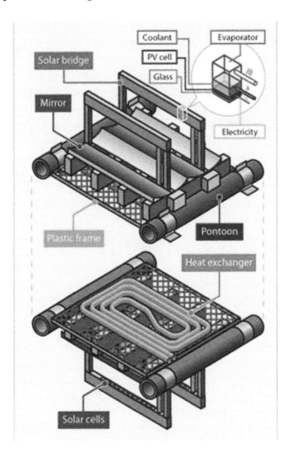

Fig. 10.32 *Schematic (Sierra, 2011) of the multi-junction solar cell application to pontoon mounting. Design of Solaris Synergy (Israel) intended for a reservoir in France. Each unit is rated at 500 Watts, and circular groups of 24 units, held with hexagonal confining wires, are provided with sensor and motor to rotate as a unit. The arrays are rated at 12 kW, and the reservoir project contains 8 arrays adding to 96 kW*

The company Solaris Synergy, mentioned in connection with Fig. 10.31, estimates that the California Aqueduct could yield 2 MW per mile, totaling up to 800 MW, and that the controlling authority is concerned that the floating solar arrays would need to be moored the banks, quite consistent with the method used by Solaris Synergy to extract the electric power.

10.7   Costs and efficiencies of photovoltaic cells and modules

A current summary of the efficiency of the leading cells and modules is given in Fig. 10.33.

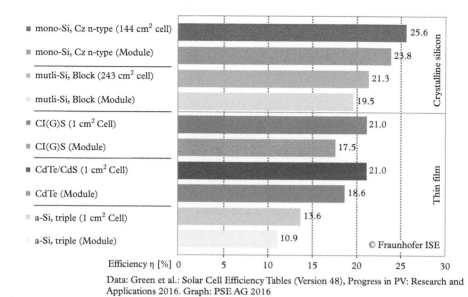

Data: Green et al.: Solar Cell Efficiency Tables (Version 48), Progress in PV: Research and Applications 2016. Graph: PSE AG 2016

Fig. 10.33 *Efficiency comparison of photovoltaic technologies in 2016. Best lab cells vs. best lab modules. (Frauenhofer Photovoltaics Report, 2016)[5], p. 24*

An important new development represented in Fig. 10.33 is the improved efficiency available in the thin film CIGS and CdTe forms. Related to this is a decrease in production of the amorphous silicon modules, as shown in Fig. 10.34.

The increase in CIGS module production is notable. The production in 2015 was about 1.1 GW, rising from less than 0.1 GW earlier than 2008. However, the whole thin film sector in 2015 accounted for less than 8 percent of module production, the predominant part being multicrystalline and single crystalline silicon.

As mentioned before Fig. 9.9, the Lazard 2015[6] report predicts an electricity cost in 2017 from utility-scale photovoltaic solar of 4.5 c/kWh, based on a mean of the costs of thin film and crystalline solar cell systems with single-axis tracking, and assuming a solar cell system (levelized) cost of $1.35/W. It appears that the recent drop in solar cell prices led the bankruptcy of firms that earlier manufactured the Dish-Stirling CSP systems.

Recent reports of even lower prices of solar electricity go as low as 2.42c/kWh.

In Chile in August 2016 a contract was agreed at auction for 2.91c/kWh. In United Arab Emirates in September 2016 a bid at auction for 2.42c/kWh was reported.

[5] Fraunhofer Institute for Solar Energy Systems, ISE (2016). Photovoltaics Report, Freiburg Germany
[6] Lazard's Levelized Cost of Energy Analysis—version 9.0, November, 2015 https://www.lazard.com/media/2390/lazards-levelized-cost-of-energy-analysis-90.pdf. (Accessed 20 March 2017)

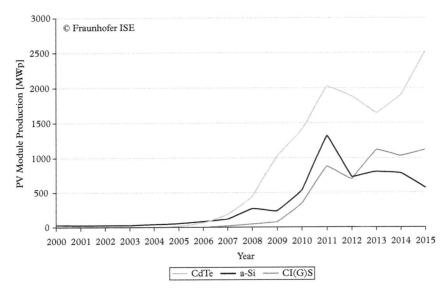

Fig. 10.34 *Thin-Film Technologies: Annual Global PV Module Production. Note the crossing of the two lower curves, such that in 2015 the production of CIGS modules exceeded that of a-Si modules. (Frauenhofer Photovoltaics Report, 2016) p. 21*

The single junction solar cell was described in Section 10.1, see Fig. 10.1 for a silicon version. The limiting efficiency of a single junction cell is 31 percent at one Sun illumination, but as high as 40.8 percent with concentrated sunlight. The Si single crystal solar cell shown in Fig. 10.1 and in Fig. 10.6 is the most common first generation solar cell. Similar "multi-crystalline" silicon cells are produced from large-grain polycrystals, that are less expensive. The fully optimized Si "PERL" cell reaching efficiency 24 percent was shown in Fig. 10.5, but the bulk of the Si cells are characterized in a range 15 percent to 20 percent. The crystalline Si solar cell is built on a wafer of Si, typically 200 to 300 μm thick, cut from a large crystal using a wire saw that wastes also one wire-width of the Si single crystal boule. The cost of wafer-based cells is higher, but in 2016 appear to be less than 1 $/watt, according to the following National Renewable Energy Lab Figure. In this figure, the bottom segment (in yellow) is the module cost and is seen to fall from around $2.50/watt in 2009 to about $0.7/watt in 2016. At the utility scale the total levelized cost per watt, about twice the module cost, is seen to be $1.42/W in 2016, close to the Lazard Figure.

The precipitous fall in prices between 2009 and 2016 shown in Fig. 10.35, for example, module cost at utility scale of about $2.50 falling to about $0.70 is largely due, it seems, to Chinese producers, at large volume with government assistance. These prices are still set largely by mono- and multi-crystalline silicon, and a great likelihood of further reduction in price is offered by the change to the less expensive thin film technology, as represented by CdTe and CIGS in Fig. 10.33. According to the

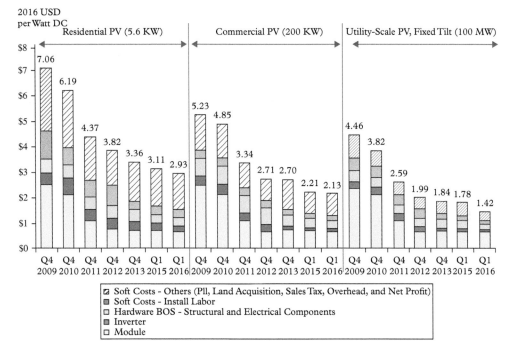

Fig. 10.35 *History of PV device/system costs. In each column, starting from the bottom, the levelized cost categories are module, inverter, hardware, labor and soft costs. The total cost in 2016 at utility scale is given as $1.42/W (see lower right portion of figure), with the module cost less than $0.70/W. (National Renewable Energy Laboratory, press release) http://www.nrel.gov/news/press/2016/37745*

Frauenhofer Photovoltaic report of 2016 (p. 18) the total solar module production in 2015 was 57 GW, and of this 7 percent was thin film. Of the 93 percent remaining, 26 percent was mono-crystalline Si while 74 percent was multi-crystalline silicon. The top ten manufacturers accounted for about 53 percent of the total in 2015. Among the top ten, the percentages ranged in order from 7 percent apiece for Trina, JASolar and Hanwha Q-Cells; 5 percent apiece for Canadian Solar, First Solar, Jinko Solar and Yingli; to 4 percent apiece for Motech, NeoSolar and Shunfeng-Suntech. All but Canadian Solar and First Solar among these appear to be in China. Concentrating photovoltaic capacity installed was about 120 MW in 2012, 100 MW in 2013, 60 MW in 2014 and 20 MW in 2015. Most of this capacity was for concentration ratios in the range 300 to 1000 and firms involved include Silevo, Panasonic, Sunpreme, CIC Choshu and Sunpower. Boeing Spectrolab, see Figs. 10.22–10.25, is not on this list but is primarily involved in the market for solar cells for spacecraft.

An overall reduction in price with manufactured volume is a studied phenomenon, called the "learning curve", that is summarized for the photovoltaic field in Fig. 10.36.

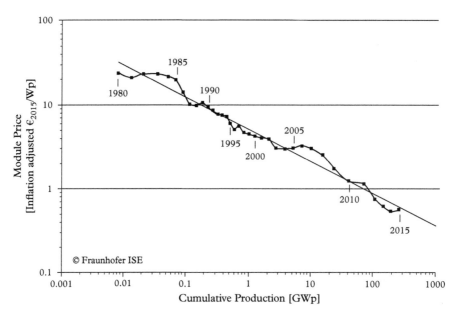

Fig. 10.36 *Learning rate: each time the cumulative production doubled the solar module price went down by 23 percent for the last 35 years. (prices in Euros) (Frauenhofer Photovoltaics Report, 2016) p. 41*

10.8   Photovoltaic research areas

What can be done to make solar cells more efficient? We have already described the main classes of solar cells that are now available. These can be characterized in terms of the number of distinct energy gaps, according to the physical form and according to use for single Sun illumination vs. concentrated illumination.

The first new concept that may alter this situation is "the intermediate band cell". The second is some form of carrier multiplication, as a means of harvesting in the external circuit the photon energy in excess of the band gap.

10.8.1   Intermediate band cells

Efficiency comes from utilizing a larger portion of the solar spectrum. The idea of the intermediate band cell is suggested by the improvement experimentally realized in the two-bandgap tandem cell. The generalization is that some homogeneous semiconductor might be found to have three characteristic energies, E_C, E_V and E_{IB}, so that the differences among these energies could represent 2 or 3 separate gaps, able to intercept a larger fraction of the solar spectrum.

A pioneering calculation along these lines is summarized in Fig. 10.37, based on a hypothetical semiconductor as illustrated in Fig. 10.38. (We will see in Section 10.8.3 that Dilute Magnetic Semiconductors may be workable examples of such a band structure.)

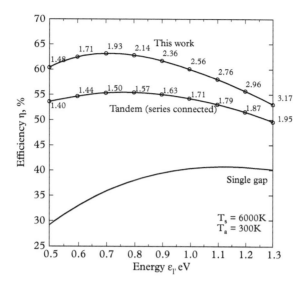

Fig. 10.37 *Efficiency of Intermediate Band Solar Cell (top curve, labeled "This work") compared with two-junction tandem cell (second curve) and single-junction cell (lower curve) at one Sun illumination. The abscissa is the value of the smaller gap, with the values for the second (larger) (labeled EH in Fig. 10.38) gap in eV printed along the upper two curves, which assume two separate gaps. These curves are calculated assuming a black body spectrum at 6000 K. Luque, A and Marti, A. (1997)*

The concept of the intermediate band semiconductor is shown in Fig. 10.38. The Fermi level lies in the *intermediate band* that allows conduction and is partially filled. In semiconductor physics this situation arises in a semiconductor with a large concentration of donor impurities, such as P in Si. At a level of doping above the Mott transition concentration the electronic states are linear combinations of electron waves centered on the impurity sites and are delocalized. The Fermi level then lies within the distribution of impurity ground state levels. In that case of large donor concentration, the energy gap between the Fermi level and the conduction band edge would be small, in the range of meV, as was discussed in Chapter 5, in relation to hydrogenic impurities. A similar situation might occur with chemical impurities that form deeper levels within the gap. In the work cited, the authors suggest the intermediate band could be formed by a regular array of quantum dots, 3-dimensionally localizing electron states, to be discussed. The small overlap between adjacent dots is assumed to form the intermediate band (referred to as mini-band) in Fig. 10.39.

A. Marti, L. Cuadre and A. Luque, (2004). "Intermediate Band Solar Cells", in *Next Generation Photovoltaics*. A. Marti and A. Luque, eds., Ch. 7. (Bristol and Philadelphia: Institute of Physics.)

A sketch of an intermediate band quantum dot solar cell device is given in Fig. 10.40. It is assumed that the metallic intermediate band results from half filling with electrons

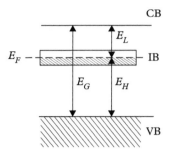

Fig. 10.38 *Sketch of hypothetical Intermediate Band semiconductor. Marti, Cuadre and Luque (2004). Fig. 7.1a, p 141.*

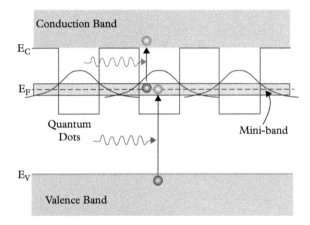

Fig. 10.39 *Two separate photon absorptions are illustrated in schematic diagram of Intermediate Band Solar Cell, as realized using coupled quantum dots to form a "mini-band". The idea is that the two smaller gaps allow utilization of part of the solar spectrum below $E_c - E_v = E_G$. In this device the usual absorption for $hc/\lambda \geq E_G$ also occurs. The upper curve shown in Fig. 10.37 assumes that carriers in the intermediate band have high mobility. (Honsberg et al, 2006)*

the quantum dot wells, as indicated. In this structure, the intermediate band is isolated from the external contacts by means of heavily doped p- and n- type emitter layers.

The idea of a quantum dot, or electron box was given in connection with Eq. 4.15, and later text. In the present context quantum dots are 3-D structures introduced locally in a junction device such as a solar cell or junction laser. The basic energy of trapping inside an empty box, a cube at potential zero for electrons is:

$$E = (nh)^2/8mL^2 \qquad (10.13)$$

where L is the edge dimension and n^2 understood to be the sum of squares of the numbers of half wavelengths of the particle's de Broglie wave $\lambda = h/p$ along the cube edges.

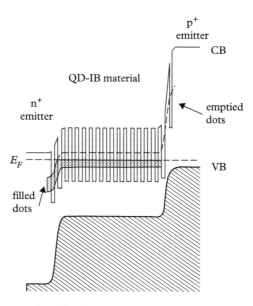

Fig. 10.40 *Sketch of Intermediate Band quantum dot solar cell device. (after Marti, Cuadre and Luque, 2004) Fig. 7.12, p 161*

A quantum dot may also be called an "artificial atom", and is characterized by discrete sharp electron energy states, and sharp absorption and emission wavelengths for photons.

Transmission electron microscope (TEM) images of such nanocrystals, which may contain only 50,000 atoms, reveal perfect crystals having the bulk crystal structure and bulk lattice constant. Quantitative analysis of the light emission process in QD's suggests that the bandgap, effective masses of electrons and holes, and other microscopic material properties are very close to their values in large crystals of the same material. The light emission comes from radiative recombination of an electron and a hole, created initially by the shorter wavelength illumination.

The energy E_R released in the recombination is given entirely to a photon (the quantum unit of light), according to the relation $E_R = h\nu = hc/\lambda$. Here ν and λ are, respectively, the frequency and wavelength of the emitted light, c is the speed of light 3×10^8 m/s, and h is Planck's constant $h = 6.63 \times 10^{-34}$ Js $= 4.136 \times 10^{-15}$ eVs. The color of the emitted light is controlled by the choice of L, since $E_R = E_G + E_e + E_h$, where E_G is the semiconductor bandgap, and the electron and hole confinement energies, E_e and E_h, respectively, become larger with decreasing L as $h^2/8mL^2$.

These confinement (blue-shift) energies are proportional to $1/L^2$. Since these terms increase the energy of the emitted photon, they act to shorten the wavelength of the light relative to that emitted by the bulk semiconductor, an effect referred to as the "blue shift" of light from the quantum dot. These nanocrystals are used in biological research as markers for particular kinds of cells, as observed under an optical microscope with background ultraviolet light (uv) illumination.

In these applications the basic semiconductor QD crystal is coated with a thin layer to make it impervious to (and soluble in) the aqueous biological environment. A further coating may then be applied that allows the QD to preferentially bond to a specific biological cell or cell component of interest. The biological researcher may, for example, see the outer cell surface illuminated in green while the surface of the inner cell nucleus may be illuminated in red, all under illumination of the whole field with light of a single shorter wavelength.

The electron and hole particles are generated by light of energy:

$$hc/\lambda = E_{n,electron} + E_{n,hole} + E_g. \tag{10.11}$$

Here the first two terms are strongly dependent on particle size L, as L^{-2}, which allows the color of the light to be adjusted by adjusting the particle size. The bandgap energy E_g is the minimum energy to create an electron and a hole in a pure semiconductor. The electron and hole generated by light in a bulk semiconductor may form a bound state along the lines of the Bohr model, described previously, called an exciton. However, as the size of the sample is reduced, the Bohr orbit becomes inappropriate and the states of the particle in the 3-D trap are a more correct description.

Experimental evidence of photocurrent from sub-bandgap light in an intermediate band solar cell structure has recently been demonstrated (Marti et al, 2006), see Fig. 10.41.

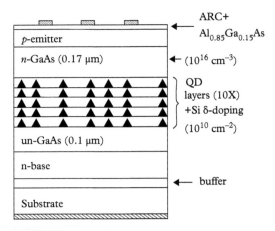

Fig. 10.41 *Example of Intermediate Band Solar Cell based on quantum dots (QD). Here "ARC" represents "anti-reflection coating", and "δ-doping" means a pure deposit, in this case of Si. The indicated quantum dots are InAs, which have also been used to lower the threshold current in versions of GaAs heterojunction lasers. The authors found extra photocurrent in this structure arising from lower energy photons, substantiating an aspect of the model depicted in Figs. 10.37–10.39, Marti, Antolin et al, (2006)*

These authors, however, state that their devices do not exhibit the main features expected for the Intermediate Band Solar Cell. The results so far obtained, thus, give little indication that a practical Intermediate Band Solar Cell device is near. Since it is difficult to fabricate a complicated structure as asked for in Figs. 10.37–10.39, the failure of the experiment may reflect the difficult fabrication rather than a failure of the essential intermediate band cell concept.

10.8.2   Impact ionization and carrier multiplication

The "Li-drifted Germanium detector"[7] is an elegant commercially available solid-state device to measure the energy of a single high-energy gamma-ray photon. A pure single crystal of Germanium is treated in a special way with Lithium, so that the mean free paths of electrons and holes are larger than the dimensions of the device. Metallic electrodes apply a small bias voltage across the germanium, and a sensitive detector of the charge flow in the circuit is provided. The germanium is pure and is also cooled to 77 K, the temperature of liquid nitrogen, so that the numbers of thermally generated free electrons and free holes is low, and only a small background current flows. A gamma ray, i.e., a photon of high energy (Mev) $E_\gamma = hf \gg E_g$, is absorbed, creating initially an electron in the conduction band and a hole in the valence band, whose kinetic energies add up to $hf - E_g$. The final result from the single absorbed gamma ray photon, as a result of internal processes in the Germanium, is n electrons and n holes flowing oppositely across the crystal and into an external charge meter, such that the total measured charge,

$$Q = n2e = 2e\left(hf/E_g\right) = \int I(t)\,dt. \qquad (10.12)$$

From this relation the gamma ray energy hf is deduced from the measured charge Q and the known value of E_g, according to:

$$E_\gamma = \left(Q/2e\right)E_g. \qquad (10.13)$$

The carrier multiplication process that occurs within the extremely pure semiconductor creates a large number $n = (hf/E_g)$ of electron-hole pairs from the initial high energy electron-hole pair. The high energy carriers reduce their kinetic energy by collisions creating additional electron hole pairs of lower energy, and in this process, loss of energy to lattice vibrations (phonons) apparently is not an important effect.

A similar process would be beneficial in the efficiency of solar cells, to capture energy from the portion of the solar spectrum in the range $hf > E_g$. This process in a solar cell could multiply the conversion efficiency by a large factor.

In this context Fig. 10.42 shows levels in a quantum dot as an element of absorber in a solar cell. The process shown is one of absorption of a high energy photon $hc/\lambda = E_g + 6\Delta$, where Δ is the energy spacing of succeeding size-quantized states in the

[7] http://en.wikipedia.org/wiki/Germanium_detector#Germanium_detector. (Accessed 20 March 2017).

quantum dot. The process depicted leads to 3 electrons compared to 1 electron, eligible to flow through the load resistance, but requires an energy threshold $E \geq E_g + 8\,\Delta$, and the further condition:

$$6\Delta \geq 2\,E_g + 4\Delta, \text{ or } \Delta \geq E_g.$$

The lower threshold to produce two electron-hole pairs would be $E \geq E_g + 4\Delta$, with:

$$4\Delta \geq E_g + 2\Delta \text{ or } 2\Delta \geq E_g.$$

The minimum excess energy released by exciton decay to give charge multiplication is $E_g + 2\Delta$. This is a high threshold: since $2\Delta \geq E_g$, the initial energy has to be more than $3\,E_g$. The further assumption is that the charge appearing at the lowest electron and hole energies will efficiently leave the quantum dot to flow in the external circuit. This may be possible but apparently has not been demonstrated in situations other than the Ge detector.

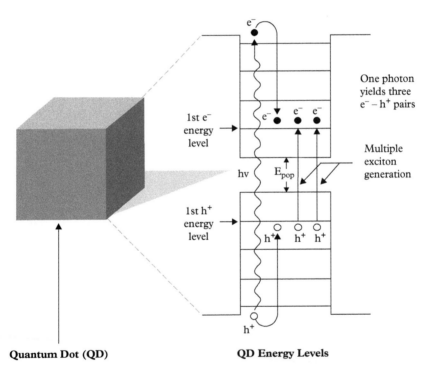

Quantum Dot (QD)   **QD Energy Levels**

Fig. 10.42 *Multiple exciton generation in a quantum dot. Because of quantum confinement, the energy levels for electrons and holes are discrete. A single absorbed photon of energy at least 3 times the energy difference between the first energy levels for electrons and holes in the quantum dot can create 3 excitons, tripling the charge in the external circuit. (Lewis and Crabtree 2005)*

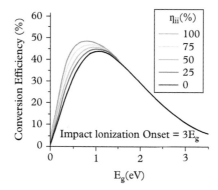

Fig. 10.43 *Predicted solar cell efficiency as enhanced by Impact Ionization. Evidence for such processes has been advanced for nanocrystals of the small bandgap semiconductor PbSe. (Schaller and Klimov 2004)*

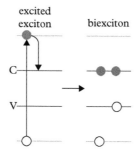

Fig. 10.44 *Basic process of Impact Ionization, as observed in PbSe nanocrystals. This process is also involved in the germanium detector, see text. (Schaller and Klimov, 2004)*

If such processes are important, the optimum solar cell would have smaller values of E_G, as shown in Figs. 10.43–44.

10.8.3 Ferromagnetic materials for solar conversion

A recent theoretical investigation (Olsson et al, 2009) has concluded that dilute magnetic semiconductors may be useful as photovoltaic materials, having the possibility of the Fermi level located in a narrow magnetic band inside a larger gap. The compound AlP:Cr has been investigated, even though it has not apparently been synthesized. The main features of the band structure are shown in Fig. 10.45. The electron band-structure diagram is split into two parts, the left for spin up and the right for spin down. The splitting of the narrow peak from the Cr ion indicates ferromagnetic order. The Cr plays the role of a "deep level" as was mentioned in connection with doping of semiconductors. The Cr level is indicated as half-filled, so this band structure is similar to that needed for the intermediate band solar cell.

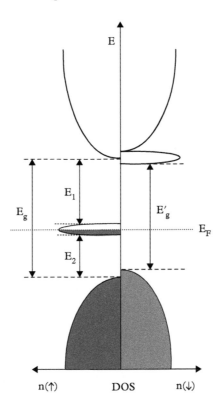

Fig. 10.45 *Sketch of electron density of states for ferromagnetic Dilute Magnetic Semiconductor AlP:Cr as suitable for a solar cell . AlP is a member of the GaAs family of semiconductors and is capable of being doped with Cr to become ferromagnetic. The narrow level in the bandgap is a d-state of Cr. The sketched bands (left side for spin up and right side for spin down) have many of the properties that were postulated for the Intermediate Band Solar Cell (Olsson et al, 2009) Fig. 1*

The authors point out that a single absorber material, rather than a multi-junction system, helps avoid optical and tunnel junction losses, and generally might be simpler to construct than a competing tandem system. On the other hand, the particular compound has not yet been made, but its properties have been surveyed theoretically as a start. The parent semiconductor is AlP, in the zinc-blende structure with an indirect bandgap of 2.43 eV. The effect of random substitution of Cr on the Al sites was estimated theoretically. The material was found to be a Dilute Magnetic Semiconductor of ferromagnetic type, with a critical temperature estimated as 160 K.

The efficiency of the resulting solar cell was found to exceed 50 percent in a suitable range of parameters, as shown in Fig. 10.46. This is an encouraging result for the proposed AlP:Cr cell, since the largest possible single bandgap cell efficiency is only 31 percent.

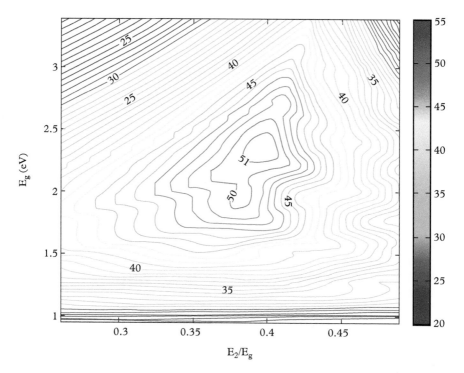

Fig. 10.46 *Photovoltaic efficiencies rise above 0.51 at one Sun in theoretical investigation of the AlP:Cr system (Olsson et al, 2009) Fig. 2*

The optimum predicted behavior for photovoltaic conversion was 52 percent efficiency at one Sun, at bandgap 2.43 eV and E_2/E_G about 0.38. This remarkably high estimate is based on an advanced calculation. The authors point to a large litera-ture related to Dilute Magnetic Semiconductors, that might be transferable to solar cell applications.

11

Energy Storage, Distribution, Use and Climate Impact

Energy in this chapter is mostly assumed to be electrical energy, with a large portion devoted to the electric grid. Later in the chapter we turn to a broader discussion of present energy uses, their emissions, predicted effects on climate and some suggestions for mitigation of the effects.

11.1   The large-scale power grid

The typically wide distance between population centers and optimum locations of renewable sources of energy makes the electrical power grid important. (The distribution problem can also be alleviated by use of an intermediate energy form such as hydrogen.) The optimum form of electric powerline depends on the distance.

11.1.1   AC vs. DC transmission

A high-voltage, direct current (HVDC) electric power transmission system (sometimes called a power super highway or supergrid) uses direct current for the bulk transmission of electrical power, in contrast to the more common alternating current (AC) systems.[1,2]

For long-distance transmission, HVDC systems are less expensive and suffer lower electrical losses. There only two conductors to put onto pylons, and the full voltage is always on the line, rather than only at the peaks of the cycle. For underwater power cables, HVDC avoids the heavy currents required to charge and discharge the cable capacitance for each cycle. For shorter distances, the higher cost of DC conversion equipment compared to an AC system may still be justified, due to other benefits of direct current links. Still, it is mandatory to invert to AC at the population center.

---

[1] https://en.wikipedia.org/wiki/High-voltage_direct_current
[2] http://web.archive.org/web/20080408011745/http://www.rmst.co.il/HVDC_Proven_Technology.pdf

Physics and Technology of Sustainable Energy. E. L. Wolf, Oxford University Press (2018).
© E. L. Wolf. DOI: 10.1093/oso/9780198769804.001.0001

HVDC allows power transmission between unsynchronized AC grid network systems. Since the power flow through the HVDC link can be controlled independently of the phase angle between source and load, it allows connection between two networks with a difference in phase. HVDC also allows transfer of power between grid systems running at different frequencies, usually 50 Hz and 60 Hz. This improves the stability and economy of each grid, by allowing exchange of power between incompatible networks.

The modern form of HVDC transmission uses technology originated in the 1930s in Sweden (ASEA) and in Germany. The state of the art in long-distance high-efficiency power lines is illustrated by ultra-high voltage direct current (UHVDC) power lines present and planned in China in Fig. 11.1. The designation UHVDC is attached to systems operating at 800 kV, larger than the 500 kV HVDC line. A famous example of

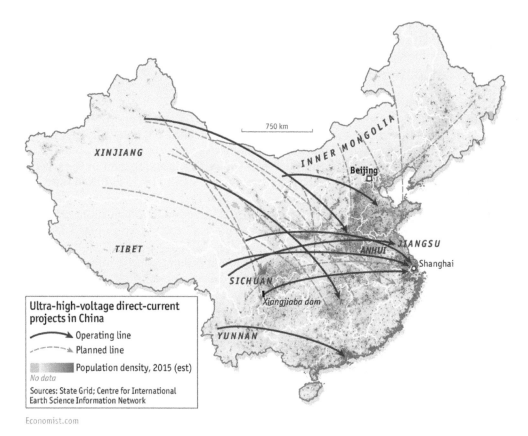

Fig. 11.1 *Map of existing and planned 800 kV HVDC powerlines in China, from The Economist Jan. 14, 2017. "Rise of the supergrid: Electricity now flows across continents, courtesy of direct current. Transmitting power over thousands of kilometres requires a new electricity infrastructure". It can be seen that the power lines arise in interior remote regions and end in regions of high population density near the coastline.* http://www.economist.com/news/science-and-technology/21714325-transmitting-power-over-thousands-kilometres-requires-new-electricity

the HVDC line is that connecting the Three Gorges Dam, providing 22.5 GW generating power, to the region near Shanghai, in China. This line, completed in 2004, runs 940 km from an AC-to- DC (rectifier) station near the dam to the DC to AC inverter station Huizhou near Guangdong. The rated power transmitted is 3 GW using bipolar lines ± 500 kV. The newer Xiangjiaba–Shanghai line, seen in Fig. 11.1 is ± 800 kV and 6.4 GW from Fulong, near the Xiangjiaba dam in Sichuan Province, to Shanghai, a distance of 2000 km, completed in 2010.

The description of the state of the art line is that the converter stations from AC to DC, and reverse, are based on thyristor switches, silicon devices that pass and shut off, alternately, the full line current that is 4000 A. The voltage drop across the individual thyristors, also known as silicon controlled rectifiers (SCRs), is limited to 8.5 kV. In HVDC terminology for this recent system, a "valve" consists of 56 thyristors in series, able totally to withstand 476 kV (Sheng et al, 2010).

The convertors are a key aspect of the HVDC system, that we discuss next.

11.1.2   AC/DC conversion methods

11.1.2.1   Silicon thyristors
The underlying device in Fig. 11.2, and in the largest UHVDC installations shown in Fig. 11.1, is the thyristor, also known as a silicon-controlled-rectifier, SCR. As shown in Fig. 11.3, it is a multi-junction NPNP Silicon device with a control electrode G that initiates turn-on of forward current, but never allows significant current in the reverse direction. It is regarded as a switch, ON only for one sign of bias, and then only if a control pulse has been received.

Recall that formation of a positive–negative (PN) junction in silicon is illustrated in Fig. 5.17 c: its I–V characteristic is described in Eq. 5.87 and shown in Fig. 5.18. The reverse current density of the PN junction is Eq. 5.85; $J_{rev} = en_p (D_p/\tau_n)^{1/2} + e\, p_n (D_p/\tau_p)^{1/2}$, where D is a diffusion constant and n and p represent the minority carrier concentrations. A desirable small value of the reverse current density J_{rev} is an attribute of a high-purity crystal with long minority carrier lifetimes, τ_n and τ_p.

Looking at Fig. 11.3, in the quiescent OFF state, the two oppositely directed junctions of the PNP regions near the anode mean that no current, i.e., only the reverse current, flows over the wide range of applied voltage between plus and minus of the reverse breakdown voltage of one or the other of these junctions. The OFF-state voltage in the reverse direction is blocked by the junction j_1 between the P-emitter and the N-base.

The off-state voltage in the forward direction is blocked at the internal junction j_2 between the P-base and N-base, see Fig. 11.3. The I-V characteristic in the off-state is shown as the dash-dot line in Fig. 11.4. The dash-dot curve for zero-control current shows that the device will go into the conducting mode at very high forward bias voltage, but for the devices used in HVDC the required voltage exceeds 8.5 kV. The ON characteristic of the thyristor is essentially the diode characteristic of Fig. 5.18. This characteristic appears and is latched into place only after a gate pulse of short duration appears in the P_2 region while the device is forward biased. The switch-on process is discussed by Streetman and Banerjee, (2009), p. 526 ff. When the SCR is biased in the forward blocking state (positive bias region of the dash-dot curve) a small current

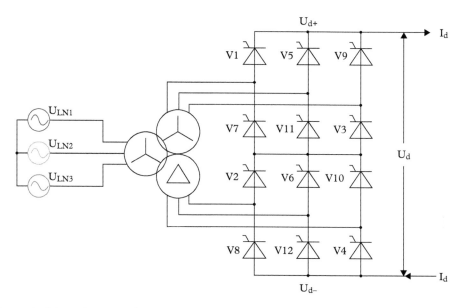

Fig. 11.2 *Schematic diagram[3] of twelve-pulse bridge rectifier that captures and adds together the 12 peak 30° sections, "pulses", of the sinusoid. The input is 3-phase AC at 60 Hz and the output approximates DC. In the UHVDC application the DC output voltage is 800 kV at about 4000 Amperes. The units V2,…V12 are called "valves" and are made up of series connections of thyristors, each with associated circuitry and deionized water cooling. Evidently, if 4 valves in series, e. g., (in the Figure) units 9, 3, 10 and 4, are seen to connect between ± 400 kV, each valve must withstand 200 kV and thus must have many thyristors per valve, if the present limit voltage per thyristor is 8.5 kV.Water cooling is needed because in the ON condition, carrying up to 4000 A, each thyristor has a voltage drop of two to three Volts and therefore dissipates several kW*

supplied to the gate can initiate switching to the conducting state. That is, a positive gate current as shown causes holes to flow into the p_2 region, that can be regarded as the base of the NPN transistor on the right (cathode) side of the structure. This added supply of holes, and the accompanying injection of electrons from n_2 into p_2, initiates transistor action in this NPN. After a transit time τ_{i2} the electrons injected by j_3 arrive at the center of the junction and are swept into n_1, the base of the PNP transistor on the left of the structure. This causes an increase of the hole injection of j_1, and these holes diffuse across the base n_1 in a time τ_{i1}. Thus, according to Streetman and Banerjee (2009), after a time delay of approximately $\tau_{i2} + \tau_{i1}$, transistor action is established across the entire PNPN structure, and the device is driven into the forward-conducting state. In most thyristors the delay time is less than a few microseconds, and the gate current required for turn-on is only a few milliamperes. Therefore, the SCR thyristor can be turned on by a very small amount of power in the gate circuit. On the other hand, in the recent devices the current I can be thousands of amperes and the power controlled by the device may be very large.

---

[3] https://en.wikipedia.org/wiki/High-voltage_direct_current.

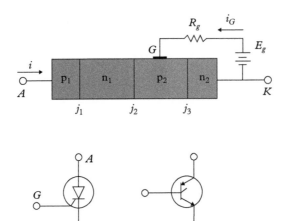

Fig. 11.3 *Schematic diagram and circuit symbols for NPNP silicon thyristor. Anode A, Cathode K and control electrode G are shown. The internal regions, from right to left are n_2, p_2, n_1 and p_1. The internal junctions from right to left are labeled 3, 2 and 1. The control gate G connects to region p_2 providing current i_G. The forward current for the device I enters at the left at A and corresponds to electron flow from right to left through the device. The control current (pulse) drives holes into the weakly doped p_2 region after Fig. 10–13a, Streetman and Banerjee, 2009*

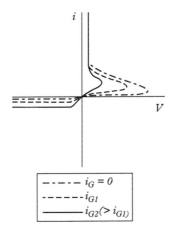

Fig. 11.4 *I-V characteristics of thyristor, schematically, with zero control current i_G (dash-dot curve), low control current i_{G1} (dashed curve) and larger control current, $i_{G2} \gg i_{G1}$ (solid curve). The dash-dot curve for zero control current shows that the device will go into the conducting mode at very high forward bias voltage, but for the devices used in HVDC the required voltage exceeds 8.5 kV, see text. The high current portion of the solid curve, latched into place with a control current pulse, represents the forward current region of Fig. 5.18, after Fig. 10–13b, Streetman and Banerjee 2009*

The device turns off if the current tries to reverse. In an AC situation the timing t of the gate pulse to turn on the thyristor is described by the "firing angle" $\alpha = \omega t$. By controlling α, the device can be changed from a rectifier to an inverter.

The state of the art in making single crystal silicon thyristors has been significantly advanced to allow their use in the kA, kV environment of HVDC, but not by changing the basic elements from the PN junctions described in Chapter 5. The advances have come from making the silicon crystals and PN junctions less defective, more ideal and larger. In the HVDC environment combinations of these devices must pass up to 4000 Amps in the ON condition and block 800 kV in the OFF condition. To meet the latter requirement, many individual devices put in series, based on the state of the art value 8.5 kV per NPNP device. The state of the art thyristors, as used in the latest UHVDC rectifier and invertor stations, are made of 6-inch diameter silicon wafers. In these devices the PN junctions operate correctly up to 8 kV reverse bias (see Fig. 11.5).

In looking at Fig. 11.5 for the high-power silicon thyristor, one sees[1] that the outer highly doped layers are the emitting zones; the weakly doped inner layers are the base zones. This suggests a view of the device as an overlap of a PNP transistor at the anode and an NPN transistor at the cathode. The control connection G is located on the P-base. As mentioned previously, the off-state voltage in the reverse direction is blocked

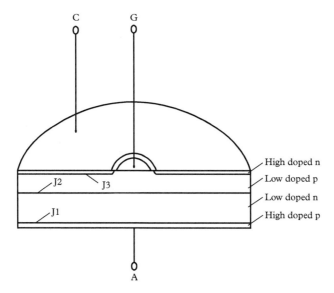

Fig. 11.5 *Schematic diagram[4] of high power single crystal silicon thyristor. From top to bottom are NPNP layers, the upper N and the lower P being heavily doped by diffusion. The upper and lower electrical contacts, not shown, are copper and molybdenum plates. The most recent version of the device is 6 inches in diameter. The control contact G is directly to the upper weakly p-type region of the single crystal. (The acronym FACTS represents Flexible AC Transmission Systems, where thyristors are also used.)*

[4] Huang, Hartmut, Markus Uder, Reiner Barthelmess, and Joerg Dorn. "Application of high power thyristors in HVDC and FACTS systems." In *17th Conference of Electric Power Supply Industry (CEPSI)*, pp. 1–8. 2008.

at junction J_1 between P-emitter and base. The off-state voltage in the forward direction is blocked at the internal junction J_2 between the P-base and N-base. The thickness of a wafer with 8000 V blocking capacity is about 1.5 mm. The extraordinary 8000 volt blocking capacity in recent silicon thyristors, attributable to the individual PN junction, was achieved[1] around 1995, and depended on improving the quality and purity of the single crystal and its diffusion doping. The recent increase of rated current to 4500 Amps is probably also related to increasing the diameter of the thyristor silicon single crystal wafer to 6 inches. The authors[1] state that improved high-purity diffusion processes, to reduce the number of undesired atoms in the wafer, and to lead to sufficiently high carrier lifetimes, have been essential to develop the high-power thyristor. In the off states, forward and reverse, currents of several milliamperes are typical and are described as "reverse currents" by Eq. 5.85. Fig. 11.6 shows recent thyristor devices.

The silicon wafer of 1.5 mm thickness, with highly p-doped bottom layer, is attached to a molybdenum carrier disk by a sintering process at 220°C. The Mo disk helps remove heat from the wafer as well as carrying the full current. The several thyristors in a "valve" are pressed together in a series connection, with supporting circuitry for each "thyristor level", or series grouping of thyristors. The supporting circuitry is illustrated in Fig. 11.7 for the case of an inverter station.

In Fig. 11.7, the auxiliary components are "snubber" capacitors and resistors, labeled C_B and R_B, DC grading resistors R_{DC} and grading capacitors C_K and the nonlinear series inductor L_{VD} labeled as a "saturable reactor". The snubber capacitor is

Fig. 11.6 *Photographs[5] of high power single crystal silicon thyristors based on 4, 5 and 6 inch diameter silicon wafers. The 6 inch version is rated for 4500 A and 8.5 kV. The upper and lower electrical contacts to these devices are copper and molybdenum plates*

[5] Huang, Hartmut, Markus Uder, Reiner Barthelmess, and Joerg Dorn. "Application of high power thyristors in HVDC and FACTS systems." In *17th Conference of Electric Power Supply Industry (CEPSI)*, pp. 1–8. 2008.

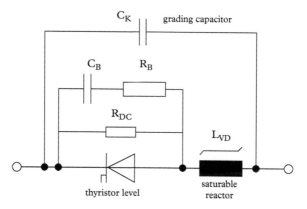

Fig. 11.7 *Supporting circuitry for "thyristor level" (series connection of thyristors) in application as DC switch, according to Huang et al[1]. Slightly different supporting circuitry is used for the rectifier station. Here the "thyristor level" in the main current flow is understood to be a direct series combination of several thyristors. The "snubber" elements C_B and R_B and saturable reactor are needed to suppress transients in switching that might damage the thyristors. These circuit elements are specially designed to operate at the extraordinarily high current, kA, and voltage, 8.5 kV per thyristor, levels. For example, the snubber resistor has wire directly immersed in flowing deionized water to allow rating of several kW per device*

needed to handle the voltage overshoot during turn-off, and the series snubber resistor suppresses oscillations. The capacitor is typically a single SF_6 unit rated for the full voltage of the thyristor, e.g., 8.5 kV, and the snubber resistor is water cooled to allow it to dissipate the full snubber current, a power level mentioned by the authors as up to 4.5 to 7 kW per resistor, with a moderate flow rate of deionized water in which the resistor wire is directly immersed.

To limit the dI/dt stress on the thyristors at turn-on, and the dV/dt stress during transients in the off state, reactors (inductors) are connected in series with the thyristor string. These reactors are non-linear to provide high inductance at the beginning of current flow but low inductance as soon as the thyristor string is turned on safely. To achieve the desired non-linearity the reactors are designed with a saturating iron core.

The components that have been described are collected together in trays arranged vertically and in series connection to allow control of the very high voltages that are being used. Fig. 11.8 (Sheng et al, 2010) shows a single Multiple Valve Unit MVU of the Xiangjiaba–Shanghai ±800 kV 6400 MW UHVDC transmission line, suspended from the ceiling of a valve hall for testing. The large white donuts are shields designed to limit corona discharges. The whole MVU unit may be at a DC level 100 kV above the ground potential on the floor of the hall. Great caution is needed to isolate the control pulse circuits, as in Fig. 11.3, that may be at 100 kV, from the operators on the floor of the hall. This is done by using an optical pulse to enable a local control bias circuit mounted in the tray, as suggested in Fig. 11.8. The Control electrode is at a small potential difference from the Cathode C or K of the thyristor, but that cathode may be at 100 kV with respect to ground and operators on the valve hall floor. According to Sheng et al (2010), the MVU as shown in Fig. 11.8 consists of two "valves" in series.

Fig. 11.8 *High-voltage MultipleValve Unit (MVU) for the Xiangjiaba-Shanghai ± 800 kV 6400 MW UHVDC transmission line, here suspended from ceiling of valve hall for testing. The thyristor strings and associated circuits are in trays stacked vertically. The rounded surfaces are designed to limit corona discharge in the air (Sheng et al 2010) Fig. 2*

A single thyristor "valve" in this installation is composed of 56 pieces of 8.5 kV 6-inch series-connected thyristors and their associated circuits, including optically-enabled pulse circuitry; and deionized water cooling. Two extra thyristors are included in the 56, to allow operation to continue if two thyristors fail. These 56 thyristor positions are group assembled in eight thyristor modules with seven thyristors in each module.

These components are among those located at the two end stations of the 2000 km bipolar line from Xiangjiaba to Shanghai as indicated in Fig. 11.1. The rectifier station, especially, needs filtering elements that are placed outside the valve hall, and are not shown here. The inverter station also needs transformers to adjust the AC output voltage to the needs of the users. The general configuration of the bipolar line is schematically illustrated in Fig. 11.9.

Increase of voltage and power capacity to 1.1 MV and 12 MW spanning 3000 km is represented in the new Changji–Guquan link in China,[6] that is shown as the longest dashed curve in Fig. 1.1. It is stated in the announcement that the thyristor valves are being provided by ABB, a major European supplier. The announcement by ABB[1] also

---

[6] http://new.abb.com/cables/references/changji-guquan-uhvdc-link

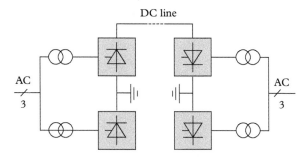

Fig. 11.9 *General configuration of bipolar DC transmission line, showing rectifier station on the left and inverter station on the right. The inverter electronics uses similarly the twelve-pulse bridge circuitry shown in Fig. 11.2 to re-establish 3-phase AC at the end of the long DC line, 2000 km in the case of the Xiangjiaba–Shanghai link[7]. Converter transformers are shown on the AC side of each station. Not shown are filtering elements that are needed especially on the AC input side of the rectifier station. In the event that one of the poles of the line fails, the ground return can still be used to get return DC for the remaining line*

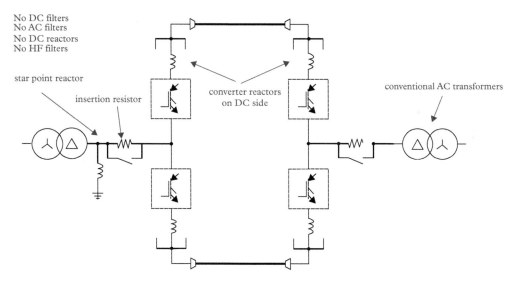

Fig. 11.10 *Modular system based on IGBT insulated gate bipolar transistor converters instead of thyristors. The converter transformers on the rectifier and inverter side are shown and are said to be simpler when the IGBT devices replace thyristors. Comment is made on reduction of the need for filtering, especially on the AC line leading into the rectifier (Friedrich, 2010)*

mentions the first 1.1 MV converter transformer, to be supplied for the new line. Because most generators at large hydropower dams have an output voltage between 6 kV and 19 kV at 50 or 60 Hz, a converter transformer is needed to increase the voltage for the HVDC line. Converter transformers are also needed on the consumer end of the line, past the inverter station, to adjust the AC voltage down to a suitable level.

---

[7] http://www.economist.com/news/science-and-technology/21714325-transmitting-power-over-thousands-kilometres-requires-new-electricity.

While thyristor systems as discussed are used in the largest present and planned installations, most of those shown in Fig. 11.1, systems based on IGBT-insulated gate bipolar transistor devices have important advantages in making more accurate AC–DC conversion and requiring less filtering. These aspects are reviewed by Chen et al (2015). It is suggested that these systems may be more widely used, as the IGBT and related semiconductor devices are improved. An example of an IGBT-based system marketed by ABB is suggested in Fig. 11.10.

11.1.3 Underground cables

The largest transmission lines are overhead lines, as suggested in Fig. 11.1. In developed countries it is much easier to use an underground cable, that offers less disruption. The cable is more viable for DC than AC, because the cable capacitance only has to be charged once, at system turn-on, rather than twice per cycle. Cables are of course needed to cross a body of water, as in the case of offshore wind turbine installations and famous inter-island connections. Chen et al, 2015 describe the types of cable in present use, with some emphasis on cross-linked polyethylene (XLPE) insulated extruded cables. A recent version of XLPE extruded cable, shown in Fig. 11.11a, is rated for 525 kV and 2.6 GW. It is said that a pair of these cables could provide power for a city the size of Paris. According to Gustafsson et al (2014), the Cu core conductor has an area $A = 3000 \text{ mm}^2$ and permits a power loss of less than 5 percent over a length $L = 1500$ km. Taking the resistivity ρ of copper as 1.68×10^{-8} Ω-m, the resistance R of the 1500 km cable, $R = \rho L/A$, comes to 8.4 Ω.

Suppose a purely DC bipolar circuit, with two of these cables, one at 525 kV and the return at –525 kV, with a resistive load R_L. If the DC power transmitted is $P = IV = 2.6$

Fig. 11.11a *Extruded HVDC cable rated for 525 kV and 2.3 GW, with XLPE (crosslinked polyethylene) insulation. The core conductor shown is Cu, the cable is also available with Al core conductor. The area of the central conductor is 3000 mm² corresponding to a diameter of 2.43 inches. Similar cables are widely used for offshore windfarms: shown is the largest available cable. http://www.abb.com/cawp/seitp 202/0cea859b5b6e1776c1257d3b002af564.aspx*

GW, with V = 1050 kV, then I = 2476 A, that is within the current rating of the 6-inch thyristors mentioned previously. (A recent statement mentions thyristors for UHVDC capable of 6250 Amperes.[8]) The equivalent load resistance is 424 Ω. The dissipated power in the two cables is $I^2R = (2476)^2 \times 16.8 = 0.103$ GW, that is 3.96 percent of the delivered power, in agreement with the manufacturer's claim.

The capacitance of the L = 1500 km cable can be estimated from the cylindrical capacitor formula,

$$C = 2\pi \, \kappa \, \varepsilon_0 \, L/\ln(b/a), \qquad (11.1)$$

where b and a, respectively, are the outer and inner radii of the cable conductors, and κ = 2.4 is the relative permittivity of the cross-linked polyethylene XLPE dielectric and b/a = 1.95 from the image in Fig. 11.11a. The result is C = 300 μF per cable. If the same cables were run at 60 Hz, the cable capacitance C would be charged twice per cycle, and the energy stored in each charged cable is ½ CV^2. The power involved in charging and discharging the cable capacitances would be 2 × 2 × ½ CV^2 × f, if the AC frequency is f. Using 60 Hz this power 2 × 4.96 GW, more than the rated power 2.6 GW. These cables are viable, then, at this length, for DC but not for AC transmission.

The underground cable is much less intrusive and easier to be permitted. Fig. 11.11b shows the comparison of the underground cable as discussed here with the HVDC conventional overhead lines, that are the type installed in China as shown in Fig. 11.1.

As suggested in Fig. 11.11b, underground cables require a much smaller right-of-way and can be installed with almost no visible impact. It is suggested in Fig. 11.11b that for 5 GW transmission a set of twelve cables would be needed, using a slightly smaller voltage rating than the one shown in the previous figure. These twelve cables are configured as six bi-poles, where only the center conductor carries current, while the outer conductor provides electrostatic shielding. There is no external electric field from the buried cable, but each bi-pole sets up an external magnetic dipole field. This magnetic field should not be a problem, and of course the alternative above-ground power line exhibits both electric and magnetic fields. The same authors, Thomas et al (2016), describe working superconducting cables that have been installed in a handful of locations. These cables are fully coaxial in that the outer conductor carries the reverse current. The superconducting cable is the least intrusive type, but is quite complex in structure and requires stations along the line to provide the needed flow of liquid nitrogen or cold gaseous helium. The extra cost of the superconducting cable is justified in reaching into urban regions where the right-of-way is difficult to obtain (Merschel et al, 2013).

A summary of recent installations of HVDC using the XLPE extruded cables is given in Table. 1, after Chen et al (2015).

---

[8] http://www.abb-conversations.com/2015/11/thyristors-the-heart-of-hvdc/

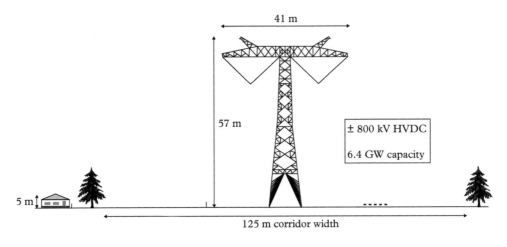

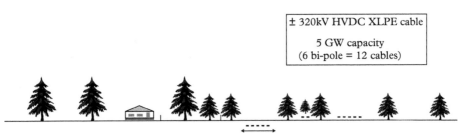

Fig. 11.11b *Comparison of the physical impact of the overhead HVDC line, with deployment of a comparable set of XLPE (cross-linked polyethylene) underground cables, one being imaged in Fig. 11.11a (Thomas et al 2016) Fig. 3*

Table 11.1 *HVDC projects with 320 KV extruded cables.*

Project Name	Commission Year	Length (km)	Voltage (kV)	Capacity (MW)
DolWin1	2012	165	320	800
SylWin1	2014	205	320	864
INELFE	2014	64	320	2000
				2 HVDCs
Xiamen	2015	10	320	1000

According to The Economist, as cited in connection with Fig. 11.1, in the US an UHVDC underground cable line will be built from Oklahoma, where there are wind-farms, 700 miles to the western tip of Tennessee, connecting to the TVA, Tennessee Valley Authority. The line will be at 600 kV and will transmit 4000 MW. It is likely that there will be more projects of this type in the US.

11.2   Large-scale energy storage

The primary forms of grid-scale storage are pumped hydro, that was discussed in Section 8.3, and compressed air energy storage.

These two forms of storage amount to 23 GW and about 0.43 GW, respectively, followed by Thermal energy storage about 0.43 GW and Batteries at about 0.3 GW. An overview of the many other forms of energy storage that are also used in connection with the power grid is shown in Fig. 11.12. This figure is oriented toward the use of storage as an augmentation of the existing grid, where the storage is filled during hours of small demand and is released during hours of peak demand. Thermal storage is not listed , even though according to Fig. 11.13 it amounts in the US to 0.43 GW. Thermal storage is fairly recent in its exploitation, and its use is tied quite closely to the renewable energy sources, solar thermal towers and wind turbines, that have wide fluctuations in basic generation that can be locally corrected with thermal storage. The time scales on which the corrections or smoothing to the grid power can be applied vary widely, from fractions of a day, as storing grid power not used at night and released the next mid-day, to very short times where fluctuations may arise from a cloud crossing the

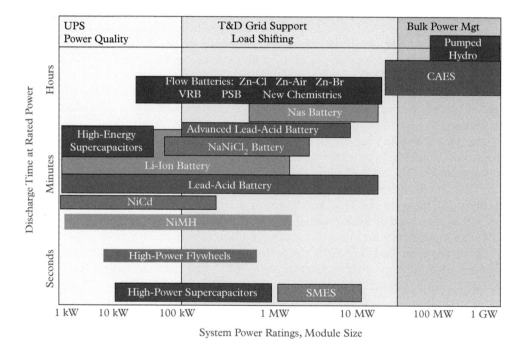

Fig. 11.12 *Qualitative overview of electrical storage technologies. In this figure the acronyms CAES and SMES respectively refer to compressed air energy storage and superconducting magnetic energy storage. Further, UPS means "uninterruptible power supply", VRB is Vanadium Redox Battery, PSB is Polysulfide Bromide Battery and T&D refers to transmission and distribution. Fig. 19, in "DOE/EPRI 2013 Electricity Storage Handbook" Sandia Report SAND 2013–5131, p. 29*

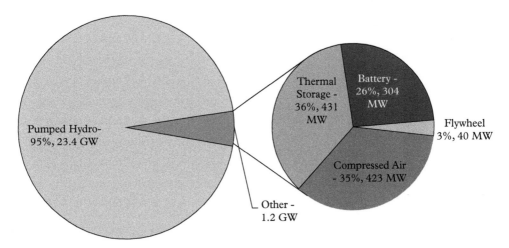

Fig. 11.13 *Rated Power of US Grid Storage projects. US Dept. of Energy Report "Grid Energy Storage" Dec. 2013. Fig. 1, p. 11, https://energy.gov/oe/downloads/grid-energy-storage-december-2013*

field of the solar tower or even shorter disruptions, are delineated in this figure, as are the power levels available. Obscure forms of storage, including superconducting magnetic energy storage (SMES), high-capacity supercapacitors and flywheel kinetic energy storage, have found application in niche cases because they can smooth out very short-term fluctuations that cannot be corrected by starting massive turbines, for example. (A non-grid application for supercapacitor storage is in short- duration storage of energy obtained from vehicle braking, to be reapplied to put the vehicle again in motion.[9])

Fig. 11.12 shows a wide variety of battery types, including flow batteries that are one approach to making larger-scale battery storage. On the other hand, large-scale battery storage can also be obtained simply by hooking together large numbers of small batteries. That would follow the pattern of the Tesla Model S, that uses 6831 small Li-ion batteries. Companies like Tesla and Samsung are also providing larger capacity Li-ion units that are intended for individual homes, to store solar cell output for use at night, or for grid correction for utility companies in place of pumped-hydro storage.

As an overview, the US in 2013 is reported to have 24.6 GW of grid-related electrical storage, with the great preponderance, 23.4 GW, being pumped hydro, described in Section 8.3. The remaining 1.2 GW include 431 MW thermal storage, 423 MW compressed air storage, 40 MW flywheel storage and 304 MW battery storage. Lists of energy storage installations are available.[10,11]

---

[9] https://www.technologyreview.com/s/415773/next-stop-ultracapacitor-buses/
[10] https://en.wikipedia.org/wiki/List_of_energy_storage_projects.
[11] http://www.energystorageexchange.org/projects/2079

11.2.1   Pumped hydro and compressed air

As discussed Section 8.3.2, pumped hydroelectric storage is limited to locations where water is available at two different heights, for example Lake Michigan and an artificial lake, some hundreds of feet higher in elevation, as shown in Fig. 8.20. An added example of a large pumped hydro installation is the Goldisthal Pumped Storage Station in Sonneberg, Thuringia, Germany.[12] This facility provides 1.08 GW for 8 hours and is the largest hydroelectric power plant in Germany. It has four pump turbines with 265 MW generating capacity, two of which can be run at variable speeds. The head, or height difference, is 301 m. The efficiency of this type of plant is in the order of 75 percent. It takes power from the grid at night when demand and prices are low and releases the power, with some loss, when the demand is high, and the power is much more valuable.

The essential idea of pumped hydro storage is that energy is stored, when grid prices are low, in the form of potential energy of elevated water, and that potential energy is turned into electricity when the need is great, and the prices are high. The efficiency, defined as power released divided by power consumed in pumping and is high, about 80 percent, from modest losses in the pumping and generating phases. The water twice goes through a Francis turbine that has an efficiency near 90 percent. An analog air-compression scheme would be to create potential energy in the form of a large volume of compressed air, and to release that potential energy later, through an expansion turbine when the need is great, and the price is high. To be adiabatic, energy conserving from the gas point of view, the temperature of the gas would greatly rise along with the pressure, and that hot gas would be released through the expansion turbine. In this hypothetical adiabatic form of compressed air-energy storage the efficiency again would limited by losses from the efficiency of the compressor, the efficiency of the expansion turbine and failure of the final exhaust gas to be again at ambient temperature. But Compressed air energy storage (CAES) in practice does not work this way. CAES in practice is much less efficient because the large amount of heat released in the compression is removed and lost, using standard practice of deploying intercoolers and after-coolers following the stages of compression. The storage caverns that are used will tolerate only air at ambient temperature. Thus, the high-pressure air has again to be heated, using natural gas. The fact that the air is already at high pressure, however, does mean that the compressor stage at the inlet of the conventional gas turbine is not needed. This means that the full power of the turbine can go to the generator, as the compressor in the standard gas turbine takes up about 2/3 of the turbine power. So, the same turbine, essentially, in the CAES will give three times the output.

CAES as practiced is based on the gas turbine as a generator of electricity, so this deserves some discussion. The conventional gas turbine[13] has three primary sections, the inlet and compressor, the combustion system and the expansion turbine, all on the same shaft that turns the generator.

---

[12] https://www.energystorageexchange.org/projects/399
[13] https://energy.gov/fe/how-gas-turbine-power-plants-work

The pressure and temperature at the inlet to the expansion turbine, where the fuel is injected, are in the ranges, respectively, more than 30 Bar and 1100–1200°C. The turbine itself is an array of stationary and rotating aerodynamically shaped blades.

In more detail, combustion (gas) turbines installed in many of today's natural-gas-fueled power plants are complex, and comprise three sections:

1. The compressor, that draws air into the engine, pressurizes it, and feeds it to the combustion chamber. The compression, to a first approximation, is an adiabatic, constant entropy, process and rapidly raises the temperature of the gas. This follows the law $T \propto 1/V^{\gamma-1}$, where $\gamma = c_p/c_v$ is 7/3 for the diatomic gases of air, as explained in Section 9.2. For a compression ratio of 3, hypothetically, this raises the temperature by a ratio $3^{1.33} = 4.31$, or to 1020°C starting from 300 K. This ideal temperature rise is reduced within the turbine because the released heat must raise the temperature of the compressor blades, that also must be able to withstand high temperature as well as large forces. It is estimated that the compression process in the turbine takes up about 2/3 of the power[14] that is generated by the gas-fed expansion turbine. So, running a gas turbine from a high-pressure air source will greatly increase its energy output, the work of compression already having been done.

2. The combustion system, is typically made up of a ring of fuel injectors that inject a steady stream of fuel into combustion chambers where it mixes with the air. The mixture is burned at temperatures of more than 1100°C. The combustion produces a high temperature, high pressure gas stream that enters and expands through the turbine section.

3. The turbine comprises an array of alternate stationary and rotating aerofoil-section blades. As hot combustion gas expands through the turbine, it spins the rotating blades. The rotating blades perform a dual function: they drive the compressor to draw more pressurized air into the combustion section, and they turn a generator on the same shaft to produce electricity.

The compression ratio is the ratio of the compressor discharge pressure and the inlet air pressure. A common form of gas turbine is derived from jet aircraft engines and operates at very high compression ratio (typically in excess of 30). One key to a turbine's fuel-to-power efficiency is the operating temperature. Higher temperatures generally mean higher efficiencies that lead to more economical operation. Gas flowing through a typical power plant turbine can be as hot as 1260°C, but some of the critical metals in the turbine can withstand temperatures only as high as 820–930°C. Therefore, air from the compressor might be used for cooling key turbine components, reducing ultimate thermal efficiency.

A simple cycle gas turbine can achieve energy conversion efficiencies ranging between 20 and 35 percent. By improved cooling technologies and advanced materials, it appears

---

[14] http://energystorage.org/compressed-air-energy-storage-caes

that higher turbine inlet temperatures may be possible, to as high as 1430°C—nearly 170°C higher than in previous turbines and achieve efficiencies as high as 60 percent.

Another way to boost efficiency is to install a recuperator or heat recovery steam generator (HRSG) to recover energy from the turbine's exhaust. A recuperator captures waste heat in the turbine exhaust system to preheat the compressor discharge air before it enters the combustion chamber. The HRSG generates steam by capturing heat from the turbine exhaust. High-pressure steam from the heat recovery steam boiler can be used to generate additional electric power with a steam turbine, a configuration called a combined cycle. In an advanced combined-cycle plant the overall energy cycle efficiency could approach 80 percent.

The CAES plant, as the Huntorf Germany plant shown in Fig. 11.14, is a modified gas turbine power plant where the power turbine does not need an initial compressor stage, since air at 40 Bar is available from the storage cavern. This means that the power turbine feeds only the generator, instead of giving most of its output power to the usually required compressor. The electrical power output available is about three times larger. Looking closely at Fig. 11.14, one sees the motor/generator at the center of the common rotating shaft, with clutches on its right and left.

With the right clutch engaged the exhaust turbine on the right drives the 290 MW generator. As shown, pressurized air at 42 Bar is fed into a combustion chamber at the inlet of the first expansion turbine, where the inlet temperature is 550°C. The exhaust gas from the first stage is reheated to 825°C and enters the second expansion turbine stage at 11 Bar and 825°C. The final exhaust gas in this installation is vented to the atmosphere, while in principle it could be used as part of the preheating of the air input to the first expansion turbine stage. While generating, the facility depletes the air pressure in the storage cavern from about 73 Bar to about 53 Bar over two hours, while a regulator is used to keep the inlet pressure to the first expansion turbine stage constant at 42 Bar.

Restoring the cavern pressure to 73 Bar is done with the left clutch engaged, taking 60 MW over 8 hours from the grid to run the central unit as a motor. Looking at the units on the left side of the diagram, one sees two compressors, each associated with a cooler, and connected by a gear reduction for the second stage compressor. As previously explained, adiabatic compression of air greatly increases its temperature and intercoolers and aftercoolers are standard engineering elements that reverse this heating, in the present case so that air entering the salt cavern is at about 50°C, comparable to the salt cavern wall temperature. The net grid energy contributed per day is seen to be $2 \times 290 - 8 \times 60 = 100$ MWh, of value ~ \$10,000, or \$3.65 M per year, taking a rate \$10 c/kWh. But the rate paid for the grid power for compression overnight is much less than the rate charged for the power generated at mid-day. The cost of the natural gas used, however, is an added major cost. This plant has been running since 1979, so it is no doubt profitable. The plant is run remotely and thus does not need on site personnel. Experts say that such plants are competitive with pumped-hydro power plants in terms of their grid applications, output, and storage capacity. However, the capital expenditure of CAES as represented in the Huntorf facility is considered less than that of pumped hydro. The turbines used in the CAES, not needing a compressor section, are adaptations of portions of conventional gas and steam turbines and are well within

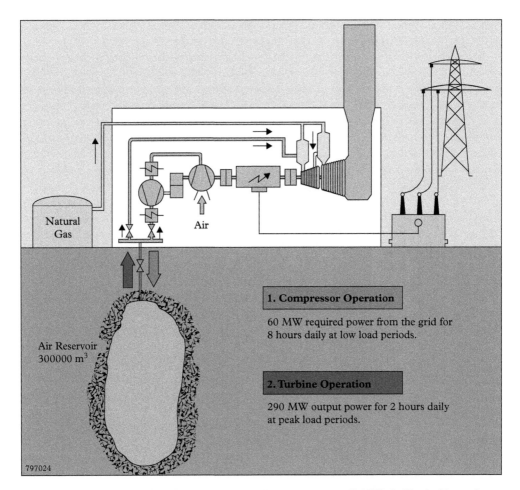

Fig. 11.14 *Diagram of cavern-based compressed air energy storage (CAES) facility in Huntorf, near Bremen, Germany. The expansion turbine runs for 2 peak hours producing 290 MW, using air pre-pressurized at up to 73 bars in the cavern. The cavern is formed by dissolving a volume of an underground salt dome. Natural gas is used to run the high and low-pressure expansion turbines, on the right, driving the generator at the center. Since the air is already pressurized, the gas turbine does not need a compressor and thus provides three times the output of a conventional gas turbine power plant. Off-peak grid power is used to compress the air in the dome*

the capability of large firms competent in these areas. The first improvement on the efficiency of this facility would be to take the exhaust from the final combustion turbine and use it to pre-heat the cool high-pressure air from the cavern before reaching the inlet of the first expansion turbine. This feature is incorporated in the McIntosh facility in Alabama,[15] that stores 110 MW for 26 hours. In this location a salt cavern of 220 ft

[15] http://www.energystorageexchange.org/projects/136

diameter, with volume 10 million cu ft, is pressurized to 1100 psi, or about 73 Bar. The released air feeds a 110 MW gas-fired combustion turbine.

This facility has a pre-heating feature, using waste heat from the turbine to heat incoming air from the cavern. This project is used for "peak shaving", providing extra capacity during the peak demand portion of the day.

A second major improvement would be to make use of the heat released in the compression phase of the operation, that could perhaps remove the need to burn natural gas to reheat the pressurized air at the inlet to the first expansion turbine. This would make the whole process adiabatic or energy conserving.

The large use of natural gas complicates the question of efficiency. Elmegaard and Brix (2011)[16] define the "electrical storage efficiency" as the product of the efficiency of compression, $\Delta E_1 / W_c$; the efficiency of storage, E_3 / E_2; and the turbine efficiency, $W_t / (\Delta E_{34} + E_f)$. The E terms are exergy, or available energy; W is work. The exergy input to the turbine in the last definition is the chemical exergy of the fuel and the exergy from the storage. The electrical storage efficiency of the first two types of facilities is thus estimated as in the range 25–40 percent, while a possible value 70–80 percent is estimated (Elmegaard and Brix, 2011) for the fully adiabatic form, including the recuperator and recovery of heat from the compressors. The third, adiabatic, type of facility has seen some planning but no construction (Zunft, 2015). This planned facility clearly would be more expensive, the author noting that only in 2010 did the projected cost fall to the level of pumped hydro.

Elmegaard and Brix give electrical storage efficiency values 29 percent and 36 percent, respectively, for the operating Huntorf and McIntosh, Alabama, facilities. The Zunft paper[17] finds the "round trip thermal efficiency" for the Huntorf and Alabama plants, respectively, as 42 percent and 52 percent, predicting the fully adiabatic plant in planning to reach "thermal efficiency", 66–70 percent.

Details of an advanced CAES system are available for a proposed, but not utilized, site near Seneca Falls,[18] NY.

Compressed air storage facilities of large size depend on a large cavity that can be filled with air. Geology provides opportunities to make such caverns, the most favorable case being the salt dome. If the cavern is produced by "solution mining", the dome the cost is estimated as $1/kWh, while if a porous rock formation or aquifer can be used for compressed air storage the cost might be as low as $0.10/kWh (Carnegie et al, 2013). Salt domes are present in large portions of the US, for example in a large area in Alabama that includes the McIntosh facility. Salt is mined in upstate New York near Geneseo and also near Seneca Falls, mentioned previously.

A planned CAES facility using an existing cavern is the Adanced Underground CAES Project with Saline Porous Rock Formation, in Kern County, CA, that is rated as 300 MW for 10 hours. The storage cavity is also described as a depleted gas reservoir.

---

[16] http://orbit.dtu.dk/files/6324034/prod21323243995265.ecos2011_paper%5B1%5D.pdf
[17] http://www.sccer-hae.ch/resources/SymposiumMay2015/Talks/SCCER2015-AdiabaticCAES-Zunft.pdf
[18] https://www.smartgrid.gov/document/seneca_compressed_air_energy_storage_caes_project

The possibility of smaller-scale above-ground compressed air storage is explored[19] (http://www.publicpower.org/files/NationalConference/Kou.pdf) in a report from the New York Power Authority. The proposal is for about 2 miles of 3-foot diameter steel pipe to form the storage chamber, a volume of about 75,000 cu ft. The proposed facility is rated as 40.5 MWh, producing 9 MW for 4.5 hours. The study concluded that such plants are likely to be less expensive and have longer life than an alternative battery plant. The same study also endorsed the large underground cavern CAES facilities as less expensive than pumped hydro facilities for large scale energy storage.

On the other hand, the CAES is an adjunct to the fossil fuel natural gas and does not seem to have a strong role in storing electrical energy in a renewable energy system based on wind turbines, photovoltaics and solar thermal facilities. The gas turbines and compressors of CAES are inherently less efficient than the water Francis turbine used in pumped hydroelectric storage. The main task of filling in extra power capacity in mid-day is admirably fulfilled with the solar facilities, such as the one shown in Fig. 10.15.

11.2.2   Flywheel kinetic and other storage

The storage considered here is not really mass storage, but an accessory to management of the power grid. It was mentioned in connection with the hydroelectric Francis Turbine generator that small adjustments in the water flow to the turbine were made by "wickets" (see before Fig. 8.18) to keep the turbine and generator rotating very close to the desired 60 Hz. The flywheel kinetic energy storage is used in a similar fashion, to rapidly compensate for changes in usage or generation that would push the generator away from 60 Hz. A large facility of this type at Beacon NY has 200 flywheels and is rated 20 MW power for 15 minutes, or 5 MWh storage capacity.[1]

The kinetic energy of a flywheel can be expressed as $\frac{1}{2} I\omega^2$ where I is the moment of inertia. A uniform cylinder of radius r and mass m has moment $I = 1/2\ mr^2$ about the cylinder axis. If the flywheel is mostly mass at radius r, ie, a cylindrical hoop, then the moment is just $I = mr^2$. The performance of flywheels is enhanced by using carbon fiber composite materials (Bender, 2015).

A schematic diagram of such a flywheel, that might be used to stabilize a power grid, is shown in Fig. 11.15. A single state of the art flywheel[1] can be rated as storing 25 kWh or 90 MJ. The well-known facility at Beacon, NY is described as having rotation speed 16,000 RPM or $\omega = 1675.5$ radians/s. The individual units are rated at 100 kW for 15 minutes. Using the previous formula, the moment of inertia can be deduced from $\frac{1}{2} I \omega^2 = 90$ MJ as 64.1 kg m². The rotor in these installations includes the motor/generator unit, as shown in Fig. 11.15. The flywheel to provide the rated 25 kWh of kinetic energy is large, it is a thick-walled hollow cylinder of length about 2 m and outer radius about 0.8 m with wall thickness around 0.3 m, roughly estimated from photographs in

---

[19] www.sandia.gov/ess/docs/pr_conferences/2010/arseneaux.pdf

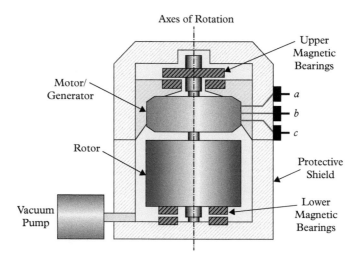

Fig. 11.15 *Diagram of flywheel kinetic energy storage device. The rotor speed and moment of inertia are maximized by using materials of great strength: carbon fiber composites. A vacuum enclosure and magnetic bearings greatly reduce friction, to extend the time of storage. The device allows rapid energy access to counter external loss in the grid. A single unit of this type can be rated at 100 kW power output and storage capacity 25 kWh. The application of a plant containing 200 such units is to reduce frequency swings caused by short term fluctuations in load or generating capacity. http://cdn.intechopen.com/pdfs-wm/6820.pdf*

the first report. Taking the mass density of carbon fiber as 1600 kg/m^3, gives an estimated mass as 3920 kg and moment of inertia I about 1660 kg m^2. The dimensions may be a bit large, but it is clear that at 270 Hz, the highest rotational frequency, the 90 MJ can be stored.[2] The decay time may be as low as 2 percent per day, with the magnetic bearing levitating the rotor and vacuum pumping removing air friction. There is no question that these are useful fast-acting storage units that can allow the grid to more easily accommodate fluctuating power from solar arrays and wind turbines. Smaller flywheel storage systems in the range 3 kW–15 kWh were offered by Velkress.[20,21]

A second type of rapid access storage is superconducting magnetic energy storage, SMES. The energy density in the magnetic field is $B^2/2\mu_0$, see (6.26). Assuming a large toroidal chamber of 500 m^3 at 5 Tesla, roughly as assumed for ITER, the stored energy is 4.97×10^9 J, about 55 times that of the single rotor mentioned previously. The corresponding magnetic pressure on the wall, that again equals $B^2/2\mu_0$ is 100 Bar. A discussion of possible applications of SMES for grid stabilization is given by Jin and Chen (2012). Both open solenoids and toroids are discussed and all are small in size compared to the ITER toroid, that was discussed following Fig. 6.7.

[20] https://web.archive.org/web/20100331042630/http://www.beaconpower.com/files/Flywheel_FR-Fact-Sheet.pdf
[21] http://www.velkess.com/velkess-now-taking-orders-for-revolutionary-energy-storage-products/.

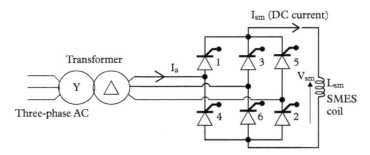

Fig. 11.16 *Diagram of superconducting magnetic storage system SMES used to smooth output from 6-pulse rectifier, used in AC/DC conversion. (Ali et al 2010) Fig. 1*

The company American Superonductor markets a trailer-based SMES system capable basically of providing 3 MW for 1 second. According to the SUFG report (Carnegie et al, 2013), pp. 59–60, three units of this 3 MW system and a dozen of a smaller version have been in use since 1992, for power conditioning applications.

Ali et al 2010, see Fig. 11.16 give an analysis of SMES applications in power conditioning, including its use to smooth output from the 6-pulse rectifier of the sort used in converting AC power from a hydroelectric generator to a DC transmission line, as discussed in Section 11.1.2.

11.2.3   Thermal energy storage, including solar salt

Storage of energy has been demonstrated in the concentrating solar troughs and solar towers, as described earlier, where oil and molten salts have been used as a heat transfer fluid (see Fig. 9.6). The largest facility with storage is the Crescent Dunes solar tower near Tonopah, NV, with capacity of 110 MW and a molten salt storage time of 10 hours. Other heat transfer fluids include water and propylene glycol. Common materials can also be used for the storage medium, including rocks and concrete. In Florida, a common sight is the solar water heater on the roof, that produces hot water stored in a tank inside the house, through a heat exchanger. The fluid in the roof panels may be an antifreeze such as propylene glycol. Planned communities exist in isolated locations where hot water storage supplies a larger group of homes. A large amount of distributed heat storage capacity is in place in individual homes, as for example in Israel, where building codes require storage, considerably reducing the demands on the electrical grids that otherwise would provide the energy for heating. This distributed storage is not included in Fig. 11.13, for example.

The firm Siemens is contracted for a 5 MW thermal storage facility[22] in the Bergdorf borough of Hamburg, Germany. The storage of heat, rated at 36 MWh, is in a constructed underground insulated container, with rocks such as basalt as the storage

---

[22] http://www.utilitydive.com/news/siemens-project-to-test-heated-rocks-for-large-scale-low-cost-thermal-ener/428055/

medium. The facility includes a 1.5 MW steam turbine and generator, that can run for 24 hours on the stored heat. The facility will be completed in 2018.

Concrete can be used as a heat storage medium, that has a lower heat capacity than water but will allow a larger temperature range than water. There are many small-scale installations of this type; individual houses can be built incorporating thermal storage of various types.[23]

A common small scale thermal storage method is the use of ice formed at night when electricity is at a low price, to provide air conditioning in the peak of day. The heat of fusion of water to form ice is 334 MJ = 93 kWh per kg. An example of such a system is the JC Penney Headquarters[24] in Plano, TX, where the storage is rated as 53.1 MWh, providing power of 4.425 MW for 12 hours. The ice is made at night and used during the day to provide air conditioning.

11.2.4 Li-ion, nickel metal hydride and other batteries

Beyond pumped hydroelectric and thermal storage, a further approach to electrical storage is in large scale batteries (Cardwell and Krauss, 2017).[25]

While batteries are more expensive than pumped hydro and have smaller energy capacity (units MWh), they are portable and can be deployed almost anywhere. While most batteries are small for computers and for electric cars, packs of small batteries, and also some large-scale batteries, are being used for grid scale electric storage.

In the emergency need for storage in Escondido, CA, the answer for San Diego Gas & Electric is large packs of lithium batteries, in fact 19,000 drawer-size packs of these batteries, as shown in the article with the name Samsung visible in the photograph. Other manufacturers of such packs are Tesla, LG Chem and Panasonic.

Sodium-sulfur batteries, on the other hand, are available in large single units, and operate at high temperature. A large-scale battery project undertaken by a US utility[1] entails "6 MW" of capacity at a total cost of $27 M, thus a capital cost of $4.5/Wp or $0.64/Wh. The basic units, rated at "1 MW" are about the size of a double-decker bus and use sodium-sulfur chemistry and operate at about 800°F. The battery is said to deliver 1 MW for about 7 hours, for a conventional rating of 7 MWh per battery. The batteries are said to be 80 percent efficient and are used to smooth power from wind turbines and make that power easier to add to a local electrical grid. The batteries[26] will "charge at night, when the wind is strong, but prices are low, and give the electricity back the next afternoon, when there is hardly any wind, but power prices are many times higher". A Sodium-sulfur NaS battery of 4 MWp, total 32 MWh capacity and cost $25 M is reported[3] in Presidio, Texas.

[23] https://energy.gov/energysaver/active-solar-heating
[24] https://www.energystorageexchange.org/projects/161
[25] https://www.nytimes.com/2017/01/14/business/energy-environment/california-big-batteries-as-power-plants.html
[26] http://inhabitat.com/bob-americas-biggest-sodium-sulfur-battery-powers-a-texas-town/

That storage cost is then \$0.78/Wh. It appears that these costs are higher than necessary, compared to \$400/kWh = \$0.4/Wh given by US Department of Energy for NaS batteries.[27] The same article estimates the cost of pumped hydroelectric power as 0.1\$/Wh.

Rechargeable batteries reversibly interchange chemical and electric energy. These are categorized by open-circuit voltage (V), and electrical energy (W–h/kg) or (W–h/l). The equivalent circuit is an electromotive force voltage, EMF, arising in the underlying chemical reaction, with a series resistance.

A battery consists of an array of individual cells that are connected in series to obtain a desired total voltage and in parallel to reach a necessary current capacity. Often, the peak rate of discharge and maximum power output are important, see Fig. 11.17.

The traditional lead-acid automobile battery consists of dense and heavy materials, lead and sulfuric acid, and is used primarily in high current applications such as starting conventional automobile engines.

Lead-acid batteries have also been used to stabilize wind turbine installations, for example one at the Tappi Wind Park in Japan, shown in Fig. 11.18, where the advanced

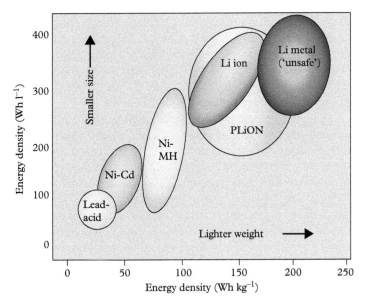

Fig. 11.17 *Comparison of energy densities for several types of batteries. Units are Watt-hours per liter (ordinate) and Watt-hours per kg (abscissa). Here the label "PLiON" refers to a class of rechargeable plastic Li ion batteries (Tarascon and Armand, 2001)*

[27] Jenny Mandel, "DOE promotes pumped hydro as option for renewable power storage", *New York Times* 15 Oct., 2010.

Fig. 11.18 *Advanced lead-acid batteries used to stabilize wind turbine installation at the Tappi Wind Park in Japan. These batteries were made by Hitachi. "DOE/EPRI 2013 Electricity Storage Handbook" Sandia Report SAND 2013–5131, Fig. 71 p. 79*

lead-acid batteries were made by Hitachi. It is reported[28] that Hitachi has provided a 10.4 MWh battery also to stabilize a 15 MW wind power installation.

Energy density is at a premium in modern applications, e.g., in cellular telephones, and also in larger-scale applications such as hybrid and electric automobiles. Li-ion batteries are used in consumer electronics, notably cellular telephones and laptop computers, because of their high energy density, in the range 160 Wh/kg and 350 Wh/l. The Li-ion battery has the largest portion of the portable battery market, about 63 percent, mostly in electronics, but not yet in most hybrid cars, but are used in the fully electric Tesla car. Ni-Cd batteries are used in power tools.

The Prius hybrid car uses NiMH batteries, presently considered a more conservative choice than the Li-ion battery, as shown in Fig. 11.19, that may present a fire hazard. Nonetheless, an excellent automobile, the Tesla Model S, is purely electric and powered by 6,831 Li-ion cells of a type that are used in laptop computers. This vehicle accelerates to 60 mph in less than 4 seconds and has a range of 210 miles.[29]

---

[28] "DOE/EPRI 2013 Electricity Storage Handbook" Sandia Report SAND2013-5131, p. 78.
[29] "The end of the petrolhead: Tomorrow's cars may just plug in", *The Economist*, 28 June 2008.

11.2.4.1 Basics of Lithium Batteries

A diagram of a lithium ion battery is shown in Fig. 11.19.

Variations on the basic Lithium battery shown in Fig 11.19 include a cathode of iron phosphate $LiFePO_4$[a] (manufactured by A123 Systems and possibly an anode consisting of Lithium titanate nano-particles (Altairnano, Inc.) that will not burn.

It appears[30] that the electrical conductivity of the iron phosphate $LiFePO_4$ cathode (LFP), that had been low, was increased by doping with metals such as aluminum, niobium and zirconium, but also probably involving nanoscopic carbon particles. These advances now make the iron phosphate cathode workable, to allow a higher discharge current, fast charging time, and stability under extreme conditions. The basic advantages of LFP cathodes over the cobalt cathodes are low cost, high abundance of iron, and freedom from overheating. Another advantage is that in the chemical reaction in the LFP case, relative to that shown Eq. 11.2 the value of x goes to zero, leaving no residual Li in the cathode when fully charged.

The chemical reaction for one form of Li-ion cell is:

$$Li_{1-x}CoO_2 + Li_xC_6 = C_6 + LiCoO_2 \qquad (11.2)$$

where the carbon is a graphite electrode. Ions, not electrons, are the current carriers. Note that lithium ions are not oxidized. In a lithium-ion battery the positive lithium-ions flow internally from the graphite anode to the cathode, with the transition metal, Cobalt, in $Li_{1-x}CoO_2$ being reduced from Co^{4+} to Co^{3+} during discharge. The performance may be as high as 160 W-h/kg.

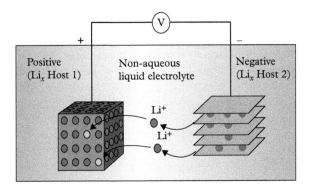

Fig. 11.19 *Operating principle of rechargeable Li ion battery. Li ions diffuse from high energy sites in the graphite anode to low energy sites in the cathode, driving charge around the external circuit. Cells using $Li_{1-x}CoO_2$ (left, in Figure) and Li_x intercalated into graphite (on right) provide 3.6V, energy densities 120–150W h/kg are widely used in portable electronic devices. Li ions reversibly enter (intercalate) and leave weakly bound positions between the graphene carbon layers of graphite (Tarascon and Armand, 2001)*

[30] N. Ravet, A. Abouimrane, and M. Armand, (2003). *Nat. Mater.* 2, 702. See also J. Gorman, (2002). "New material charges up lithium-ion battery work-Bigger, cheaper, safer batteries", *Science News* 28 Sept. 2002.

The performance of Li-ion batteries can be improved by nanostructuring the electrodes, to enlarge the effective surface area. For example, it is stated that both higher power and higher storage capacity can be obtained by applying the active anode and cathode materials in a thin film to copper nano-rods anchored to sheets of copper foil. Enlarged electrode surface area by nanostructuring can increase the short-circuit current of a battery, assuming good electrical conduction to the nanostructured areas.

11.2.4.2 NiMH (nickel metal hydride) batteries

A nickel metal hydride battery, abbreviated NiMH, is a type of rechargeable battery, with 1.2-volt nominal voltage, similar to a nickel-cadmium battery, but has an anode of hydrogen-absorbing alloy, instead of cadmium. The anode reaction occurring in a NiMH battery is:

$$H_2O + M + e^- \leftrightarrow OH^- + MH. \tag{11.3}$$

The battery is charged in the forward direction of this equation and discharged in the reverse direction. Nickel (II) hydroxide forms the cathode. The metal M in the anode of a NiMH battery is typically an intermetallic compound. Several different intermetallic compounds have been developed, which mainly fall into two classes. The most common form of anode metal M is AB_5, where A is a rare earth mixture of lanthanum, cerium, neodymium, and praseodymium and B is nickel, cobalt, manganese, and/or aluminum. Other batteries use higher-capacity negative electrodes based on AB_2 compounds, where A is titanium and/or vanadium and B is zirconium or nickel, modified with chromium, cobalt, iron, and/or manganese. Any of these compounds M serves the same purpose, reversibly forming a mixture of metal hydride compounds. When hydrogen ions are forced out of the potassium hydroxide electrolyte solution by the charging voltage, it is essential that hydride formation is more favorable than forming a gas, allowing a low pressure and volume to be maintained. As the battery is discharged, these same ions are released to participate in the reverse reaction.

NiMH batteries have an alkaline electrolyte, usually potassium hydroxide. A NiMH battery can have two to three times the capacity of an equivalent size NiCd battery. However, compared to the lithium-ion battery, the volumetric energy density is lower. The specific energy density for the NiMH battery is approximately 70 W-h/kg. It is reported that a NiMH Prius car battery contains 12 kg of Lanthanum. An advantage of the NiCd battery over the NiMH is typically higher short circuit current.

Batteries used in fully electric vehicles EV (Nissan Leaf) and plug-in electric vehicles (PHEV) (Chevy Volt) are in the range 20 kWh to 50 kWh in capacity. The batteries in hybrid electric vehicles HEV like the Toyota Prius are generally in the range of 2 kWh.

The broader question of energy storage in the power grid can be discussed in terms of many small batteries. The total sales of the Toyota Prius are about 1 million, half in the US. The storage capacity of 0.5 million Prius cars, assuming 2 kWh per car, is seen to be 2 GWh. Comparing to Fig. 11.20 for the State of California, the peak power is 50 GW, it is seen that the application of all the Prius batteries in the US. to the grid in California at this peak power would not be appreciable, it would meet demand for only 2/50 h = 2.4 minutes.

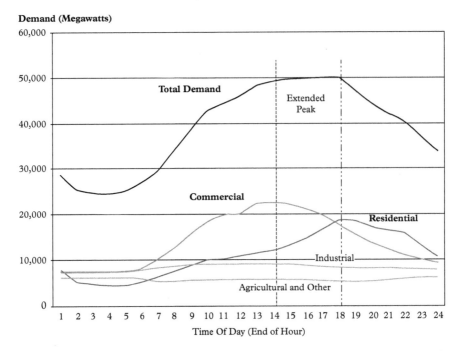

Fig. 11.20 *The total load profile[31] for the State of California US on a hot day in 1999. The demand during the day is about twice the demand at night*

If the cars were of the EV or PHEV category, assuming 50 kWh battery capacity, and if the number of cars is 0.8 of the population of California (37 Million), thus 29.6 Million, then the storage capacity would be 29.6 × 10^{6-} × 50 kWh = 1480 GWh which is enough to supply peak demand in Fig. 11.20 for about 3 hours.

One estimate of the cost of this storage capacity could be based on $30,000 per PHEV car, that would give $0.89 trillion.

The cost of 1480 GWh of sodium-sulfur battery storage at $0.4/Wh is $0.592 trillion, and would be $0.148 trillion if provided by pumped hydro storage. These numbers exceed the State of California budget, on the order of $100 M. It appears that these high storage costs exceed the cost of generating capacity. For the case of California at peak demand in Fig. 11.20 as 50 GW, the nominal cost of that electric generating capacity at $1/Wp would be $50 billion. Reducing the cost of battery storage is a goal of ongoing research and development in the US Deptartment of Energy and in the auto industry worldwide.

Larger-Li-based batteries have been described for buses and vans.[32]

---

[31] http://Www.mpoweruk.com/electricity_demand.htm (Data from Lawrence Berkeley National Laboratory, U.S. Dept. of Energy.)

[32] Jim Witkin, (2011). "With battery from Texas, E.V. bus begins its rounds at Heathrow". *New York Times* 23 Sept. 2011.

The batteries described are 85 kWh, and are based on lithium-iron phosphate technology.

11.2.4.3 *The vanadium redox flow battery*

An example of a flow battery, adaptable to larger size than Li-ion batteries, is the vanadium redox battery. The principle of this device is shown in Fig. 11.21.

A utility application of the vanadium redox VRB-flow battery is in Oxnard CA where Prudent Energy has installed 200 kW modules for a total of 600 kW capacity providing 3600 kWh of electrochemical storage.

A possible approach to the electric energy storage might be "one battery per person". The average power per person in the US is about 1.48 kW. (This figure includes industrial and other electricity uses and is higher than the needs of the typical individual.) So, an individual owning one of the 85 kWh batteries would have a buffer well- exceeding the nominal 85 kWh/1.48 kW = 57.4 h = 2.4 days in case of a power outage. (This

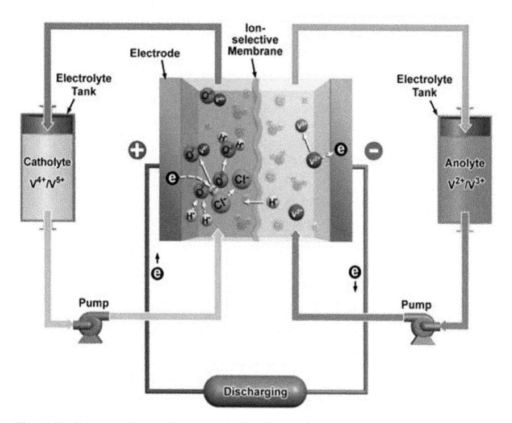

Fig. 11.21 *Schematic diagram illustrating the Vanadium Redox Battery. Adjusting the size of the storage tanks allows increased electrical storage capacity. "DOE/EPRI 2013 Electricity Storage Handbook" Sandia Report SAND 2013–5131, Fig. 48, p. 55*

Table 11.2 *Grid Storage Technology Cost Estimates, 2011 and 2020.*

Technology	2011 Range		2020 Range	
	$/kWh	$/kW	$/kWh	$/kW
Na-S	257–491		181–331	
Li-ion	850–1000		290–700	
Pumped Hydro	10	1500–2300	10	1640–2440
Compressed air	3	850–1140	3	500–1140
Flywheel	148	965–1590	81–148	200–820
Redox flow battery	173–257	942–1280	88–173	608–942

From Viswanthan et al, (2013) (Table E.1, p. iv).

battery would not replace external facilities, that contribute to the high average usage, 1.48 kW, that might be shut down.) In urban areas it is common practice for each apartment building to have a water storage tank on its roof. Adding a set of batteries to the roof would not be so different. If the price per Watt–hour is $0.40/Wh, then the 85-kWh battery cost would be $34,000. This is expensive, and it might be more economical to buy an auxiliary generator. An apartment building with 100 persons might need 30 kW, that can be purchased for about $8000.

The lowest-cost form of energy storage does seem to be pumped hydro power. However, pumped hydro is a large-scale grid solution, limited in its availability, and will not help the off-grid individual or small wind or solar producer. The modern form of flywheel kinetic energy storage is available at any location and is modular, allowing it to be scaled to any size above 100 kW or so.

A report comparing the costs of grid storage methods is available[33] from (Viswanathan et al, 2013). Table 11.2 predicts a reduction in cost per W for flywheel storage to a range between 0.2 to 0.82 $/W by 2020.

11.3 Hydrogen for energy storage and fuel cells

Energy from wind and solar can also be stored and distributed by converting it to molecular hydrogen gas. This is most conveniently done by electrolyzing water. The gas can be distributed in pressurized containers or pipelines. The energy can be reclaimed by fuel cells or even by burning H_2 in a nearly conventional automobile engine. Hydrogen is also a marketed commodity with many uses including making ammonia and processing heavy hydrocarbons to make kerosene or gasoline.

---

[33] http://energyenvironment.pnnl.gov/pdf/National_Assessment_Storage_PHASE_II_vol_2_final.pdf.

11.3.1   Efficient catalytic dissociation of water, electrolyzers

Inert electrodes such as stainless steel or platinum, immersed in water (containing a small addition of ions to promote conductivity) will evolve hydrogen gas at the cathode and oxygen gas at the anode if ~ 1.9 volts is applied. The chemical potential energy associated with a molecule of hydrogen is 1.23 eV, so the efficiency of the electrolyzer is stated as 65 percent. According to (Turner, 2004) the efficiency of commercial electrolyzers is in the range 60 percent to 73 percent. Actually, only about 4 percent of commercial hydrogen production is by electrolysis, and about half of that is by electrolysis of brine (NaCl plus water) with chlorine gas the primary desired product, the hydrogen sometimes being abandoned. The primary source of hydrogen presently is natural gas, methane, followed by a high temperature process using steam. This is influenced by the presently low price of natural gas.

One form of electrolyzer is based on the proton exchange membrane (PEM), typically Nafion-PFSA (perfluorosulfonic acid), mass-produced by DuPont, in a configuration similar to that of Fig. 11.27, operated in an inverse fashion. That is, water is fed in and hydrogen and oxygen gases exit, given a supplied external voltage, V. Platinum catalysts are needed in this device. A commercial device of this sort is the Hogen Series C[34]. This device produces pure hydrogen at high pressure. The HP 40, a single home-size unit, produces 2.28 kg of hydrogen at 2400 psi (160 Bar) in 24 hours, using deionized water, air and electricity as inputs. Much larger units, typically using an alkaline electrolyzer bath, are needed for hydrogen filling stations for fuel cell cars. An example of a large unit is the Norsk Atmospheric Type No. 5040, rated at 1046 kg per day (Ivy, 2004).

If such an electrolyzer is connected to a photovoltaic array (PV) of efficiency 12 percent, connecting enough PV cells in series to achieve a working voltage greater than 1.9 volts, then the overall conversion efficiency: sunlight energy into chemical energy in the form of hydrogen, is 0.65 × 0.12, that gives 7.8 percent. This is an off-the-shelf approach with commercial products available at present. A tandem solar cell is needed to reach the minimum 1.9 volts, that is available, for example, in amorphous silicon tandem cells, providing 2.2 Volts. Of course, a series connection of conventional solar cells can easily provide any requisite voltage. We will see an example of a monolithic PV-hydrogen electrolytic converter with efficiency 12.4 percent. For comparison, a high temperature "solar-thermal" water-splitting process is predicted (Fletcher 1977, Weimer 2005) to provide 24 percent efficiency in hydrogen production, but requires a large facility.

11.3.1.1   *Photo-catalytic dissociation of water into hydrogen and oxygen*

One element in this category is a cell for splitting water to produce hydrogen with sunlight, called photo-catalytic splitting of water. Direct photo-catalytic water splitting is

---

[34] http://pdf.directindustry.com/pdf/proton-energy-systems/hogen-hp-high-pressure-electrolyzer/22808-19084.html

not easy to accomplish. Efficiency is needed in the several steps: absorption of photons to create electron hole pairs, in their transport to the water interface, in dissociating the water, and in separating and collecting the hydrogen. The large area needed for a useful amount of hydrogen production, stemming from the low-energy density in sunlight is a difficulty, which could be addressed with focusing mirrors or Fresnel lenses. We consider a device in Fig. 11.22 (Khaselev and Turner, 1998) based on a tandem solar cell.

This device can be regarded as a water electrolysis cell in series with a tandem solar cell, to supply the needed voltage to split water. Crudely, one can think of the semiconductor layers as a (light activated) battery, the right-hand cathode terminal is in the electrolyte and the left-hand terminal ("ohmic contact") is insulated from the electrolyte, but connected to the platinum (foil or gauze) anode through the external ammeter. The electrolyte in the cell is 3 molar H_2SO_4, a strongly acidic solution conductive by H^+ ions, also known as solvated protons or "hydronium ions" H_3O^+.

Consider first the electrolytic cell aspects. The cathode was coated with a thin layer of platinum particles, in the nature of "platinum black", by an electroplating procedure. The cathode reaction (reduction) is:

$$2H^+ (aq.) + 2e^- \rightarrow H_2 (gas). \tag{11.4}$$

The anode reaction (oxidation) is:

$$2H_2O(liq) \rightarrow O_2 (gas) + 4H^+ (aq.) + 4e^- \tag{11.5}$$

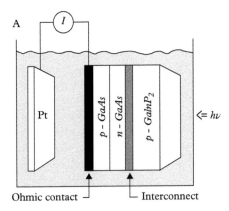

Fig. 11.22 *Schematic diagram of illuminated monolithic photovoltaic-photo-electrochemical device for hydrogen production by water splitting, with 12.4 percent efficiency in converting solar light energy into chemical energy. In this cell, with no external connection, hydrogen bubbles appear on the illuminated surface at the right (cathode), and oxygen bubbles appear at the left (platinum anode). Mass spectroscopy revealed the evolved gases to be pure hydrogen and oxygen in ratio 2 to 1. In this device, electrons flow to the illuminated contact where they act to release hydrogen gas. This cell is built on the GaAs multijunction tandem solar cell technology as described in Chapter 10 (Khaselev and Turner, 1998) Fig. 1A*

Taking twice the first reaction and adding to the second reaction, we get:

$$2H_2O(liq) \rightarrow O_2(gas) + 2H_2(gas). \tag{11.6}$$

The standard potential for this reaction is 1.23 Volts. To make the reaction work, a larger voltage is needed, the excesses being termed cathodic and anodic overvoltages.

Solar cell output voltages are limited by the bandgaps of the underlying semiconductors, that are 1.83 eV on the right, and 1.42 eV on the left, in Fig. 11.22. The sum of these voltages provides an upper limit of voltage from the tandem cell, that well exceeds the voltage, nominally 1.23 V, to decompose water.

The photo-current flows along the wire at the top of the diagram (an ammeter was shown in the previous diagram) and the Fermi levels of the ohmic contact and the platinum electrode are aligned. So, the voltage developed in the semiconductor layers is dropped across the electrolyte, driving the decomposition of water. In the two semiconductor layers, it is seen that the Fermi level is flat (no voltage drop) across the

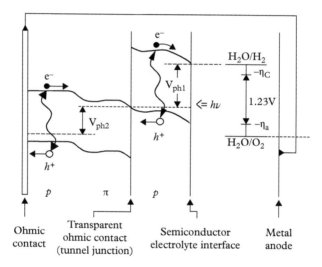

Fig. 11.23 *Schematic band diagram (electron energy increasing vertically) of illuminated monolithic photovoltaic-photo-electrochemical device for hydrogen production by water splitting, with 12.4 percent efficiency. (This is the same device as shown in Fig. 11.22, but the platinum anode has been moved to the right side of the picture.) In this device, electrons flow to the illuminated semiconductor-electrolyte interface, where they act to release hydrogen gas. Energy levels for water reduction (release of hydrogen) and water oxidation (release of oxygen) are sketched at right side of diagram. An alignment is needed between conduction band edge and water reduction level, at the semiconductor interface. Alignment is also needed between the lower water oxidation energy and a level providing holes from the anode. So, on the right, the metal Fermi level (dashed line on right) has to be pushed down by 1.23 eV + 2 η to allow a hole to be injected into the water (i.e., platinum anode accepts an electron from the water) and thus to release oxygen (Khaselev and Turner, 1998) Fig. 1B*

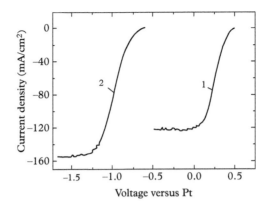

Fig. 11.24 *Current voltage diagrams of tandem cell (right, curve 1) and single right-hand cell (p-GaInP$_2$ cell) under white light illumination. A voltage source has been inserted in the upper connecting wire as shown in Fig. 11.23, to obtain curve 1. The voltage source has been inserted between the platinum electrode and the inner Ohmic junction (labeled "transparent ohmic contact") to obtain curve 2. Note that the zero of current is at the top of the figure. At zero voltage a current of 120 mA flows through the tandem cell, but to get a current from the (illuminated) single cell (curve 2) a voltage has to be provided externally. The efficiency of the cell is calculated from curve 1 assuming that energy 1.23 eV is associated with each hydrogen molecule (Khaselev and Turner, 1998) Fig. 2*

connecting "tunnel diode interconnect", so that the photo-voltages of the two junctions are additive. The light enters on the right, and the right-hand junction has the larger bandgap, so that lower-energy photons that are not absorbed on the right pass to the left to the second junction where they are absorbed.

In such a tandem cell the electrical current has to be the same in each junction, so that the photocurrent of the series tandem cell is less than the short-circuit current of the weaker junction. The voltages add directly, allowing a higher conversion efficiency than a single junction photocell. The efficiency was calculated as the chemical energy in the resulting hydrogen divided by the radiation power input. The chemical energy was calculated from the measured current multiplied by the voltage 1.23 V that is related to the hydrogen molecule, see Fig. 11.24.

The efficiency of water splitting was calculated from curve 1, with no external voltage applied, as power out divided by power in, or 0.12 A × 1.23 Volts/(1.190 W) = 0.124, for one square centimeter. The light intensity was about 11 suns for this experiment. One can infer that adding an additional junction in the cell to make a 3– or more junction tandem cell could result in excess cell output voltage beyond that needed to release hydrogen. Such a cell might be configured to simultaneously split water and supply power to an external load.

The situation observed in curve 2 (zero photocurrent at zero applied voltage, i.e., the water splitter does not work) is usually the case for a single junction. The reason is that the conduction and valence band energies must be separated by 1.2 eV or so

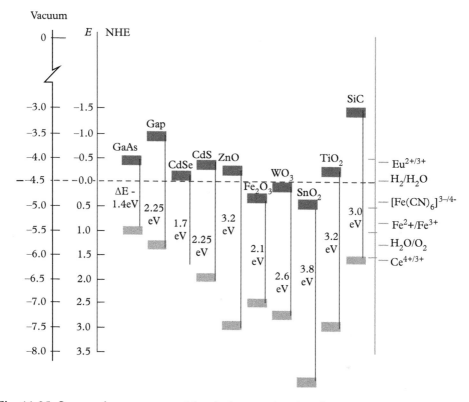

Fig. 11.25 *Survey of energy gaps and band-edge energies of semiconductors and oxides of use in solar cells and water splitters. Note that the reduction energy level of water, taken as zero on the hydrogen electrode energy scale of electrochemistry, is about 4.5 eV below the vacuum energy. On the right are shown energy levels for some additional reduction/oxidation reactions in water solution (Graetzel, 2001)*

for efficient absorption of solar spectrum photons. At the same time, the conduction band edge has to be above the water reduction energy level and the valence band edge has to be below the water-oxidation energy level (these two levels, separated by 1.23 eV + 2 η ~ 1.6 eV, are shown in Fig. 11.23 on the right). It is common to reduce the overvoltage losses η at the semiconductor/electrolyte interfaces by depositing platinum particles as catalysts. The platinum deposition is usually done electrolytically from a solution containing platinum ions. These conditions cannot usually all be satisfied at once without inserting an external voltage, as has been done in the present example by the extra PN junction. A survey of the energy levels is given in Fig. 11.25.

According to the compilation of Fig. 11.25, GaAs, GaP and TiO_2 are suitable to act as photo-cathodes to release hydrogen from water, in that their conduction band edges all lie above zero on the "NHE" normal hydrogen electrode scale.

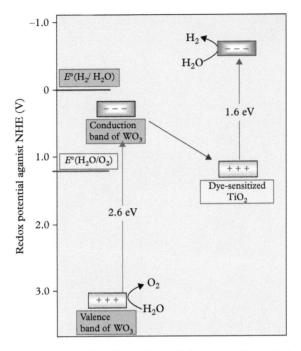

Fig. 11.26 *Proposal for inexpensive thin-film tandem cell for water splitting. In its realization, the two cells are stacked, just as in Fig. 11.22. The downward sloping arrow in the center of this Figure represents the "transparent ohmic contact" connecting the two cells in Fig. 11.22. The "Dye-sensitized TiO₂" cell on the right is as described in connection with Fig. 10.17, and the 1.6 eV arrow shown here corresponds to the dye absorption energy rather than the titania band gap (Graetzel 2001) Fig. 6*

11.3.1.2 *Practical tandem-cell water-splitting device*

Shown in Fig. 11.26 is a schematic of a potentially low-cost tandem device that achieves direct cleavage of water into hydrogen and oxygen by visible light. This is based on two photosystems in series (tandem), with electron flows as shown in the figure. The top cell (exposed to light) is a thin film of tungsten trioxide that absorbs the blue portion of the solar spectrum. The valence band holes oxidize water directly to oxygen. The conduction band electrons are fed into the second photo-system, the dye-sensitized nanocrystalline TiO_2 cell. This is placed directly under the tungsten trioxide film and captures the green and red part of the solar spectrum. The photoelectrons in the conduction band of the titania reduce water to produce hydrogen gas.

This device perhaps can be produced as a wide-area membrane with water on both sides, generating oxygen on one side and hydrogen on the other side. Unlike the expensive GaAs-based semiconductor system discussed earlier, these oxide surfaces are inert with respect to water exposure and do not involve single crystals nor vacuum-processing steps.

11.3.1.3   Dual-purpose thin-film tandem cell devices

One might imagine a monolithic dual purpose self-regulating photo-electrochemical device comprised of a multi-junction tandem solar cell, electrolytic cell and power sharing circuitry. The tandem cell structure under illumination will be efficient inherently by using several different gaps. Its series structure will give an open-circuit voltage larger than that necessary to decompose water. The excess photo-voltage with respect to water splitting can be diverted to an external load. The cell will generate primarily electric power under heavy load (small load resistance) and will generate primarily hydrogen gas at light load (high load resistance). The structure might be built up, e.g., as a CIGS cell on aluminum foil, with for example a tungsten trioxide blue light cell deposited on top. The upper surface of the tungsten trioxide cell could be metallized with a transparent electrode, and fed through a load resistor to a second aluminum foil below the original aluminum foil, the inter-space between the two foils being filled with water to be electrolyzed.

11.3.2   Hydrogen fuel cell status

A fuel cell takes hydrogen gas as input and produces electricity, with efficiency in the range of 50 to 70 percent, greater than the internal combustion engine, ~ 30–35 percent. A proton exchange membrane (PEM) fuel cell element is sketched in Fig. 11.27 (Lister and McLean, 2004). The input gas channel on the left takes in hydrogen, and the input gas channel on the right takes in air. This is to be regarded as the nth unit in a stack making up a practical fuel cell. The central membrane PEM is often Nafion, or Nafion-PFSA (per-fluorosulfonic acid), a sulfonated fluoropolymer, invented by Dupont and related to Teflon, that allows motion of protons but blocks electron flow. In the diagram shown, protons, H^+ ions, cross the PEM from the left to the right, causing equivalent electron current to flow in the external circuit. Catalysts are needed to cause the cathode oxygen reduction reaction and the hydrogen oxidation reaction (releasing the protons), that are written into the figure. Platinum particles are the traditional catalysts, but active work proceeds to find less expensive catalysts (Debe, 2012). The present situation is that fuel cells remain expensive as components of automobiles, but are in wide use, with the expectation of lower costs in the future.

Automaker Honda in 2008 (Packler, 2008) released its hydrogen-fuel cell powered sedan, the FCX Clarity. This vehicle has a 280-mile range, using pressurized hydrogen. Pressurized hydrogen is available only in a few places, for example at about $5/kg at a series of stations in California. This vehicle is expensive, largely due to the cost of the fuel cell. It accelerates rapidly and can reach 100 mph. The manufacturer is said to have a production line available, that it could run if the infrastructure of hydrogen stations expands. At the moment the manufacturer is leasing these cars at a low introductory rate. This car has a fuel cell "the size of a desktop PC that weighs about 150 pounds" and has a power rating of 100 kW (134 HP, at 746 W/HP). It is anticipated that the cost of the fuel cell unit will fall, so that the cost of producing the car will fall from "several hundred thousand dollars" to below $100,000. The manufacturer says that the car is efficient, "equivalent of 74 miles a gallon of gas".

More recently, Toyota markets its 2015 Mirai hydrogen car at $57,500, with three years of free fuel. It is said that the range per filling for the Mirai is 312 miles, and that

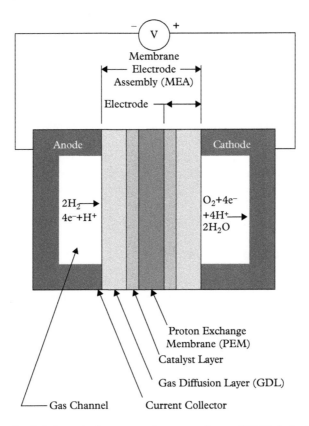

Fig. 11.27 *Schematic of single typical proton exchange membrane (PEM) fuel cell. A "fuel cell" is a series-connected stack of such devices. Such devices, "run backwards", are run as one form of electrolyzers of water to give free hydrogen for energy storage, to run fuel cell cars, or for production of ammonia (Litster and McLean, 2004) Fig. 1*

filling takes five minutes. This feature is more convenient than electric cars, that take longer for re-filling. Other manufacturers sell cars that simply burn hydrogen gas in an internal combustion engine, an expedient that loses the efficiency inherent in the fuel cell/electric motor combination, and ignores the release of carbon dioxide.

11.3.3   Storage and transport of hydrogen as a fuel

Gaseous hydrogen can be piped as is natural gas, with preference for plastic pipe to avoid questions of hydrogen embrittlement of steel. It is commonly stored at high pressure in steel tanks.

In its liquid form hydrogen has a higher density. However, liquid hydrogen requires the low temperature of 20.3 K. Liquid hydrogen has high energy density per unit mass, but on a volume basis the energy density is about a factor 4 less than gasoline or diesel fuel. Storage tanks for liquid hydrogen, that are vacuum vessels with "superinsulation" in the form of 30 to 300 layers of aluminized mylar, are available in a variety of sizes, including

100 liters and 190 liters. These tanks accommodate hydrogen liquid at atmospheric pressure, without internal refrigeration and have losses in the range of 1 percent per day, decreasing with increasing tank volume. A line of sedans has been developed and tested by manufacturer BMW, that use such liquid storage tanks, the gas powering internal combustion engines. (It is more common in hydrogen cars to use pressurized hydrogen gas.)

This schematic[1] of a proposed process cools hydrogen at atmospheric pressure in a series of three heat exchangers, each heat exchanger block being cooled by expansion of a mixture of neon and helium gas. The proposed cycle would require only one pass of hydrogen through the system, while existing cycles require multiple passes to liquefy all the gas. The energy and capital expenses in liquefaction are not negligible. Commercial liquefaction schemes usually involve compression of the hydrogen gas. Most commercial hydrogen is presently processed from natural gas.

Tank trucks are on the highway transporting liquid helium, and other high value cryogenic liquids including liquid oxygen and liquid nitrogen. Liquid natural gas is transported on a larger scale by ship and railroad tank car. On a large scale, hydrogen transport as a liquid is economic for some applications.

On the smaller scale, to replace the gas tank of an automobile, liquid storage may be a less viable option, since the fuel would evaporate over relatively short storage times. Presently, as mentioned, the most common hydrogen storage is in a high-pressure tank.

11.3.3.1 Surface adsorption for storing hydrogen at high density

Alternative possibilities for dense hydrogen storage are by reaction of hydrogen to form chemicals such as metal hydrides and by temperature-dependent adsorption of hydrogen onto lightweight high-area substrates, that may include decorated graphitic surfaces. Hydride storage tanks have been tested in connection with autos and trucks. A figure of merit of such a storage medium is the weight density of hydrogen that can be achieved. A practical benchmark of 6.5 percent by weight of hydrogen has been chosen by the US Department of Energy.

A hydrogen storage model based on graphene layers is shown in Fig. 11.28. The modeling associated with Fig. 11.28 suggests that graphene layers optimally spaced at 0.8 nm, may, with some uncertainty, accommodate hydrogen at the 6.5 wt percent level at room temperature and moderate pressure. This structure could be described as graphite intercalated with hydrogen. The calculations of Patchkovskii et al (2005) indicate the hydrogen molecules will be delocalized (i.e., able freely to move, between the graphitic layers) that would promote rapid filling and emptying of the hydrogen charge. On the other hand, the spacing of the graphene layers is a critical parameter that has to be arranged independently. No obvious route to obtain the desired spacing of graphene layers is known, although there is a large literature of "graphite intercalation compounds". Any intercalation compound will have additional mass and will block free volume for the intended hydrogen molecules.

The attraction of hydrogen to transition metal atoms, including Ti, is stronger than the attraction to graphene, that was discussed previously. According to (Durgun et al, 2006) that bonding strength is about 0.4 eV, compatible with adsorption/desorption at room temperature. The interaction can be reliably calculated with advanced methods, leading to predictions, indicated in Fig. 11.29, of hydrogen storage

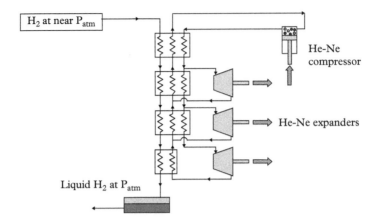

Fig. 11.28 *Possible single-pass low pressure hydrogen liquefaction cycle, proposed by Gas Equipment Engineering Corp[35]. In this diagram the resistor-like symbols are heat exchangers, and the cooling occurs as the He-Ne gas mixture, initially pressurized, is allowed to expand*

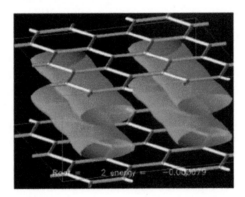

Fig. 11.29 *Nanometer-scale schematic diagram of a hydrogen storage configuration examined theoretically. The tubular regions, located between graphite planes of variable spacing, are regions of high probability density for hydrogen molecules. The modeling is based on van der Waals attraction (Patchkovskii et al, 2005)*

in certain simple organic molecules with transition metal adsorbates, in the range of 14 wt–percent.

Durgun et al (2006) find that the hydrogen molecules in the right panel of Fig. 11.29 are bound with energies varying from 0.29 eV to 0.49 eV. These energies are considered suitable for room temperature storage of hydrogen. This is a reliable theoretical estimate,

---

[35] Innovative hydrogen liquefaction cycle, Gas Equipment Engineering Corp., in DOE 2007 Merit Review, May 15, 2007.

but synthetic chemistry is needed in order to create the underlying molecule, $C_2H_4Ti_2$. The molecules would need to be dispersed, in a state more like a gas than a liquid, if hydrogen is to be readily inserted and extracted upon demand. This leaves the problem that the molecules will not reliably remain in the storage tank, but might flow out as the hydrogen fuel is withdrawn. The authors suggest that the molecules might be imbedded in a nanoporous matrix. The expected means to bring hydrogen gas in or out of the storage sites is to adjust the temperature. A nanoporous matrix might facilitate this, if it were continuous and could be uniformly heated by passing current through it.

11.3.3.2   *Storage of hydrogen in ammonia NH₃*

Hydrogen is also "stored" in anhydrous ammonia, NH_3, a colorless gas with a pungent odor, that is produced in great quantities principally for making fertilizer and other chemicals. Worldwide production of ammonia in 2015[36] was about 146 million metric tons, where a metric ton or tonne is 1000 kg with market prices ca $470/short ton (2000 lb, or 907 kg) in 2015. So, the cost is low, $0.518/kg.[37]

At normal conditions anhydrous ammonia is a gas but becomes a liquid at 240 K (or at 300 K at around 10 Bar, 1500 psi). It is reasonable to say that hydrogen is stored in ammonia, because the hydrogen is easily released, while saying hydrogen is stored in water does not make sense, because the hydrogen in water cannot be easily released. Ammonia will burn, with some difficulty, in air, and can be used in automobile engines with some modification. Hydrogen actually forms about 18 percent by weight of ammonia. Ammonia is much easier to liquefy than hydrogen and there is already an extensive pipeline network in the Midwest for ammonia, principally serving its agricultural uses. To form ammonia, from hydrogen and nitrogen, the final reaction is direct stoichiometric combination of hydrogen and nitrogen, to form NH_3 in a catalytic process called the "ammonia synthesis loop" (Haber–Bosch process), a well-known but difficult reaction carried out at pressures on the order of 100 atmospheres. (The energy cost of the Haber-Bosch process is high, since it involves four passes through a bed of catalysts, at about 400°C, each time releasing pressure from 100 atmospheres and then re-pressurizing.) Ammonia is the second-most produced chemical worldwide, although its production is much less than that of hydrocarbons. So, if hydrogen or ammonia were to become the energy carrier in a future carbon-depleted (or carbon-forbidden) world economy, the production of ammonia would need to be increased substantially, and the present reliance on natural gas for the hydrogen would need to be shifted back to electrolytically produced hydrogen. There is no fundamental reason why these steps could not be taken, there is enough water and enough nitrogen, and electricity for electrolysis could come from wind and solar sources.

---

[36] US Geological Survey, Mineral Commodity Summaries, January 2016, p. 119.
[37] The low price is because natural gas prices are low in the present era. Ammonia presently is made by the Haber Bosch process using hydrogen derived from steam treatment of natural gas, methane.

The hydrogen for ammonia production, in the present regime of very low natural gas prices, comes principally from steam reforming of methane (natural gas). We can speak of the (nonrenewable) hydrogen production in ammonia, worldwide, as obtained from the ammonia output multiplied by the fraction of the mass of anhydrous ammonia represented by hydrogen. This fraction is $(3 \times 1.008)/[(3 \times 1.008) + 14.0067] = 0.1776$, where the mass of nitrogen is 14.0067 in atomic mass units. Therefore, the worldwide production of hydrogen (specifically the part associated with production of ammonia) in 2013 was 146 million metric tons $\times$ 0.1776 = 25.9 million metric tons. (The proportion of this produced in the US was 8.17 percent, or 2.12 million metric tons of hydrogen.) Commercial hydrogen production is also done from "town gas", methane plus carbon monoxide, that is a result of gasification of coal. These latter processes are carbon-dependent (non-renewable) processes that release greenhouse gas. An independent estimate is world production of hydrogen (2004) is 50 million tons, about half for ammonia and about half for petroleum industry; "hydro-cracking" to produce light hydrocarbons from heavy hydrocarbons.

11.3.4   Further aspects of hydrogen and ammonia

One scenario for future distribution of energy is in the form of hydrogen, compressed or liquefied. The "hydrogen economy" scenario is promoted along with the idea that automobiles may switch to electric motors driven by fuel cells run on hydrogen gas stored in the car's fuel tank. A scenario of this sort is being tested in Iceland, that has renewable energy resources and no oil, coal or gas.

Hydrogen as a fuel is attractive because it burns to produce water, with no adverse effect on the atmosphere. Hydrogen is not abundant in its free form, for example, the availability of hydrogen as a byproduct of commercial liquefaction of air to produce nitrogen and oxygen is limited. Commercial (non-renewable carbon-based) production of hydrogen, as we have said, is extensive and is typically based on methane (natural gas). Approximately half is used make ammonia for agriculture, the other main use being in oil refining,

http://www.hydroworld.com/articles/hr/print/volume-28/issue-7/articles/renewable-fuels-manufacturing.html.

Commercial storage and transport of hydrogen is conventionally at high pressure in steel cylinders, or aluminum-lined containers reinforced with carbon fibers, or as liquid in a cryogenic container. The network of hydrogen "filling stations" in California dispenses hydrogen at 5000 psi and at 10,000 psi ($\sim$ 330 Bar to 660 Bar). There are significant energy costs in pressurizing hydrogen and also in liquefying it. (In principle, hydrogen can also be stored at high density adsorbed to high surface area substrates, such as graphitic planes decorated with hydrogen adsorbing centers, such as Ti atoms, as mentioned previously.) Finally, ammonia, NH_3, is easily liquefied at 1500 psi, and then contains hydrogen at high density. Liquid ammonia, in spite of its toxicity, is routinely piped over large distances in connection with its use as a fertilizer.

It is reported that Iceland produces 2000 tons of hydrogen per year by electrolysis, using spare electrical capacity. Most of it is used indeed in ammonia production, some in running fuel-cell buses and boats. Eventually the price of natural gas may rise enough that a renewable route to ammonia production will be competitive.

11.4 Hydrogen and ammonia as potential intermediates in US electricity distribution

As an upper limit on the hydrogen that might be needed, we can calculate the hydrogen production equivalent in energy to the US. consumer electricity usage per year, about 460 GW in 2013, according to Fig. 1.2. The total electric energy in 2013, taking a year as 3.154×10^7s, was 1.45×10^{19} J. If the energy carried per H_2 is 1.23 eV = 1.968×10^{-19} J, and the mass per H_2 is $2 \times 1.008 \times 1.6605 \times 10^{-27}$ kg, then the equivalent mass of hydrogen per year is 2.47×10^{11} kg = 2.47×10^8 metric tons, or 247 million metric tons. (Another estimate (Turner, 2004) is that the total annual transportation energy need of the US could be met (2004) by 150 million tons of hydrogen.)

This estimate of 247 million metric tons of hydrogen is 9.5 times the worldwide production of hydrogen presently in the production of ammonia, principally for fertilizer. (Ammonia production is 146 million tonnes, and hydrogen is 17.8 percent by weight of ammonia.) The price per metric ton of ammonia in 2015 was $0.518/kg. If the cost of ammonia is attributed entirely to its hydrogen, that is 17.8 percent of its weight, then the cost of hydrogen is $0.518/(0.178) = $2.91/kg. It is clear that this is an overestimate, because the costs associated with fractionating air to get nitrogen, and with the Haber reaction to make ammonia, would be avoided in the pure hydrogen production. An independent estimate of cost of hydrogen from natural gas is $2.70/kg. It is also stated that retail prices per kg for hydrogen at filling stations in California and in the Washington DC area range from $5 to $6 per kg. Again, according to Fig. 1 in Turner (2004), electrolysis of water would match $2.70/kg for producing hydrogen at electricity costs in the vicinity of 4 c/kWh. It is also stated that the lowest-price hydrogen presently is produced by electrolysis at wind farms.

As the price of natural gas rises it might occur that ammonia (fertilizer) production would represent a market for renewably produced hydrogen, for example from electrolysis using electricity from wind farms or PV farms, or thermo-chemical splitting of water. Water and nitrogen are regarded as renewable resources, effectively infinite in supply, while carbon is not renewable. Turner (2004) also states that potential electricity produced by solar and wind could (if the capacity were built) be more than adequate to produce all needs for hydrogen, based on electrolysis of water, and also that there is enough water for this purpose.

Commercially available also, as we have mentioned, are fuel cells, with efficiency in the range of 50–70 percent, that produce electrical current from hydrogen gas. So, a remote location might obtain continuous electrical power by the combination of PV (solar cells), electrolyzer, hydrogen storage, and fuel cell. This is a presently viable approach, an alternative to the simpler combination of PV and batteries.

A case can be made that ammonia itself is better suited to be the energy intermediary than hydrogen. As mentioned it liquefies at lower pressure and can be quite easily converted to free hydrogen to run a fuel cell. It is poisonous and pungent, but not strongly flammable or explosive. It is more amenable to pipe transport than hydrogen, and an extensive ammonia pipeline network is already present. Ammonia has been used to power buses and automobiles (usually in conjunction with a small amount gasoline or natural gas). It has a lower energy per unit volume than gasoline, thus gets lower mileage, but it is less expensive by a significant factor. It is completely free of carbon emission. If a rupture occurs it quickly disperses because it is lighter than air. There is some progress in developing fuel cells that will convert anhydrous ammonia to energy and nitrogen, but it is established (Lan and Tau, 2014) that the proton exchange membrane PEM cells will not do this.

It is reasonable to consider a long-term future where hydrogen or ammonia would be the energy intermediary in a portion of the transportation sector. If gasoline and diesel are priced out of use for transportation fuels, electric vehicles powered by batteries or by fuel cells will be more important. Charging batteries from an electric grid supported only by solar and wind, with more extensive storage, is one option. The other option is using the intermediary, hydrogen or ammonia, to run the vehicle's electric motor by a fuel cell. The hydrogen or ammonia would be available from dedicated stations. Alternatively, hydrogen might be produced at home with a unit like the Hogen HP 40 hydrogen generator mentioned previously. (That would be analogous to the electric car owner with solar cells on his roof and a battery pack, to charge his car.) On the other hand, it seems that ammonia, based on the advanced Haber Bosch process, could not be produced in home size units. The transportation power use in the US in 2014 was about 27 percent of the total, that was 3.29 TW for 2014, thus about 904 GW, compared to 460 GW for consumer electricity.

11.5   Energy use and climate impact

We start this section with discussion of energy use in transportation, principally autos. The burning of fossil fuels for cars, leading to severe smog situations in large cities like Los Angeles, CA was an early event leading to awareness of alteration of the atmosphere by human activity. In 2017 smog is not a huge problem in cities in the US and Europe, although the problem remains in Beijing and Delhi. Portions of the problems were fixed by removing lead from gasoline and by mandating catalytic convertors in gasoline autos in the US. The future of the auto regulations in the US may be in some doubt in the short term, but certainly cars are much better than they used to be, and consumers may be reluctant to go back to inefficient cars spewing pollution.

In the longer term the rational view is that burning of carbon will be taxed to reduce its impact on climate, that causes economic loss to everyone. Failing that, it will eventually be depleted in supply, become expensive, (see Section 1.4) and transportation will shift to non-carbon vehicles. Such a shift has already started. At present hybrid electric vehicles (HEVs) like the Toyota Prius (also available as PHEV, plug-in version) or Chevy Volt reduces gasoline consumption and fully electric cars as the Chevy Bolt, Nissan Leaf and Tesla Model S use carbon only in a portion of the charging electricity. Four million Prius cars have been sold, with the total hybrid count at 11 million as of 2017. Owners of fully electric cars may use solar cells and batteries to charge their electric cars, giving freedom from the grid on a local basis. This PHEV fleet is much smaller (Neil, 2017), only 159,000 were sold in the US in 2016. These can be very impressive automobiles; Neil describes the large Chrysler Pacifica PHEV van that can run 33 miles solely on its 16 kWh lithium-ion batteries, plug-in rechargeable, driving the two high-torque electric motors combined to 198 hp. After the first 33 miles the larger gasoline engine starts, at 268 hp, moving the car and charging the batteries. (The owner of this car has zero chance of being stuck with a dead battery in a remote spot, as can happen to the owner of the purely electric Tesla Model S.) The fuel mileage over the first 100 miles is 84 miles per gallon, for this large fully equipped station wagon. It will go up to 75 mph even in the silent electric-only mode. For many drivers who go less than 33 miles typically per day, the vehicle will be running only electric, with no exhaust fumes and virtually no sound emitted. The base price for this vehicle is $45,000, only about $2000 more than the purely gasoline version. (The same virtues at half the price roughly are available in the smaller Prius cars.)

Turning to the much smaller hydrogen powered car fleet; the Honda FCX of 2008 and the Toyota Mirai of 2015 require only electricity and water to generate hydrogen. The filling stations for these cars are based on electrolysis of water, using units like the Norsk Atmospheric Type No. 5040 mentioned in 11.3.1. There are a few instances of running vehicles on ammonia, that will remain as an important commodity with the eventual scarcity carbon, as ammonia can be produced from electrolytic hydrogen and nitrogen through the Haber-Bosch process and will always be available for fertilizer. In Belgium between 1943 and 1945 a fleet of buses was run on a combination of ammonia NH_3 (liquefied in a tank) and coal gas, that is 50 percent hydrogen. Truck fleets now may be run on methane (even derived from cow manure) in the form of compressed (CNG) natural gas. For example, a large dairy farm in Indiana (Kowitt, 2016), with 35,000 cows, and 46 semi-trailer trucks to deliver the milk, runs its truck fleet entirely on farm-produced compressed natural gas CNG (methane) processed from manure.

11.5.1 One billion autos, with rising demand from seven billion persons

There are 1.2 billion cars now globally, and the number in 2035 is predicted to be 2 billion. In the US, transportation uses between 25 percent and 30 percent of power and also represents 28 percent of the greenhouse gas emission, second only to 34 percent from electric power generation, much from coal power plants.

A broader view of vehicle power systems has been considered by Zamfirescu and Dincer, 2009: a portion of the results are shown in Fig. 11.30. The performance aspects considered here are the size of the fuel tank needed and the cost per mile of travel for each power system. The great advantage of gasoline is the high energy density that

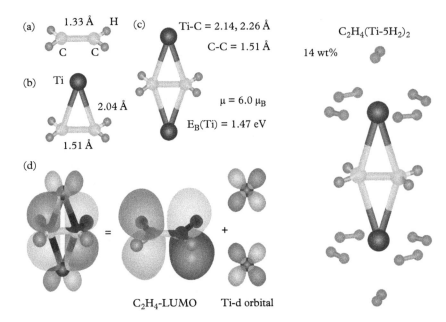

Fig. 11.30 *Diagrams showing molecular hydrogen storage schemes. (a) Structures based on ethylene C_2H_4. Panel (b) indicates a single Ti atom with bond-length 2.04 Angstrom from carbon, followed by adsorption of two Ti atoms at binding energy 1.47 eV. Panel (d) shows how this structure can be regarded as a covalent bond between the Lowest Unoccupied Molecular Orbital of the ethylene molecule (large lobes in panel d) with the 3d orbital of Ti (small lobes in panel d). Right-hand side of the Figure shows binding sites of 10 hydrogen molecules that are provided by the two titanium atoms in this structure. This corresponds to 14 wt % of hydrogen. (Durgun et al 2006)*

allows a small tank (or long range) for the vehicle. The compactness is best for the liquid fuels gasoline, LPG, methanol and ammonia. The considered storage for hydrogen is in a hydride form (along the lines discussed in Section 11.3.3.1) rather than as pressurized gas but still requiring a large volume tank. The results for operation cost in Canadian dollars for 100 km (62.1 miles) is surprisingly low for the ammonia cases, apparently due to the low cost of commercial ammonia already noted. (In 2009 the Canadian dollar was about 0.95 US dollars.) It appears that estimated capital cost for the vehicle power system is not part of the analysis.

11.5.1.1 *Efficiency of autos*

In these estimates it was assumed the gasoline, CNG and LPG cars operate at 28 percent efficiency, with 40 percent for methanol. For the H_2 vehicle a proton exchange membrane (PEM) fuel cell system of 50 percent efficiency, with hydrogen stored in metal hydride tanks, was assumed. For the ammonia vehicles the choice of internal combustion engine vs. fuel cell was left open, but overall system efficiencies of 35 percent and 45 percent were assumed.

One approach to an ammonia vehicle is to modify a hydrogen fuel-cell vehicle like the Honda FCX with an ammonia tank and a converter to break the ammonia into hydrogen and nitrogen. "Thermal cracking" of ammonia is a well-known technology (Ganley et al, 2004) to release hydrogen.

Such a vehicle, apart from its running cost, would not be cheaper than a hydrogen vehicle, that is expensive at present on the basis of the fuel cell.

(a) Fuel tank compactness

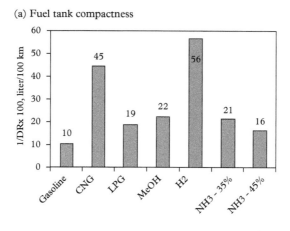

(b) Specific driving cost

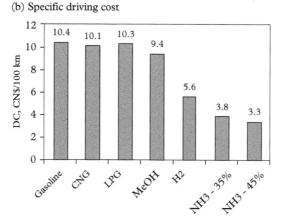

Fig. 11.31 *Comparative performance analysis of several vehicular power systems. Here CNG stands for compressed natural gas, LPG is liquefied petroleum gas, MeOH is methanol (wood alcohol). In the estimates it was assumed the gasoline, CNG and LPG cars operate at 28 percent efficiency, with 40 percent for Methanol. For the H2 vehicle a PEM (proton exchange membrane) fuel cell system of 50 percent efficiency, with hydrogen stored in metal hydride tanks, was assumed. For the ammonia vehicles the choice of internal combustion engine vs fuel cell was left open but the overall system efficiencies of 35% and 45% were assumed (Zamfirescu and Dincer, 2009) Fig. 5*

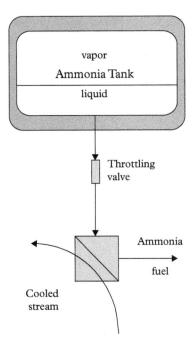

Fig. 11.32 *Schematic diagram of ammonia vehicle fuel tank, with liquid ammonia at nominal 1500 psi, showing expansion to gaseous state through throttle valve allowing cooling of other parts of the engine, in indicated heat exchanger. (Zamfirescu and Dincer, 2009). Fig. 3a*

The AmVeh is a small ammonia-fueled vehicle sold in South Korea.[38] This small car uses a fuel ratio 70 percent ammonia to 30 percent gasoline in a conventional gasoline engine. The design allows conventional cars to be converted to partial ammonia use. These cars significantly reduce carbon emissions as ammonia has no carbon. It is noted that in South Korea there are about 0.31 cars per person. A second ammonia car is a converted Toyota called the Marangoni Toyota GT86.[39] It uses also a dual fuel supply, and it is stated that with a 30-liter ammonia tank (at 20 cents per liter) the car will go 180 km. At low speeds, as in city driving, it can run only on ammonia, while at higher speeds a mix of gasoline and ammonia is used. These two cars use schemes similar to that in Fig. 11.31, with liquefied ammonia at about 1500 psi.

11.5.1.2 *Carbon emission in auto production and use*

According to Fig. 11.32 the transportation sector in the US in 2014 derived its power mostly from petroleum and natural gas, 92 percent and 5 percent respectively, and only 5 percent from renewable energy. Thus, 95 percent of the transportation power led directly to carbon emissions, contributing to global warming. Also, the industrial sector, involved

---

[38] https://nh3fuelassociation.org/2013/06/20/the-amveh-an-ammonia-fueled-car-from-south-korea/
[39] https://nh3fuelassociation.org/2013/04/25/ammonia-fuel-marangoni-eco-explorer/

in manufacture of autos and trains is shown to get 38 percent of its power from petroleum and 44 percent of its power from natural gas. This figure also shows that 91 percent of coal was burned directly for electric power, contributing 42 percent of the power for electricity. Note that 23 percent of petroleum and 34 percent of natural gas go to the industrial sector including the formation of hydrogen in the steam forming process. The long-term challenge, based on the predictions of Yergin (see Section 1.4), and others, that fossil fuels will be unavailable on a hundred-year timetable, will be to boost the lowest two sources in this figure, Renewable Energy and Nuclear Electric Power now totaling only 18 percent of US power sources, to replace the present 82 percent supplied by petroleum, natural gas and coal. These fossil fuels are guaranteed to be transformed into atmospheric CO_2, discounting the chance that a meaningful fraction of the CO_2 can be "sequestered" (buried) on a long time scale. (The climate change aspects of this will be considered in Fig. 11.33.)

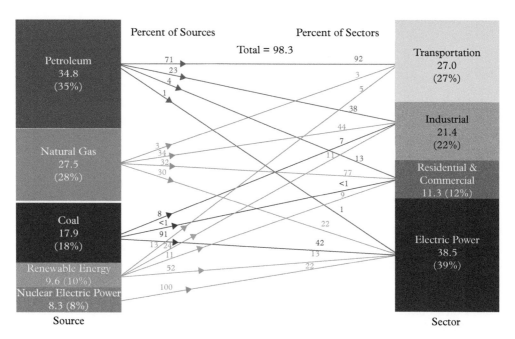

Fig. 11.33 *Transportation sector use of fossil fuels, shown in US Primary Energy by Source and Sector for 2014, in Quadrillion BTU units. In this Figure "Renewable Energy" contains Hydroelectric, Geothermal, Solar/Photovoltaic and Biomass. Here "Petroleum" does not include biofuels that have been blended with petroleum, that are in "Renewable Energy". One "Quadrillion BTU" is an energy unit equivalent to 1.055×1018 J. The "Electric Power" value 38.5 Quads per year corresponds to 1.29 TW average power for US in 2014. This value exceeds the 460 GW consumer electric power value, stated earlier, by including a large entry for combined-heat-and-power (CHP) plants. All sector entries: Industrial, Commercial and Residential, and Electric Power contain CHP portions and commercial-only and industrial-only electric power. These data have been obtained from US Energy Information Administration, Monthly Energy Review (March 2015, Tables 1.3, 2.1–2.6. (2017 US Energy and Jobs Report_0–1.pdf), US Department of Energy, January 2017, p. 83*

This gradual transformation in the powering of transportation, and other aspects of the economy, will proceed on the individual consumer level as well as on the economy-wide level. The individual consumer can become independent of fossil fuel to a large extent by solar power and batteries for household electricity, heating and for transportation; or by hydrogen and fuel cells if an electrolyzer can be purchased. The consumer cannot avoid reliance on the larger economy for equipment including cars, fuel cells, electrolyzers, electronics, air conditioners and chemicals such as ammonia and pharmaceuticals. The larger economy can shift more to electricity from renewable sources with a more flexible grid with more storage, to fertilizer production using ammonia derived from electrolytic hydrogen rather than natural gas hydrogen. Some applications, such as the hybrid diesel-electric locomotives, require the high-energy density of diesel, but biodiesel is possible for a fraction of these.[40]

The gas turbine can also run on bio-fuel, as described for jet engines by Handler, 2016. It is said that all United Airlines flights from the LAX Los Angeles Airport are running on a mix of kerosene and biofuel.

11.5.2   Diesel-electric trains, natural gas trucks

Diesel-electric locomotives are advanced hybrids with power ratings up to 4500 hp and 6000 hp (3.36 MW and 4.48 MW). Freight trains over a mile long with several locomotives or pairs of locomotives spread out in the length, are assembled to haul coal, petroleum, iron ore and manufactured goods in an efficient manner. Six- to twelve-cylinder diesel engines are used to run electric motors. In some passenger locomotives regenerative braking is used to power onboard electric services. Specialized smaller diesel electric locomotives used in switching yard sometimes, but not often, use regenerative braking in conjunction with large onboard arrays of lead acid batteries (Austen, 2009).

Very high-speed trains are already fully electric, with no carbon emissions, but it is not likely that freight trains will be changed to electric. It is conceivable that a fraction of freight trains might use biodiesel fuel, but the diesel electric locomotive is unlikely to be supplanted. General Electric has recently agreed to provide 1000 of their diesel electric locomotives to India with long-term service contracts.[41] It is also possible to refit a diesel electric locomotive to run on CNG compressed natural gas and this has been done in several places.

The well-known oil man T. Boone Pickens has long advocated changing the US trucking fleet from diesel to compressed natural gas (CNG) or liquefied natural gas (LNG). (CNG is pressurized to 3000 to 3600 psi; there are about 15 million CNG

[40] http://www.thehindu.com/news/national/karnataka/train-running-on-biodiesel-flagged-off/article7947215.ece
[41] http://www.railwaygazette.com/news/business/single-view/view/ge-wins-1000-locomotive-indian-railways-deal.html

vehicles worldwide, about 250,000 in the US and about 900 CNG filling stations.) Apart from his business interests, Mr. Pickens sees a US CNG trucking fleet as a means to reduce dependence of the US on foreign oil.[42] It is noted that the large metro bus service of Los Angeles has adopted buses running on CNG. Diesel buses in the large Los Angeles fleet have been replaced by basically similar units, still using Cummings engines and Allison transmissions, but running on compressed natural gas. Gas tanks on the tops of buses are a familiar sight in New York City. In a longer term there might be a business opportunity for a competent manufacturer to produce a large truck using a hydrogen- or ammonia- fuel cell driven electric motor. If an electric motor at 6000 hp can haul a freight train, a smaller electric motor can certainly be configured to haul an 18-wheel semi-trailer, with expected savings in operating costs, considering the efficiencies of fuel cells and electric motors and the low price of ammonia. Such a fleet would be superior to that envisioned by Pickens.

11.6   Carbon emissions and temperature change

An interesting insight into the expected effects of the various energy uses on the global temperature is offered by work reported in the IPCC 2013 report and shown in Fig. 11.34. The calculated results, shown for various sources of greenhouse emissions, are the expected temperature change after 100 years for an initial one-year dose of the specific emission. For example, in the upper (a) section of the Figure is shown that "Household fossil fuel and biofuel" burning (assuming a one-year dose now) will raise the temperature in 100 years time by about 2.1 mK. This effect from home furnaces, etc., is seen to be relatively large, not much smaller than the 2.5 mK approximately that is predicted for "Road" transportation (automobiles, etc), and even about half that predicted for "industry". The main contributor, shown by the red color coding, is CO_2, from the gas burning in the home and apartment furnaces, the gasoline burning of the cars, and the various fossil fuels used in industry. Turning to the shorter-term 20-year prediction in (b), one sees a much larger temperature rise from "animal husbandry", in color brown for methane, that decays more rapidly in the atmosphere than does the CO_2. The methane (CNG) derived from cow manure was used by the Indiana dairy farmer mentioned to run his fleet of 46 semi-trailer trucks, that would have the effect of transferring the emission from "animal husbandry" to "road (transportation)" in this figure, and give it a longer lifetime in the atmosphere. On the 20-year scale one sees a larger negative, cooling, effect from the SO_2 sulfur dioxide emissions (green), in "Industry" and "Energy". The methane is also noticeable from "Waste/landfill" entry and suggests that landfills, as well as dairy, pig, and cattle farmers, could make money by collecting and selling the methane, if not obviously to

---

[42] http://oilprice.com/Energy/Energy-General/Boone-Pickens-On-Natural-Gas-Its-The-Way-To-Defeat-OPEC.html

(a)

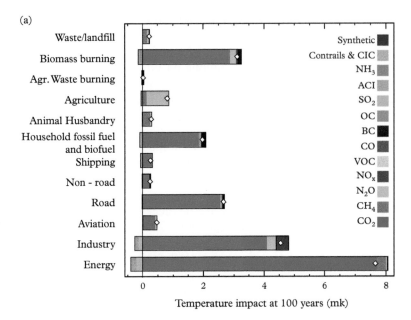

(b)

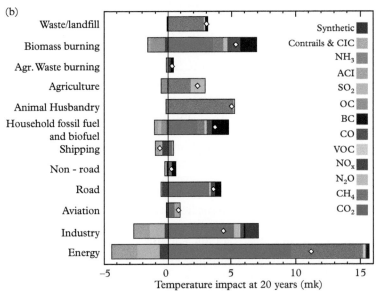

Fig. 11.34 *Net global mean temperature change by source sector after (a) 100 years and (b) after 20 years (for a 1-year pulse of 2008 emissions). Emission data for 2008 are taken from the EDGAR database. IPCC 2013, Fig. 8.34 p. 720*

the benefit of the future global temperature. It is well known that there are more pigs than people in Iowa, and also well known that pig farms are smelly. The pig manure, rather than polluting the rivers,[43,44] could better be turned into CNG compressed natural gas for sale, following the lead of the Indiana dairy farmer (Kowitt, 2016, see 11.5). If persons in West Virginia can no longer mine coal, perhaps they could learn to raise pigs or cattle, and perhaps even learn to manufacture and sell CNG on the side.

In looking at these two Figures, one sees that the home furnaces (household fossil fuel) and biomass burning are really large contributors, and that sources that one might worry about, including aviation, shipping and trains (that are in "non-road" transportation), are not so large. To change the large greenhouse gas effect from building heating, predominantly the gas furnace, the most immediate step is to improve efficiency by putting in more insulation and double-glazed windows, etc. It is conceivable to shift to electric heating or solar heating for new homes, but retrofitting seems unlikely. Heat pump systems are also more efficient but are really only an option for new homes and buildings.

Biomass burning occurs in forest fires that are not controllable, but it is stated that 90 percent of biomass burning is human initiated. It appears that biomass burning is a leading source of heat and cooking in less-developed areas of the world,[45] and that the best hope is that the stoves may become more efficient. The large entry on biomass burning appears to reflect in part the great increase in global population, say since 1900, (it was 1.6 billion then and 7 billion now) and the fact that many live in poverty. According to the World Bank[46] in 2013 10.7 percent of the world's people lived on less than \$1.90 per day. That is still 760 million persons, nearly twice the US population, who will be lucky to be able to cook over a wood stove.

It is also seen from the 20-year plot that biomass burning also produces some organic carbon OC (blue) that, like SO_2, tends to block sunlight and act as a negative emitter.

The temperature scale in Fig. 11.34 is in mK, or 0.001 K, seemingly small. A crude estimate of the temperature change in 20 years can be made from Fig. 11.34b. If we take the predicted net change from "Energy" to be 15.5 mK–4.5 mK = 11 mK, we can be quite sure that for every year between now and 20 years out, the change will be greater than 11 mK, since the effect of the one-year dose now decays each year. If we suppose that we continue to add in each of the next 19 years the same dose, following the likely scenario of little change in emissions, and reasonably add the resulting temperature changes, we will get from "Energy" a predicted 20 × 11 mK change, or 0.22 K.

[43] According to David Pitt, in an article entitled "Polluted Iowa waterways rise 15 % in 2 years", (*Associated Press*, 14 May, 2015, "Two of the most frequently cited problems for rivers and streams are ... largely the result of manure spills or waste storage leaks from large scale hog or cattle operations".

[44] http://www.desmoinesregister.com/story/money/agriculture/2015/05/14/polluted-iowa-rivers-lakes/27318299/

[45] http://www.worldatlas.com/articles/20-countries-burning-the-most-waste-and-biomass-for-energy.html

[46] http://www.worldbank.org/en/topic/poverty/overview

Adding the contributions from all the sources shown, this predicts a global temperature rise of 0.9 K or 1.6°F in 20 years' time, i.e., by 2037.

If we take the same approach to the 100-year prediction, in (a) of Fig. 11.34, we get 1.48 K or 2.7°F, assuming no change in emissions, by 2117. This is a crude estimate only, a severe underestimate, ignoring the fact that the temperature change will be larger in the earlier years of the intervals, and still assuming the business as usual scenario of unchanging emissions for the full-time intervals.

It is appropriate to look into the methods used to provide the predictions in the Fig. 11.34.

The predicted temperature changes are calculated using the concept of the Absolute Global Temperature Potential (AGTP) that is tabulated in units of Kelvins/kg in IPCC 2013 on p. 731 ff. The background for the concept is discussed, e.g., by Shine et al, 2005. "Alternatives to the global warming potential for comparing climate impacts of emissions of greenhouse gases". *Clim. Change* 68, 281.

The AGTP gives the temperature change, after a specified time delay, typically 20 y or 100 y, for a specific emission, in kg. The algorithm for use of this concept is:

$$\Delta T(t) = \sum_i \int_{0-t} E_i(s) \, AGTP_i(t-s) \, ds \qquad (11.7)$$

namely, a convolution of the emission scenarios and $AGTP_i$ (in K/kg), where E_i is the emission component in kg, t is time, and s is time of emission (Berntsen and Fuglestvedt, 2008). "Global temperature responses to current emissions from transport sectors". *Proc. Natl. Acad. Sci. USA* 105, 19154.

In this framework some important AGTP and radiative efficiency values are given in Table 11.3, data from IPCC 2013 p. 731.

The behavior of carbon dioxide in the atmosphere is anomalous. A study (Joos et al, 2013), of the unusual assimilation of CO_2 by the atmosphere and ocean, finds that a pulse of emission into the atmosphere shows a rapid decline in the first few decades followed by a 1000-year-scale remaining concentration. A single time constant cannot be ascribed to this behavior. For example, if a 100 Gt-C emission pulse is added to a constant CO_2 concentration of 389 ppm, 25 ± 9 percent is still found in the atmosphere after 1000 years; the ocean has absorbed 59 ± 12 percent and the land has absorbed the remaining 16 ±14 percent. The response in global mean surface air temperature is an increase by 0.20 ± 0.12°C. within the first 20 years, thereafter and until

Table 11.3 *Temperature potential parameters of key emitters.*

Emitter	Lifetime, y	Radiative Eff., $Wm^{-2}ppb^{-1}$	AGTP 20-y, K/kg	AGTP 100-y, K/kg
N_2O	121	3×10^{-3}	1.9×10^{-13}	1.3×10^{-13}
CH_4	12.4	3.6×10^{-4}	4.6×10^{-14}	2.3×10^{-15}
CO_2	(~ 1000)*	1.4×10^{-5}	6.9×10^{-16}	5.5×10^{-16}

*The decay is complex, see text.

year 1000, the temperature decreases only slightly, while the ocean heat content and sea level continue to rise.

11.7   The large city as sustainable energy leader

City governments can be very forward-looking and stable, with competent bureaucracies able to plan on long time horizons. Cities also have a great deal of autonomy.

As an example, the City of New York carried out a careful mapping assessment of the solar power capacity of its one million-plus rooftops (Navarro, 2011), concluding that 66.4 percent of the buildings of the city had roof space suitable for solar panels. It was concluded 5.847 GW could be generated putting solar panels on hundreds of thousands of buildings. It was concluded that 49.7 percent of the city's peak power usage could be generated from solar power and 14.7 percent of the city's annual electricity use, taking into account typical weather conditions. David Bragdon, director of the *Mayor's Office of Long-Term Planning and Sustainability* said the city could realistically add "thousands of megawatts" in solar power. The city would likely establish a uniform approach and presumably would negotiate with a company like First Solar to do the installations. The budget of the city was $65.7 B for 2012. Nominally the installation would cost around $5.8 Billion, which, over 10 years, would be 0.9 percent per year of the city budget.

Such a project would be a large-scale deployment of small installations, and would be in the power of the city to mandate. The article (2011) states that US nationwide installed solar capacity is 2.3 GW, less than half the roof top potential of New York City. (The US installed capacity is now about 35 GW, but the 5.9 GW still is a very large number.) The existence of a *Mayor's Office of Long-Term Planning and Sustainability*, able to obtain city funding of $450,000 for the aerial survey, is in itself encouraging, and suggests that large cities may be an avenue toward rational implementation of renewable energy systems.

11.8   The possibility of carbon capture at power plants

The concept of sequestering carbon dioxide from exhaust gases originated (Rao and Rubin, 2002) with the idea of selling the separated carbon dioxide, particularly for its use in enhanced oil recovery operations, in which the gas is pressurized into the old oil field to push out the remaining oil. A working system for separating carbon dioxide, however, was first set up by Statoil of Norway, in response to a carbon tax. Statoil of Norway, since 1996, has been storing carbon dioxide from its Sleipner West gas field in a sandstone aquifer 1000 m beneath the North Sea. This facility is smaller than a typical coal–electric plant. Such a plant captures CO_2 with processes based on chemical absorption using a monoethanolamine (MEA)-based solvent. MEA is an organic chemical belonging to the family of compounds known as amines. MEA was earlier developed as a general non-selective solvent to remove acidic gas impurities such as H_2S from natural gas streams.

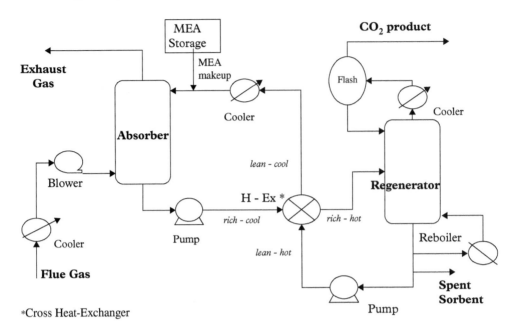

Fig. 11.35 *Removal of carbon dioxide from coal plant exhaust flue gas (enters at lower left), using monoethanolamine-based absorber (MEA), resulting finally in separated output of pure CO_2, compressed to 136 Bar, at upper right. Other flue gas components exhaust at upper left. (Rao and Rubin, 2002) Fig. 3*

A careful and detailed design study of MEA-based carbon dioxide scrubbing has been carried out by Rao and Rubin, 2002. These authors analyzed 500 MW pulverized coal burning electric power stations with and without the added MEA based carbon dioxide scrubbing. The essential features of the system are an Absorber, where the carbon dioxide attaches to the MEA, and a Regenerator where the CO_2, now at much higher concentration, is stripped from the solvent. These features, with further details, are shown in Fig. 11.35. The reference power station of the design study is modern and incorporates all the emission control features now required by US law. The comparison study adds only the carbon sequestration feature, but this is a large step, as will be seen.

The complex system illustrated in Fig. 11.35 removes about 90 percent of the carbon dioxide from the exhaust gas and results in highly pure CO_2 at 136 Bar, potentially for sale. The pumps, blowers, coolers, reboilers and compressor in this plan actually consume about 27 percent of the plant's electric power output. Of this power need, 49 percent is for the solvent regenerator, right side of the figure, and 34 percent is to compress the output carbon dioxide.

It is seen from Table 11.4 that the carbon dioxide-scrubbed plant, nominally at 500 MW, has reduced output and needs a higher price of electricity, 9.7 c/kWh vs. the control plant with electricity cost 4.92 c/kWh. It is also seen that the SO_2 sulfur dioxide emission is completely removed from the plant with the MEA scrubbing system.

The reference plant is designed to operate on low-sulfur US coal, 0.48 percent sulfur (typically from Wyoming), and meets requirements of the US Clean Air Act of 1990 for emissions of sulfur dioxide and other noxious byproducts, regulated in connection with the formation of acid rain downwind from power plants. So, the reference plant already has flue gas desulfurization FGD, electrostatic precipitation EPP, selective catalytic reduction SCR and low nitrogen emitting burners LNB. The capital cost of the MEA scrubbed plant is increased from $571 M to $705 M. Including the reduced output, the capital cost referenced to capacity is increased to $2.16/W from $1.24/W. (These are both quite low numbers, in fact.) The performance of the MEA system would be degraded if higher-sulfur coal typical of the eastern US were used. The output carbon dioxide is compressed to 136 Bar and assumed to be transported 65 km for disposal. The price for the coal delivered at the plant was assumed to be $23.19 per tonne; a recent price for Wyoming coal at the mine is about $10 per ton, and the shipping cost is comparable. This analysis does not address what is done with the compressed carbon dioxide, that only in select cases has been purchased and used to pressurize ageing oil fields to push more oil out. Unfortunately, the ultimate fate of such carbon dioxide is quite open to question.

There is a literature, e.g., (Huijgen et al. 2007), devoted to the methods for getting carbon dioxide to react to form carbonate minerals, that would provide a stable seques-tration. One possibility is to spread powdered olivine $(Mg,Fe)_2 SiO_4$ on farmland; olivine is abundant and costs about $25/ton. Pulverized olivine is known to "weather" relatively quickly, taking carbon dioxide from the air, and forming stable carbonate minerals. The "weathering" reaction is:

$$(Mg,Fe)SiO_4 + 2CO_2 \leftrightarrow 2 \ (Mg,Fe) \ CO_3 + SiO_2. \tag{11.8}$$

Table 11.4 *Comparative performance and costs for new 500 MW Coal Power Plants.*

parameter	units	reference plant	w/CO_2 control
net plant capacity	MW (net)	462	326
CO_2 emission rate	g CO_2/kWh (net)	941	133
SO_2 emission rate	g SO_2/kWh (net)	2.45	0.0003
NO_x emission rate	g NO_x/kWh (net)	0.45	0.58
CO_2 sequestered	10^6 tonne CO_2/yr		2.58
cost of electricity	$/MWh (net)	49.2	97.0
CO_2 mitigation cost	$/tonne CO_2 avoided		59.1

From Rao and Rubin, (2002), Table 4.

This reaction is exothermic but is known to be slow. In the roughest estimate, a ton of olivine at \$25 might eventually stably remove a ton of CO_2. The cost of removing a ton of CO_2 in the MEA process, according to Table 11.2, is \$59, even before turning the gas into a stable compound. The possibility of using olivine directly in reacting with flue gas as in a coal plant exhaust has been considered by Kwon et al, 2011; Huijgen, Comans, and Witkamp (2007), *Energy Conversion and Management* 48, 1923; Kwon, Fan, DaCosta, and Russell (2011) p. 23.

Their laboratory experiment showed that the rate of the reaction of $MgSiO_4$ with CO_2 was increased by raising the temperature to about 200°C, but that adding water vapor to the carbon dioxide did not have much effect. A chemical engineering analysis of this possibility using a fluidized bed reactor would seem appropriate. The likely result is that the rate of the reaction even at raised temperature is so slow that an impractically large bed of heated pulverized olivine would be needed to react the bulk of the CO_2.

On the other hand, consider the situation if such a carbon mineralization process could be made practical. The result Eq. (11.8) basically is that two molecules of CO_2 are removed per one of olivine, leaving three new molecules of carbonate and quartz. Looking at Table 11.4, a single 500 MW power plant calls for sequestering at least 2.58 million tonnes[47] (2.64 million short tons) of CO_2 per year that otherwise goes up the chimney. The successful mineralization remedy would require approximately the same number of railroad cars bringing in pulverized olivine as bringing in the Wyoming coal, and then again fully loaded taking away the resulting carbonate residue. Such a massive amount of residue might find some use in paving or replacing gravel from gravel pits, but also might end up being dumped in the ocean. This brings to mind the enormity of the discharge to the atmosphere from coal power plants. Roughly, the 2.64 million short tons, at 115 tons per car and 115 cars per train is 200 trainloads per year, for one coal plant. The mineralization remedy, if successful, would more than double the train loads, to add the mineral input and output, and create new disposal questions and expenses.

It is evident from Figs. 11.32–33 that coal is a major source of energy for electricity and also of potential global temperature rise. Fig. 11.32 shows that coal is the source of 17.9 Quads of energy in the US in 2014 and that coal was the input power for 39 percent of electricity generation that totals 39 Quads, where a Quad is 1.055×10^{18} J. (Total US energy use in 2014 was 98.3 Quads.) Fig. 11.34 shows that energy leads to the largest predicted global surface temperature rise. The discussion here of carbon capture shows that such efforts appear to double the cost of electricity from a modified coal plant, although the final cost is not impossible. (The 9.7 c/kWh is less that the figure usually quoted for New York City, 14c/kWh.) However, such installations have not been adopted except in one or two isolated situations, and appear to be greatly limited by plausible and credible geological locations to permanently dump underground vast amounts of carbon dioxide gas. We have seen previously that the mineralization-on-site possibility, even if successful, would be extremely awkward to carry out. In large part these possibilities all seem unrealistic and the remedy in more practical terms is to switch to other sources of electric power.

[47] One tonne = 1.023 short tons (that are 2000 lb).

11.9   Localized self-sufficiency vs. the large grid

Local assembly of several elements of renewable energy, including solar and wind, with battery and/or hydrogen fuel cell storage, is technically possible but requires competence or outside management, beyond the capital expense. This might become part of the expertise of future home or subdivision builders, but is not so at present. The diversity of sources and storage in technically available systems will allow a continuous source of power completely independent of the grid. Alternatively, such a system, if the local government and utility are open to it, could exchange power with the grid and actually form a part of the grid as a "virtual power station". In either case, a system such as shown in Fig. 11.36 (Wang and Nehrir, 2008) would be suitable.

This system was designed for a cluster of five homes in the Pacific northwest. It could be reduced in size and complexity, for example by removing the fuel-cell portion in favor of more batteries. It would also appear that such a system could be expanded easily with a larger turbine and/or more solar cell area, to provide, for example, power for a ranch or the dairy farm in Indiana that was mentioned in Section 11.5, or a ski resort or a small municipality. Operating such a system would require technical competence and depends on the larger economy to provide the essential high-technology parts including the wind turbine, solar cells, electrolyzer, compressor and fuel cells. The assumed 50 kW wind turbine might cost $150,000, a 100 kW unit $350,000, while a 2000-Watt turbine for a single home is $2,499 with free shipping from Home Depot. An analysis of costs in such a system was given by Nelson et al (2006). One conclusion at that time was that the fuel cell cost was very large, leading to a recommendation that all storage would better be done with batteries. That would simplify the construction and maintenance of such a system, especially for a smaller installation as for a single home.

The larger part of this literature is aimed at freedom from the grid in the interest of lower cost and the need of the isolated home-owner or rancher. A second interest in the "stand-alone" power system is for augmenting the power grid, a recent article in the *Wall Street Journal* (Sweet, 2017).

The article mentions the utility Pacific Gas and Electric as looking to batteries to solve a supply-demand mismatch, storing excess solar power and feeding it as needed to the grid. Shown is a large combined solar and wind facility in Palm Springs, CA, with comment that California's solar farms create so much power during daylight hours that they often drive real-time wholesale prices in the state to zero. Meanwhile, the need for electricity can spike after sunset, sometimes sending real-time prices as high as 100 c/kWh. (An image of a large photovoltaic plant in California, with capacity 0.55 GW, is shown in Fig. 10.15.) The article states that US solar installed capacity in 2016 is 34.1 GW, with 45 percent of it in California. The conventional solution to the need for extra power is the natural gas "peaker plant" that is turned on when needed. According to this article, such a plant leads to electricity costs between 15.5 and 22.7 c/kWh, while a competing lithium battery plant would give electricity at a cost between 28.5 and 58.1 c/kWh. Nevertheless, the utilities expect the battery plant cost to fall, and

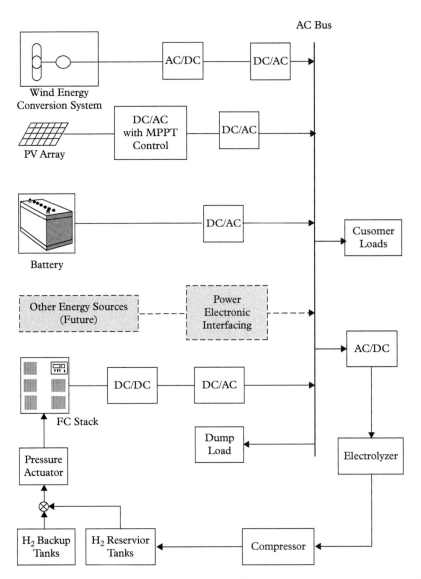

Fig. 11.36 *Schematic plan of stand-alone power system, originally designed for a cluster of 5 homes, with an average peak demand of 14.6 kW. In this figure MPPT stands for maximum power point tracker, to orient the solar array toward the Sun. The wind energy control system (WECS) includes a control over the pitch angle of the wind turbine blades. The "dump load" unit is provided to take up the power if the hydrogen tanks are full, and would not be needed if the system were tied to the local grid that would accept excess power. The "customer loads" output could include an interface with a local utility grid, making the stand-alone system a "virtual power station". (Wang and Nehrir, 2008) Fig. 1*

Edison International in the Los Angeles area is going forward with the "virtual power plant" suggested in Fig. 11.37 that will consist of more than 100 office buildings and industrial properties. The Tesla Li-ion batteries in this virtual power plant will be able to deliver more than 360 MWh to the grid on short notice, enough to power 20,000 homes for a day. At other times the batteries will allow those buildings hosting the solar arrays to get power at lower rates. The project is being managed by Advanced Microgrid Solutions, mentioned as source of Fig. 11.37. A smaller similar project is described for Consolidated Edison in New York, with a statement that by using the "virtual power station" the company can put off for some years a much larger and more costly addition to its generating capacity. The "virtual power station" plans shown in Fig. 11.37 and mentioned for Consolidated Edison in New York could be improved by adding wind turbine capacity, as in the plan of Fig. 11.36. In an urban situation, as on the roof of the handsome new 388 Bridge Street apartment building in Brooklyn, New York, vertical axis wind turbines are very suitable, interesting to look at, easy to operate without

What is a Virtual Power Plant?

Utilites use solar panels, batteries and software to create 'virtual power plants,' or networks that store excess power and release it to buildings or the grid when needed.

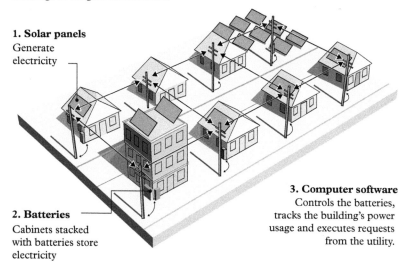

1. Solar panels
Generate electricity

2. Batteries
Cabinets stacked with batteries store electricity

3. Computer software
Controls the batteries, tracks the building's power usage and executes requests from the utility.

THE WALL STREET JOURNAL.

Fig. 11.37 *Artist's depiction of a "virtual power plant", where a community agrees to a common solar power and battery storage "stand alone" power system with emphasis on its active connection to the local utility grid. This system is crude compared with that of the previous figure, but represents a welcome and unexpected idea of cooperation with a power utility. (Sweet, 2007)*

Source: Advanced Microgrid Solutions

need for altering the yaw angle or the blade pitch as would be needed for a horizontal axis wind turbine (see Section 8.1.6).

In the state of Nevada things have been quite different, as the local utility backed away from allowing consumers with roof-top solar to sell power back to the utility. However, it appears[48] that the Nevada utility has restored an earlier more favorable agreement on accepting power from "virtual" sources, see Fig. 11.37.

There is likely to be continuing dispute over such things in some states, even opposition in legislatures, as the US government has, for the moment, turned back to favor the conventionally conceived economic interest of businessmen who would rather not worry about pollution and who see renewable power as a threat. Recent news items have included that more employment now occurs in renewable energy than in conventional energy, and that the person of job description "wind turbine technician" is most sought after in today's US economy.

---

[48] http://fortune.com/2016/09/13/solarcity-berkshire-solar-settlement/

12

Prospects for Sustainable Power and Moderate Climate

12.1   A brief summary

To summarize briefly, we have found in earlier chapters of this book that on a hundred-year time scale the supply of fossil fuels is brought into question, in particular the supply of easily harvested oil. On shorter time scales the sharply rising level of carbon dioxide in the atmosphere is tied both to the modest temperature rise of the planet and to industrial use of fossil fuels, notably coal burned to produce electricity and oil used in transportation, especially automobiles. It is well documented that the excess carbon emitted globally is approaching one trillion tons, and that carbon dioxide molecules absorb in the infrared, trapping the earth's outward radiation. Historically such changes have led to rising sea levels that are indeed presently being seen. With notable dissenters, scientific and enlightened world opinion broadly accepts the logical conclusion that carbon fuels should be phased out and energy obtained from sustainable and non-polluting sources including wind, solar and hydroelectricity. This is considered more urgent because the carbon dioxide has a lifetime of nearly a century after emission into the air. Such changes to renewable fuels are indeed in progress.

The preceding chapters have addressed the physics and technology underlying these sustainable energy sources, and have also explained the mechanisms by which carbon fuel burning changes the balance of energy flow from the sun and back to outer space, that sets the earth's temperature. It does not take too much physics to realize that higher temperature of the globe means more water in the ocean and higher sea level evidenced by the gradual inundation of Tangier Island in the Chesapeake Bay and the need to relocate city streets to higher ground in Miami Beach, Florida. The centuries-old record extracted from ice cores in Greenland and Antarctica clearly tie together higher carbon-dioxide concentration, global temperature and sea level, and present surveys of Arctic and Antarctic ice show a potential of 80-meters sea-level rise in the remaining ice, now likely on schedule for melting. The connection between slightly higher temperature and wider swings in local weather is harder to quantify, but one basic principle is that higher sea-water temperature means higher vapor-pressure of water in the atmosphere and the condensation of water vapor back to liquid drops releases huge amounts of kinetic energy powering hurricanes, typhoons and tornados.

Physics and Technology of Sustainable Energy. E. L. Wolf, Oxford University Press (2018).
© E. L. Wolf. DOI: 10.1093/oso/9780198769804.001.0001

The wide disgust with polluted air has also led momentum to curb dirty fuels. These issues have been settled in Los Angeles and New York but are still in play in Delhi and Beijing. A competition exists between these concerns and the growing energy needs of the increasing world population. The historical records show many ice-age/interglacial (hot, high sea level) eras in the earth's past, involving many 100-meter sea-level changes, but these global changes exhibit time-constants on the order of 5000 years (see Fig. 1.11) The 80-meter sea-level rise may be already set into motion,[1] by the trillion-ton dose of carbon, but there is plenty of time to move away from Miami Beach, for example, and to build new cities on higher ground. The massive loss of value of coastal real-estate might be mitigated by a tax on the carbon emissions paid to the citizens of Miami Beach, if not to the less mobile citizens, e.g., of Dhaka, Bangladesh. Such a measure is presently proposed by a group led by former US Secretaries of State George P. Schultz and William Baker (Schwartz, 2017).

The proposal is for a tax of $40/ton on carbon dioxide by emitters. The proceeds are suggested to be distributed to citizens at large, leading to an estimated dividend of $2000 per US family. A provision of the proposal is that companies that emit greenhouse gases should be protected from lawsuits over their contributions to climate change. The proposal is supported by leaders including Michael Bloomberg, former mayor of New York, Steven Chu, former US Secretary of Energy, and William Summers, former US Secretary of the Treasury, and Stephen Hawking, eminent physicist.

12.2   The present situation, with emphasis on jobs and the economy

On the near term and on the time-scale of hundreds of years these changes toward sustainable energy and away from flood-prone locations provide opportunities for economic activity, for example building apartments on higher ground that are better insulated, need less heating and cooling, and have rooftop solar installations and batteries to lessen dependence on a power grid.

The American workforce has already filled many new jobs in sustainable energy sectors, as indicated in Table 12.1.

As seen in Table 12.1, more persons, 373,807, in the US in 2016 were employed in power production by solar energy, than in all of fossil-fuel power generation,[1] listed as 187,035, and attributed as Coal, 86,035, Gas, 52,125 and Oil and Petroleum, listed as 12,840. By comparison, 68,176 were employed in nuclear power generation.

The many solar jobs are concentrated in states including California, North Carolina, Arizona, Nevada and Utah. A summary of utility-scale solar-electric generation capacity in the US is shown in Fig. 12.1.

It was reported (Stewart and Higgins, 2017) that Tesla Corp, led by Elon Musk, has agreed to build the world's largest lithium battery system to back up a wind farm in

[1] http://www.forbes.com/sites/niallmccarthy/2017/01/25/u-s-solar-energy-employs-more-people-than-oil-coal-and-gas-combined-infographic/#a003ecc7d27f.

Table 12. 1. Generation and Fuels Employment by Sub-Technology.

	Electric Power Generation	Fuels	Total
Solar	373,807	–	373,807
Wind	101,738	–	101,738
Geothermal	5,768	–	5,768
Bioenergy/CHP	26,014	104,663	130,677
Corn Ethanol	–	28,613	28,613
Other Ethanol/Non-Woody Biomass, incl. Biodiesel	–	23,088	23,088
Woody Biomass Fuel for Energy and Cellulosic Biofuels	–	30,458	30,458
Other Biofuels	–	22,504	22,504
Low-Impact Hydroelectric Generation	9,295	–	9,295
Traditional Hydropower	56,259	–	56,259
Nuclear	68,176	8,595	76,771
Coal	86,035	74,084	160,119
Natural Gas	52,125	309,993	362,118
Oil/Petroleum	12,840	502,678	515,518
Advanced Gas	36,117	–	36,117
Other Generation/Other Fuels	32,695	82,736	115,431

Table 12.1 Employment in 2017 shows more workers in solar energy electric power generation than in all fossil-fuel power generation. In this chart CHP refers to "combined heat and power".
United States Department of Energy United States Energy and Employment Report, 2017.

Australia, to be built by the French firm Neoen. This arrangement with the Australian government was prompted by an earlier blackout, during a heat wave, of a large electric grid in Australia.

The project will provide 129 MWh of storage, will use 500 to 600 of the Tesla "Power Pack" lithium-ion battery modules, and will cost about 80 M$. It will provide power for 30,000 homes, about the number of homes that lost power during the long blackout. The article mentions an estimate that the grid storage market in the US will reach 3 Billion $ by the beginning of the next decade, up from 320 M$ in 2016.

Another anecdotal report is that the auto-maker Volvo, now owned by the Chinese firm Geely, plans to cease production of gasoline engines in favor of electric engines for its future cars.

Utility-Scale PV Installed Capacity, Top 10 States, as of August 2016

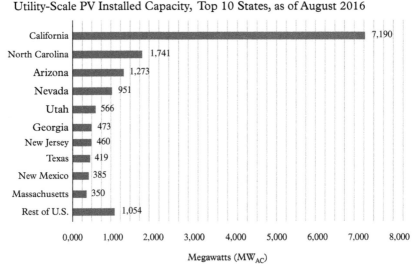

Fig. 12.1 *Installed utility scale photovoltaic capacity in the US as of August 2016. The sum of all these numbers is 14.862 GW*

Source: Energy Information Administration, 2016.

A trend toward renewable energy is clearly in place. A prediction of how this may play out in the future is given by Bloomberg New Energy Finance, shown in Fig. 12.2.

12.3   Approaches to a carbon-free economy

The changes toward renewable energy have been driven in large part by government rules related, e.g., to mileage for autos, catalytic converters for car exhausts, and portions of power to be provided by renewable sources. The changes also represent entrepreneurship of small companies, as they recognize that they can make and sell such new products as wind turbines and solar panels. We have seen in the previous section that the economy is indeed changing toward renewable energy including the prediction in Fig. 12.2 that by 2040 the leading producers of electricity will be wind and solar in the US.

The transformation of the entire US economy, not just electricity, entirely away from fossil fuels oil and coal, however, is not going to happen under the present rules. Deep de-carbonization of the whole economy would require a strong government role. While there is no early likelihood of such a rule at the national level, it is quite possible at the state level. For example, a bill has been introduced into the Senate of California[2] requiring all energy in the State of California to be provided by renewable sources by 2045.

---

[2] http://sd24.senate.ca.gov/news/2017-05-02-california-senate-leader-introduces-100-percent-clean-energy-measure.

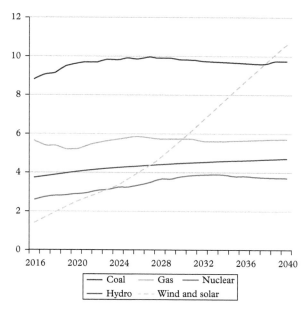

Fig. 12.2 *Annual Electricity output by technology, units of 1000 TWh. Ten on this scale corresponds to average power 10,000 TW/24 × 365 = 1142 GW. In 2040 the leading producers, in order, are predicted to be:*[3] *wind and solar, coal, gas, hydro and nuclear*

Is it possible for the US to essentially cease using oil and coal and remain prosperous? Before answering that this will never happen, one should consider the situation of Sweden. As suggested in Fig. 12.3, starting around 1970 that country started to change from a situation similar to the US, to a new era with increasing prosperity co-existing with greatly reduced carbon emission per person. The prosperity of this small country is due in large part to its thriving business community, with well-known global companies including IKEA, the merchandizer; Volvo, the auto maker, and Nokia, a pioneer in wireless telephones. This is a capitalist economy, rather than a planned economy. At the same time the country offers its citizens substantial benefits including free health care and free university education. It has been suggested that the "safety net" of Sweden stimulates small firm entrepreneurship, innovation and prosperity, allowing the essential risk-taking to be undertaken reducing the chance of impoverishing the family of the entrepreneur. This small country is proof that fossil fuel use is not needed for prosperity in a modern way of life.

One of the energy efficiencies that underlies Fig. 12.3 is "district heating". Common steam heating is provided for sections of a city, as is presently the case in portions of Manhattan, New York City. The common heat source for Swedish district heating includes burning of garbage and other biofuels, as well as "secondary" or "waste" heat from industrial processes, that can include nuclear power plants. (It can also allow

[3] https://about.bnef.com/blog/coal-and-gas-to-stay-cheap-but-renewables-still-win-race-on-costs/

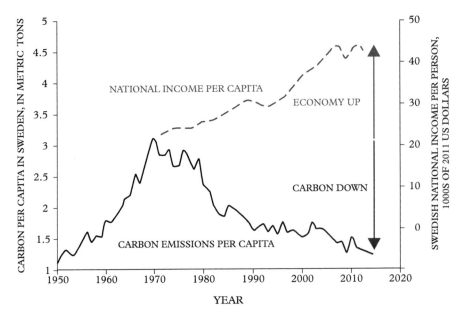

Fig. 12.3 *Carbon emission per capita and income per capita in Sweden from 1950 to 2020, a period of deep de-carbonization of the local economy. (Pierrehumbert 2016) Fig. 1, http://dx.doi.org/10.1080/00963402.2016.1145908*

direct sun heating, of course.) A high temperature is not needed to provide building air- and water heating, so that exhaust energy that usually goes to waste is suitable. An additional efficient source of low-temperature heat is a heat pump bringing heat from underground rock that is significantly warmer than the air temperature all through the winter. This can be done on individual homes, but is probably better managed and cheaper as part of a larger system. If there is excess electricity, e.g., from a wind turbine, it can also be used, in the absence of batteries, to heat the common reservoir in a district heating system. A complete description of such a district heating system is given by Sibbitt et al (2012).

12.3.1 The Civil Engineering approach of M. Z. Jacobson and collaborators

A series of brilliant detailed papers on the organization of carbon-free energy economies has been written by Prof. M. Z. Jacobson in the Civil Engineering Department at Stanford University, and several collaborators.

The most recent paper that we will focus on gives a detailed plan for an entirely carbon-free energy system for the US and demonstrates that the planned system provides enough storage to support the electric grid with no outages over a period of six years, from 2045–2050.

The choice to leave nuclear power, bio-fuels, and carbon-sequestered fossil fuel energy out of the energy plan is explained in a longer companion paper (Jacobson et al, 2015a), and earlier references therein.

Very briefly, nuclear power is omitted on basic concerns of potential for weaponizing, the much longer times needed to bring plants to the grid, and the great difficulty with storing wastes. Bio-fuels are rejected on a general argument that on a per square meter basis solar cells can produce 20 percent efficiency while photosynthesis is only on the order of 1 percent, plus the fact that solar cells produce zero emission. Carbon sequestration is excluded, even though it is part of the assumption of the IPCC reports, on the basis of costs and more realistically on the absence of any workable technology, as was concluded in Section 11.8. These exclusions can be looked on also as a confident demonstration that purely renewable energy is robustly available in non-unique forms, even those excluding nuclear reactors and any vestige of fossil fuel power.

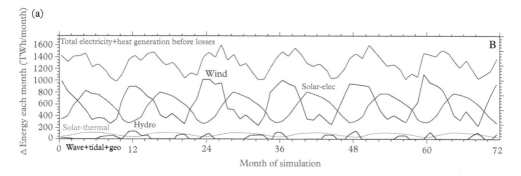

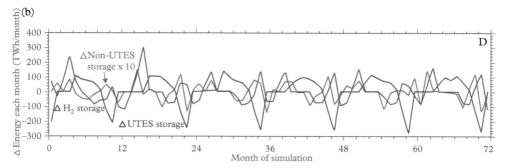

Fig. 12.4 *Simulated performance of proposed carbon-free US energy economy over six years (upper). Total power (TWh/month) as sum of solar-electric, wind, solar-thermal, hydro, wave, tidal and geothermal. Peak power on this graph, 2200 TWh/month comes to 2.22 TW. It is seen that main contributions from wind and solar electric tend to complement each other on an annual basis. Jacobson et al 2015, Fig. 2b (lower). Changes of energy in storage over 72 months of simulated carbon-free US energy economy. Main forms of storage are UTES (underground thermal energy storage) typified by district heating, and Hydrogen storage. Non-UTES storage, shown on expanded scale, includes pumped-hydro storage, that is assumed by the model to be expanded in capacity by addition of extra turbines to existing dams and modest expansion in number of hydroelectric dams in Alaska only. Jacobson et al 2015, Fig. 2d*

The important work of Jacobson et al (2015) has been attacked by Clack et al (2017), ostensibly for its exclusion of nuclear, biofuels and carbon-sequestered fossil fuels, omissions suggested to raise costs. The rather strident and uninformative attack has promptly been refuted line by line by Jacobson et al (2017), see also[4].

The author agrees with the assessment (Jacobson et al, 2017) that the analysis of Clack et al (2017 is "is riddled with errors and has no impact on Jacobson et al's 2015 conclusions".

The proposal is for an entirely new carbon-free energy economy to be completed by 2050, with detailed plans state-by-state provided by Jacobson et al (2015). The idea is to go to the bank for a loan to build the new system, with the bank making money in the end from repayment with interest as the new economy flourishes. The entire transportation system is proposed to be run on electric motors, fueled by batteries or by hydrogen (using fuel cells), or directly on hydrogen. This includes aviation, with the known ability to power a mildly modified aviation turbine with liquefied hydrogen, or to use the hydrogen to turn a propeller via a fuel cell and an electric motor.

The other main change is a vast increase in the amount of storage, so that the grid is always supplied with electricity while the sun and wind power vary. Heat energy is also stored to be used later primarily as heat energy, although conversion of some heat energy to later electricity is also allowed, as in the concentrating solar CSP plant such as described in Fig. 9.7. The direct use of sun to heat fluids and store energy in underground rock (UTES) decouples heating from the electric grid. Low-cost electricity at night is used to freeze water to ice that is melted in the next day for air conditioning. Enlarged hydropower storage capacity is planned with extra turbines on existing dams to increase the available short-term electric capacity. It is pointed out that there are many un-utilized dams, as suggested by our description (text following Fig. 8.18), of the Cannonsville Dam on the Delaware River north of New York City, including a whole set of dams in existence on the major waterways of the US. It is further proposed to build four new hydropower dams in Alaska.

In more detail, following Jacobson et al 2015, p. 15065, the key elements in the plan, used in the simulation mentioned in Fig. 12.4, are "(i) UTES (underground thermal energy storage) to store heat and electricity converted to heat; (ii) PCM-CSP to store heat for later electricity use; (iii) pumped hydropower to store electricity for later use; (iv) H_2 to convert electricity to motion and heat; (v) ice and water to convert electricity to later cooling and heating; (vi) hydropower as last-resort electricity storage and (vii) demand response (DR)". DR is an acronym for a smart grid system, or advanced metering infrastructure, using smart meters, smart thermostats, energy management and storage systems and dynamic lighting controls, to enhance efficiency. Further, "PCM-CSP" is a variation on use of molten salt heat storage (see Fig. 9.7) replaced by

---

[4] https://www.ecowatch.com/pnas-jacobson-renewable-energy-24446. (Accessed 30 Jan. 2018.)
http://sd24.senate.ca.gov/news/2017-05-02-california-senate-leader-introduces-100-percent-clean-energy-measure.
https://www.ecowatch.com/pnas-jacobson-renewable-energy-24446.

a "phase change material" PCM heat storage system, believed to be of higher efficiency than the molten salt system.

The costs of the new system include extensions to short- and long-distance power lines. The long-distance costs are taken as 1.2 c/kWh for 1200 to 2000 km lines. The basic assumption is that 30 percent of all wind and solar electric power generated are subject to long-distance transmission line costs. Assumed costs for hydrogen are 4 c/kWh-to-H_2 for the electrolyzer, compressor, storage equipment and water.

The proposed new installed renewable power capacity, along with estimates of the numbers of jobs created, are summarized in Table 12.2.

The time-frame of the proposed transition from present to future carbon-free energy economy is suggested in Fig. 12.5, according to Jacobson et al (2015).

The important work of Jacobson shows clearly that it is possible to plan a carbon- and nuclear-free energy economy, calling for greatly enlarged energy storage, that will provide a black-out-free electric grid at reasonable cost. The opposition that such plans face by fossil fuel and nuclear advocates is suggested by the strident and misleading attack mounted by Clack et al (2017).

Table 12.2 *Summary of proposed carbon-free power capacities with estimated numbers of created permanent jobs in the US. In this Table, "CSP plant" refers to concentrating solar power plant. (A companion Table 11 estimates a loss of 3,859,000 jobs from eliminating energy generation and use from the fossil fuel and nuclear sectors.) Table 10 of Jacobson et al (2015).*

Energy technology	Installed MW	Jobs per installed MW		Number of permanent jobs	
		Low	High	Low	High
Onshore wind	1 639 819	0.14	0.40	229 575	655 927
Offshore wind	780 921	0.14	0.40	109 329	312 368
Wave device	27 036	0.14	0.40	3785	10 814
Geothermal plant	20 845	1.67	1.78	34 811	37 103
Hydroelectric plant	3789	1.14	1.14	4319	4319
Tidal turbine	8823	0.14	0.40	1235	3529
Residential roof PV	375 963	0.12	1.00	45 116	375 963
Com/gov roof PV	274 733	0.12	1.00	32 968	274 733
Solar PV plant	2 323 800	0.12	1.00	278 856	2 323 800
CSP plant	363 640	0.22	1.00	80 001	363 640
Solar thermal	469 008	0.12	1.00	56 281	469 008
Total	6 288 375			876 275	4 831 206

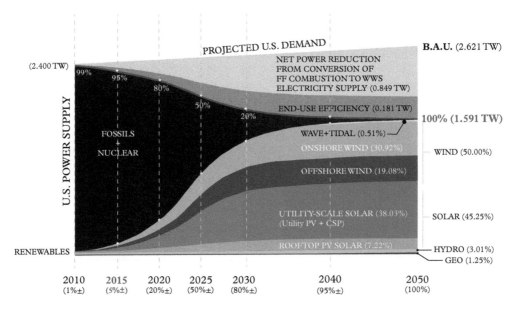

Fig. 12.5 *Schematic diagram of proposed transition to carbon-free energy economy as described by Jacobson et al 2015. In this diagram "B.A.U." means business as usual. (Jacobson et al 2015, Fig. 5)*

12.3.2   The renewable energy future

A summary and suggestion for practical advance on the renewable energy front was offered by Sachs (2016).

Sachs suggested that all sources of energy will be needed, and that pragmatic measures will become more mainstream. Thus, this view is of continued expansion of population and demand for energy, provided by a mix of renewable energy with increasing contributions from solar and wind, nuclear power, perhaps with innovation in reactor design such as integral fast reactor technology that uses its own nuclear wastes as energy inputs in a closed-fuel cycle. Such reactors have also passive safety features that would make them safer in case of a shutdown. Sachs suggested that continued contributions of fossil fuel will be made possible by development of carbon sequestration technology, in which the carbon-dioxide emissions are captured and stored underground. An added working example of this is the Boundary Dam coal-fired power plant in Saskatchewan, Canada. It is suggested by Sachs that experience with this technology will lead to diminishing costs. The Boundary Dam CCS facility[5]

---

[5] http://www.saskpower.com/our-power-future/carbon-capture-and-storage/boundary-dam-carbon-capture-project/.

follows the general plan described in connection with Fig. 11.34, suggested in Table 11.2 to give electricity at 9.7 c/kWh, up from 4.9 c/kWh with no sequestration at a coal fired power plant.

The situation at the Boundary Dam location is favorable in that an oil field is nearby, purchasing the pressurized CO_2 for Enhanced Oil Recovery EOR. The prices available from the Boundary Dam CCS facility are similar to those quoted. The Saskatchewan Power Utility speaks of future disposal of the CO_2, beyond the needs of the depleted oil field, in an underground aquifer. Unfortunately, these conditions are rare and are not at all likely to be widely utilized. On-site conversion of CO_2 to stable carbon-bearing minerals such as olivite is extremely awkward (and has never been utilized in any test situation) as was discussed at the end of Section 11.8.

12.4 The climatic future

The future of the earth's climate is quite difficult to assess. We have seen in Fig. 1.11 a 420,000-year history of temperature and carbon dioxide concentration, including four cold periods, glaciations, and five interglacial warm periods, as at the present. (According to the terminology in the literature, this whole 420-ky period lies in an "ice age", since the polar ice caps have been retained over the whole era. The four cold periods shown in Fig. 1.11 are referred to as "glaciations".) In that history the carbon dioxide level never exceeded 300 ppm, while it is presently at 407 ppm. (In the more distant past however, higher pressures and temperatures have occurred, including relatively short "hyperthermal" heating events) The recent global temperature has risen only about 1°C since the industrial age, as shown in Fig. 1.7a, but the present temperature seems to be below what the historical record would suggest for 407 ppm. So further warming is clearly expected, but there is little guidance on the extent and time scale of the expected further warming.

12.4.1 The role of the oceans in past "hyperthermal events"

The oceans cover 72 percent of the earth and have a volume of 1.3×10^9 km^3. It is estimated that in recent years the oceans are taking up about 2 Gt C/y, and that they have absorbed about 39 percent of the total anthropogenic carbon emitted over the past 200 years. This dissolved carbon is almost all in the top layers of the ocean and has changed the pH so far by about 0.1 at the surface with almost no change of pH in the deep sea (the average ocean depth is 3800 m.) Here pH is defined as the negative of $\log_{10}$ of the hydrogen ion concentration, in moles per liter. Pure water is neutral at pH 7, with pH values less than 7 being acidic. Sea water contains dissolved carbon according to the reaction:

$$CO_2 + H_2O \leftrightarrow HCO_3^- + H^+. \qquad (12.1)$$

A recent article[6] has drawn attention to the most extreme climatic possibilities that have been suggested for the global climate as a result of the industrial age addition of about one trillion tons of carbon, raising the carbon dioxide level from about 280 ppm to 407 ppm. The recent article points to the accepted scientific fact that very long ago, 252 million years, at the end of the Permian era, a "hothouse earth" condition occurred, leading to the extinction of 90 percent of existing life, on land and in the oceans, at that time. (Brand et al, 2012; Kaiho and Koga, 2013). That era was also one of extreme volcanic eruptions, over millions of years, leaving for example vast areas of Siberia covered with basaltic rock, the result of extensive lava flows. This event is also referred to as having a "Superanoxic" ocean, greatly oxygen deficient, and it is suggested that not only was the sea depleted of oxygen, but also (Kaiho and Koga) that the whole atmosphere was depleted by 14 percent of its oxygen. (Another mentioned present estimate[1] is of 1.8 trillion tons of carbon that is held in permafrost eligible for melting, that might be released in further heating as the potent greenhouse gas methane, creating further heating.) It is estimated that the equatorial ocean in this extinction era was as hot as 40°C, that is 104°F. This event is evidence for the instability of the earth's climate. On the other hand, there is no present evidence for extensive loss of ocean oxygen nor present indication of volcanic eruptions. The only thing that has happened in our era is the rapid emission of about a trillion tons of carbon in the form of carbon dioxide.

To be sure, such warming has not occurred in the past 420,000 years as clearly shown in Fig. 1.11. But the extinction event did occur at the end of the Permean 252 million years ago, and plausibly could happen again as a consequence of the present sharp increase in carbon dioxide levels. A discussion of this event also brings in the important role of the ocean in producing extreme climate change.

This extreme warming and extinction event of 252 My ago, developed slowly over millions of years, and is suggested by Brand et al 2012 to have led to tropical ocean surface temperature of 39°C or 102.2°F, a second estimate. The extinction event is interpreted as being driven by release of greenhouse gases carbon dioxide and methane, possibly from volcanism, plus poisonous hydrogen sulfide generated in the ocean. It appears (Grice et al, 2005) that hydrogen sulfide can be generated in the ocean as a response to low-oxygen (anoxic), high-sulfur conditions, called "euxinic".

12.4.2 The euxinic ocean

This technical term refers to an ocean deficient in oxygen and rich in sulfur, a condition believed to have occurred in past "hyperthermal" events of the earth climate. The euxinic ocean is stratified, with a small oxygen-rich layer at the top and a deep anoxic layer reaching to the bottom. The layers are suggested to invert in the hyperthermal events, bringing carbon and sulfur to the top, to leak massively to the atmosphere.

The ocean as we know is too salty to drink, so it surely contains Na^+ and K^+ ions. Fish have gills to get oxygen from the sea and fresh water, and run-off of fertilizer into rivers and lakes is associated with fish-kills from lack of oxygen, as well as algal growths.

---

[6] http://nymag.com/daily/intelligencer/2017/07/climate-change-earth-too-hot-for-humans-annotated.html.

Seawater has for billions of years received chemicals from volcanic vents in the sea floor, to allow iron, sulfur, phosphorus and many other elements to slowly accumulate. The weight percents of impurities in seawater are listed as Cl, 1.84; Na, 1.08; Mg, 0.129; S, 0.091; Ca, 0.04; K, 0.04; Br, 0.0067 and C, 0.0028. But the ocean is vast, with volume mentioned as 1.3×10^9 km³. So, the mass of carbon in the ocean, starting from the weight percent 0.0028, is 1.3×10^{18} m³ $\times$ 1000 kg/m³ $\times$ 2.8×10^{-5} = 3.64×10^{16} kg. Since a gigatonne is a billion times a thousand kg, this estimated mass of carbon in the ocean is 3.64×10^4 Gt C. This 36,400 GtC is mostly[7] in the form of bicarbonate ion, HCO_3^- as seen in Eq. (12.1).

There are suggestions that during "hyperthermal" high-temperature events, lasting typically 100 ky, ocean chemistry changed to allow some of the carbon to come to the surface as CO_2 and to massively escape into the atmosphere, reinforcing the greenhouse warming. Sulfur in the form of H_2S is also a possible release from the "euxinic" ocean. The oceans at present are not at all euxinic. The best example of a small euxinic region in the present world is the Black Sea. A cartoon describing such an ocean is shown in Fig. 12.6. The ocean conditions shown in Fig. 12.6a are referred to as euxinic conditions. There is present evidence for small amounts of hydrogen sulfide generation[8] near the coast of Namibia, in Africa. "Hot Sulphur Springs" in the State of Colorado reminds us that sulfur can get into surface water from deeper layers of the earth. The rapid warming and extinction of 252 Myr ago is said to have been driven by greenhouse gas from volcanic activity and possibly also by methane from melting of permafrost or release from methane hydrate (clathrate).

The euxinic ocean model is shown in panel a of Fig. 12.6. The upper oxygen-rich layer supports photosynthesis with acceptance of CO_2 from the atmosphere and releasing oxygen. Growth of organic matter in the upper layer, represented chemically as CH_2O, takes part in reducing sulfates to sulfide, according to the reaction:

$$2\ CH_2O + SO_4^{2-} \rightarrow H_2S + HCO_3^- \tag{12.2}$$

Certain bacteria (chlorobiacene, green sulfur bacteria) can reduce sulfate, to sulfide.

The modern ocean is not stratified as suggested in Fig. 12.6, but is now absorbing fossil-fuel carbon dioxide.

This absorption has been modeled by Archer et al (1997), who estimate that the oceans sequester 70–80 percent of the emitted atmospheric carbon on a time scale of several hundred years. Chemical neutralization of CO_2 by reaction with $CaCO_3$ on the sea floor:

$$CaCO_3 + CO_2 + H_2O \rightarrow 2HCO_3^- \tag{12.3}$$

accounts for another 9–15 percent decrease in the atmospheric concentration on a time scale of 5.5–6.8 kyr. (Calcium carbonate, one form being calcite, appears in ocean sediments and coral reefs. Calcite, along with Dolomite, $CaMg(CO_3)_2$, forms limestone.)

---

[7] https://en.wikipedia.org/wiki/Oceanic_carbon_cyclehttps://en.wikipedia.org/wiki/Oceanic_carbon_cycle.
[8] https://earthobservatory.nasa.gov/NaturalHazards/view.php?id=18791.

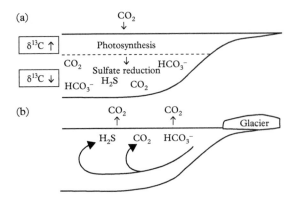

Fig. 12.6 *Schematic of layered sulfur-rich ocean such as believed to have led to massive carbon dioxide and hydrogen sulfide release at the end of the Permian period, killing most of existing life on land and in the sea. (a) Euxinic ocean in the million years preceding the event the deep oceans are suggested to have come to a stratified condition as shown. (b) Inversion of the strata lead to leading to huge release of carbon dioxide and toxic hydrogen sulfide gas when the "Anoxic Deeper Ocean" is pushed up into the top layer, as shown. This is referred to in Grice et al 2005 as "Photic Zone Euxinia": the Photic Zone being the upper oxygen rich sea surface, and "Euxinia" being the anoxic sulfur-rich condition. (Knoll et al 1996) Fig. 1*

(Reaction with $CaCO_3$ on land accounts for another 3–8 percent, with a time scale of 8.2 kyr.) The final equilibrium with $CaCO_3$ leaves 7.5–8 percent of the emitted CO_2 still in the atmosphere, even after 8.2 ky. According to Archer et al (1997), final consumption of all the atmospheric carbon dioxide excess comes only after 200 kyr, as the carbon dioxide is consumed by weathering of basic igneous rocks .

Chemical weathering of rocks can be illustrated with the dissolution of calcite into bicarbonate, starting with (12.1), followed by:

$$H_2 CO_3 + CaCO_3 \rightarrow Ca(HCO_3)_2. \tag{12.4}$$

After this weathering reaction, the soluble bicarbonate, $Ca(HCO_3)_2$ might be washed away in the rain.

The upper portion (a) of Fig. 12.6 indicates that the sulfidic deep portion of the euxinic ocean has reduced isotope 13 of carbon, compared to the oxygen-containing surface, showing the symbol "$\delta^{13}C\downarrow$". This is based on the definition:

$$\delta^{13}C = \left[\left(^{13}C/^{12}C \right)_{sample} / \left(^{13}C/^{12}C \right)_{standard} - 1 \right] \times 10^3 \tag{12.5}$$

This measured isotope ratio is an empirical tool to identify the origin of the carbon. Positive carbon isotope excursions (CIE), relative to the agreed standard, are found with organic carbon, originating in vegetation, while negative excursions are observed in inorganic carbon.

A review of the end-Permian extinction, at the Permian-Triassic boundary, (Payne and Clapham, 2012) concludes that the largest forces driving the event were the two episodes of "flood basalt" volcanism, vast eruptions of lava, the second of which occurred in Siberia. This massive event is estimated to have released 30,000 Gt C that led to great greenhouse warming and made the ocean anoxic. The chronology of these events is summarized in Fig. 12.7.

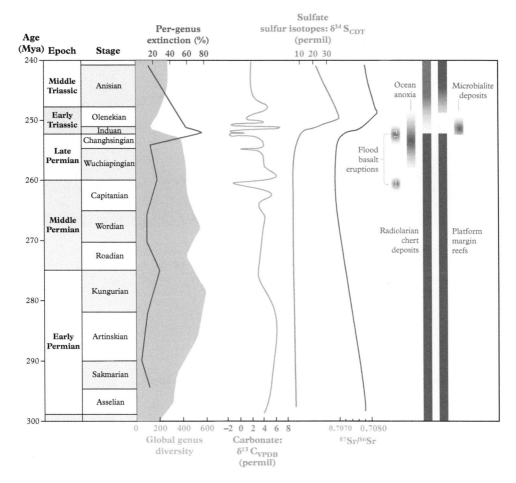

Fig. 12.7 *Summary of indicators in period containing the Permian extinction, a hothouse-Earth event at the Permian-Triassic boundary, at about 252 Myr ago. The left-most curve shows the percent of species going extinct, peaking at about 80 percent, while the left-most shading indicates the diversity of biological species showing a minimum after the event, during the Induan stage of the Early Triassic. The right-hand of the figure shows duration of events that accompanied the extinction: two stages of extreme volcanic activity (flood basalt eruptions) and periods of anoxic ocean behavior as described in text above. The second curve from the left shows oscillations in the carbon isotope ratio with peaks that approximately coincide with the two peaks in the volcanic activity. The large positive excursion in the sulfur isotope ratio following the extinction suggests a role of sulfur from the volcanic activity, perhaps indicating hydrogen sulfide in the atmospheric changes that occurred. (Payne and Clapham, 2012) Fig. 1*

The massive volcanic events that occurred in Siberia have been elucidated by the work of Svensen et al (2009). The photograph in Fig. 12.8 is of a 3000 m pit that remains at the top of a "pipe" that extends down through two blocking rock layers into a region that contained molten lava.

Fig. 12.9 (Svensen et al, 2009) sketches a suggestion as to how the "volcanic pipes" that greatly enhanced lava and greenhouse gas emissions, evolved in a region already

Fig. 12.8 *Photograph of an exemplary "volcanic pipe" pit, 3000 m wide, in eastern Siberia. The dotted line outlines the dimension of the pipe that led directly down into the Earth's magma allowing vast release of lava and heating a surrounding adjacent "metamorphic areoule" region, to release carbon-containing volatile chemicals acting as greenhouse gases. (Svensen et al 2009) Fig. 4a*

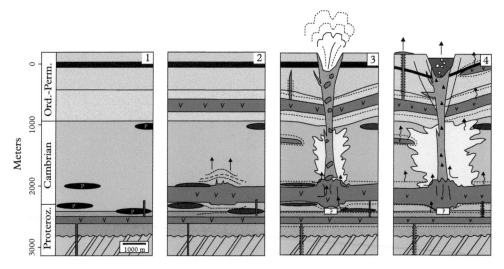

Fig. 12.9 *Schematic of origin and function of "volcanic pipe", up to 1000 m wide, and anchored at depths 2 to 4 km, as found in many locations in eastern Siberia. Regions marked V represent volcanic lava and P represents petroleum accumulation. (Svensen et al 2009) Fig. 6*

rich in petroleum deposits. In this Figure symbol P represents petroleum and V represents lava.

In summary, Payne and Clapham indicate that the scenario that best explains the Mass Extinction of 252 My ago invokes the massive Siberian volcanism as the primary trigger of extinction. Releases of volcanic carbon dioxide as well as volatilized sedimentary organic carbon led to global warming, ocean acidification and perhaps destruction of atmospheric ozone. Enhanced weathering and nutrient runoff increased pre-existing ocean anoxia. These earth system changes caused global population declines and resulted in mass extinctions.

Coming back to the present situation, where a short dose of carbon, totaling about 1000 GtC has been released since 1850 by the industrializing world economy, we see no suggestion of an onset of volcanism, and can be optimistic that there will be no repeat of the 252 My "hothouse earth" and extinction events.

12.4.3 The PETM event as a possible model for the climatic future

A weaker hyperthermal event that may resemble the present situation is the Paleocene-Eocene Thermal Maximum (PETM). This occurred 55.2 My ago, and is believed to have been triggered only by release of greenhouse gas, with no indication of large volcanic activity. A broad survey of the global temperature over the past 80 My is shown in Fig. 12.10.

In Fig. 12.10, the isotope ratio of oxygen 18 vs. oxygen 16 is used as an indicator of the temperature, and the oxygen samples were formed at the bottom of the ocean at depths that can be reliably correlated with their age. The conclusion is that the earth has generally been warmer than presently, over the long interval back to about 52 My ago, with a spike in temperature at 55.5 My ago, as labeled in Fig. 12.10. Not shown here are indications that the carbon dioxide concentration in the atmosphere was also considerably higher than at present, perhaps as much as 900 ppm over very long time periods.

The actual PETM event, marked in Fig. 12.10, but not well revealed, is known to have been a short (on geological scales) peak of 5–8°C lasting only about 100 ky. This is shown

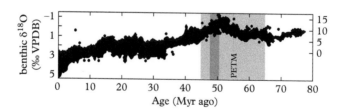

Fig. 12.10 *Estimate of global temperature over 80 Million years, based on oxygen isotope mass, 18 relative to 16, shown in parts per thousand. ("VPDB" denotes choice of an agreed reference sample, see (12.4), following common recent practice."Benthic indicates ocean floor, and temperatures inferred are shown on right, in Centigrade.) The Paleocene-Eocene Temperature Maximum at 55.5 Myr is indicated but is too brief an excursion to be seen clearly in this plot. (Sexton et al 2011) Fig. 1*

with better time resolution in Fig. 12.11, that again is based on isotope ratios of materials deposited at the bottom of the ocean, forming sediment layers recently extracted and studied in the Ocean Drilling Program. The solid line, indexed to the left side of the figure, is the carbon isotope ratio (12.5) in parts per thousand taken from bulk carbonate samples at a particular boring site, shown as ODP1266. (Ocean Drilling Program Site 1266 is on the Walvis Ridge, in the southeastern Atlantic Ocean, and is reached by a drilling ship sailing from Rio de Janeiro, Brazil.) This curve says that the kind of carbon in deposited carbonate sediments at the sea floor changed very noticeably over about 20 ky. The dashed curve, indexed to the right, indicates that the ocean temperature, as represented by the oxygen isotope ratio in a particular organism, *N. truempyi*, located at site ODP1263, changed by nearly 10°C on this same 20 ky time scale. (*N. truempyi* is a particular small calcite-shelled benthic (ocean bottom) foram creature that went extinct, leaving its shells as a sediment layer, because of acidic conditions during the PETM.) To change the temperature of the whole ocean obviously takes some time. 20 ky is short on the global ocean time scale, but of course a long time on the scale of human activity. Our whole civilization, dating back to the Pyramids of Egypt has endured for only 5 ky, one quarter of the time period we are inferring for the onset of the PETM hyperthermal event. The question is whether 1850, near the start of the industrial carbon-burning era, will be the analog of time zero on a new plot like Fig. 12.11, taking the world 10°C. warmer and killing all the coral reefs and fish in the sea?

A separate set of data showing the PETM as a short-lived hyperthermal event correlating with loss of calcium-carbonate-containing animals is shown in Fig. 12.12 (Zachos et al, 2005). Again, the ocean floor is formed in large part by the shells of the myriad species of foraminifera, including planktonic forms that float in the water near the sea floor and benthic forms that live on the ocean floor. Dying planktonic foraminifera continuously rain down on the sea floor adding to the remains of benthic foraminifera, with their calcium-carbonate-rich shells. Thus the sediment at the bottom of the ocean is conventionally rich in carbonate. Zachos et al (2005) surveyed $CaCO_3$ content and

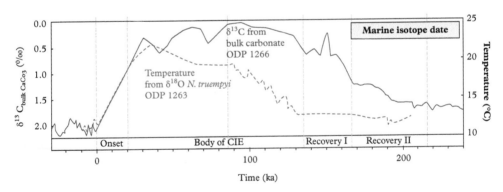

Fig. 12.11 *Short term estimates of Carbon Isotope Excursion CIE and proxy temperature over the PETM hyperthermal event 55.5 Myr ago. During this event, not shown, the seas became very acidic and corals and other $CaCO_3$-based organisms died. (McInerney and Wing, 2011). Fig. 4c*

carbon isotope ratio at southeastern Atlantic Ocean Drilling Program sites (ODP) 1262, –63, –65, –66 and –67, with labeled depths including 1500, 2600 and 3600 m. (See right-hand panel of Fig. 12.12.)

At each site, with a drilled column extracted and analyzed, the P-E boundary (55My ago) was characterized by an abrupt transition from carbonate-rich ooze to a dark red "clay" layer that then graded back into ooze. (A typical clay is kaolin, that is a mixture of $2SiO_2$, Al_2O_3 and $2H_2O$.) The carbonate content was typically less than 1 wt percent in the clay layers and 80–90 wt percent in the underlying and overlying ooze layers. The thickness of the clay layers increased with depth, from 5 cm at the shallowest site (labeled 1500 m) to 35 cm at the deepest site. The benthic foraminiferal extinction occurred at the base of the clay layer at each site. There is always a decrease in $\delta^{13}C$ at the base of the clay layer followed by a gradual recovery. Further, a large amount of

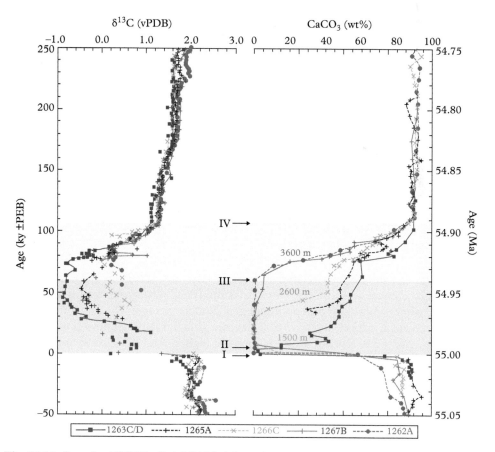

Fig. 12.12 *Records of PETM of brief (100 ky) hyperthermal event at 55 Ma before present. Left trace, change in carbon isotope ratio, suggests different origin of deposited carbon during the event. Right trace: carbonate content of deposits, inversely related to acidity of the water. (Zachos et al, 2005) Fig. 3*

calcite at the ocean floor was dissolved during this event, following Eq. (12.3), referred to as neutralization of CO_2. During the PETM event thus there were no living calcium carbonate shelled creatures to add to the ocean floor sediment, these creatures were killed by the ocean acidity. Zachos et al noted that although the benthic foraminifera, including the *N. truempyi* mentioned previously, died out at the beginning of this event, most plankton species, small-shelled animals floating above the sea floor, survived. So this event was by no means as destructive as the end-Permian event 252 My ago. They suggest that release of more than 4500 Gt C is consistent with the temperature rise of this PETM event. Regarding the future, these authors suggest that if the full present fossil fuel reservoir is burned, about 4500 GtC, then the impacts on the deep sea pH and biological species will be similar to the PETM event. However, because the present full burning of fossil fuels is expected to be on a short time scale, say 300 years, and this time is less than the mixing time of the ocean, the impacts on surface ocean pH and biota will probably be more severe.

In a review of the PETM, McInerney and Wing (2011) suggest that warming associated with carbon release implied about two doublings of atmospheric CO_2 during the event. Although there was a major extinction of benthic foraminifera, most groups of organisms did not suffer extinction. Geographic distributions of most kinds of organisms were radically rearranged by 5–8°C of warming, with tropical forms moving poleward in both marine and terrestrial realms. It is clear that coral reefs were destroyed in the PETM by the temperature and acidity, as well as in the earlier end-Permian extinction event (see Fig. 12.7, upper right, where "platform margin reefs" were replaced by "microbialite deposits"). It is known[9] that coral reefs, such as the Great Barrier Reef of Australia, that has already been severely bleached, are very temperature sensitive, and will die with a temperature rise of 1.5°C, and thus may, in the present era, be gone by the year 2050.

12.5 Near-term predictions

More recent warming events related to sea-level rise are noted[1] by Dr H. Portner, a distinguished climate scientist and co-chair of the IPCC (Intergovernmental Panel on Climate Change) working group on impacts of climate change: "We know that during the last interglacial period, with 0.7 to 2 degrees global warming above pre-industrial levels,….we had a sea level of around 7 meters higher. And if we compare and look back at the last period in Earth's history where we had 400 parts per million CO_2, and that's where we are now, we had a similar degree (7 meters) of higher sea level."

Confirmation of points about the sea and temperature rise is offered by Brannen 2017, who says that the sea has become 30 percent more acidic (i.e., a pH change of about −0.114) since the start of the Industrial Revolution, one evidence being the presently observed pitting of shells of sea-dwelling planktonic snails. It is predicted that by 2050

---

[9] http://www.dw.com/en/climate-change-a-simmering-threat-to-our-ocean/a-39160034. (Accessed 2 Aug. 2017).

the Southern Ocean will no longer be able to host those creatures, that form a critical part of the diet of salmon; again is predicted the loss of coral reefs on the same time scale. Finally Brannen (2017) suggests, without a reference, that the present expected global temperature rise by 2100 is now about 4°C, that exceeds the earlier prediction of about 2°C. The expected temperature rises have been addressed in the reports of the IPCC (International Panel on Climate Change). An early prediction[10] for the temperature-change relative to 1990, for the year 2100 is 1.4–5.8°C, with predicted CO_2 concentrations in the range 479–1099 ppm. An enlarged recent set of predictions from the IPCC 2013 report is shown in Fig. 12.13.

This figure shows on the left that the global mean temperature has risen about 0.7°C since 1850, about the start of the industrial revolution. These data are similar to

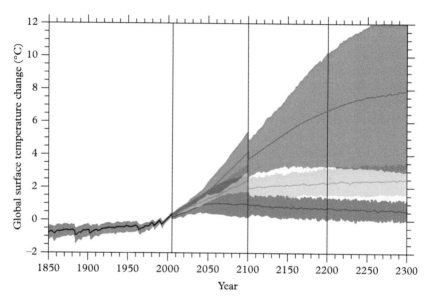

Fig. 12.13 *Temperature change, relative to period 2005–2012, predicted out to year 2300 AD. Black curve on left, with error band, is observed average global temperature change. Four curves, top to bottom, in second panel (2010–2100), with error bands, are projections on cumulative emission scenarios for period 2012–2100, respectively: 270 (140–410) GtC, 780 (595–1005) GtC, 1060 (840–1250) GtC, and 1685 (1415–1910) GtC. For the last scenario (here considered unlikely, see text), with 1685 GtC total emitted, the mean predicted temperature change by 2300 is about 8°C. (Intergovernmental Panel on Climate Change IPCC Report 2013, Fig. 12.5 on p. 1054, see also Table SPM.3 on p. 27). http://www.climatechange2013.org/report/full-report/*

[10] http://www.ipcc.ch/ipccreports/tar/wg2/index.php?idp=29. (Accessed 30 Jan. 2018.)

http://www.dw.com/en/climate-change-a-simmering-threat-to-our-ocean/a-39160034. (Accessed 2 Aug. 2017.)

http://www.ipcc.ch/ipccreports/tar/wg2/index.php?idp=29. https://en.wikipedia.org/wiki/Oceanic_carbon_cyclehttps://en.wikipedia.org/wiki/Oceanic_carbon_cycle.

https://earthobservatory.nasa.gov/NaturalHazards/view.php?id=18791.

http://www.dw.com/en/climate-change-a-simmering-threat-to-our-ocean/a-39160034. (Accessed 2 Aug. 2017.)

http://www.ipcc.ch/ipccreports/tar/wg2/index.php?idp=29.

those shown in Fig. 1.7a. It is clear that the observed temperature rise is small: most of Fig. 12.13 is the result of modeling. The four curves shown in the second panel of the Figure are based on four assumptions (see caption) on the amount of carbon emitted into the atmosphere during the 88-year interval 2012–2100: the amounts assumed range from 270 GtC to 1685 GtC. The latter number can be expressed also as 1.685 trillion tonnes or 1685 pg (petagrams). The upper predictions for the year 2100 from this figure are close to 4°C, but are contingent, within the modeling, on an emission rate of 1685 GtC over 88 years, thus 19.1 GtC/year. The present rate of carbon burning, however, is reported as 9.8 GtC/year, only half that assumed in the model, so that the 88-year estimate of release is 862.4 GtC. This is close to the 780 GtC related to the second curve from the bottom, that approaches 2°C by 2100 and 2.5°C by 2300 AD. From the perspective of the modeling and current rate of carbon burning the temperature rise by 2100 will be in the vicinity of 2°C. As noted previously this is well above the temperature rise to kill all the coral reefs by 2100. However, it does not appear to be enough to lead to a hyperthermal event of the same magnitude as the Paleo-Eocene Thermal Maximum (PEMT) as described in Fig. 12.11.

It is clear that there will be need for economic activity to respond to the slow changes that surely are coming in the climate and sea level and to adapt power sources to those that do not clog the atmosphere. Engineering will certainly be in demand in this process.

Bibliography

Allen, M., Frame, D., Huntingford, C., Jones, C., Lowe, J., Meinshausen, M., and Meinshausen, N. (2009) "Warming caused by cumulative carbon emissions toward the trillionth tonne", Nature 458, 1163.

Alferov, Zh. I. and Rumyantsev, V. D. (2003). "Trends in the development of solar photovoltaics". In Marti, A., and Luque, A. (Eds.), Ch. 2 in *Next Generation Photovoltaics*. (Bristol and Philadelphia: Institute of Physics.)

Algora, C. (2004). "The importance of the very high concentration in third-generation cells". In Marti, A., and Luque, A. (Eds.), Ch. 6 in *"Next Generation Photovoltaics: High efficiency through full spectrum utilization.* (Bristol and Philadelphia: Institute of Physics Publishing.)

Angulo, C., Arnould, M., Rayet, M., Descouvemont, P., Baye, D., and Leclercq-Villain, C. (1999). "A compilation of charged-particle induced thermonuclear reaction rates", *Nucl. Phys.* A 656, 3.

Archer, D., Kheshgi, H. and Maier-Reimer, E. (1997). "Multiple timescales for neutralization of fossil fuel CO_2". *Geophysical Research Letters* 24, 405.

Armaroli, N. and Balzani, V. (2006). "The future of energy supply challenges and opportunities" *Angew. Chem. Int. Ed.* 46, 52.

Arrhenius, S. (1896). "On the influence of carbonic acid in the air upon the temperature of the ground". *Phil. Mag.* **5**(41): 251.

Atzeni, S. and Meyer-Ter-Vehn, J. (2004). *The Physics of Inertial Fusion*. (Oxford: Oxford University Press.)

Austen, I. (2009). "Hybrid Locomotive Maker Loses Steam". *The New York Times*, 5 Feb. 2009.

Bahcall, J., Pinsonneault, M. and Basu, S. (2001). "Solar models: current epoch and time dependences, neutrinos, and helioseismological properties". *Astrophysical Journal* 555, 880.

Barnett, A., Honsberg, C., Kirkpatrick, D., Kurtz, S., Moore, D., Salzman, D., Schwartz, R., Gray, J., Bowden, S., Goossen, K., Haney, M., Aiken, D., Wankass, M., and Emery, K. (2006)."50% efficient solar cell architectures and designs". Record of the 4th IEEE Worldwide Conference on Photovoltaic Energy Conversion, Hawaii, Vol. 2, p. 2500.

Bender, D. (2015). "Flywheels" SAND2015-3976 (Albuquerque NM: Sandia National Lab) May 2015.

Benitez, P., Miñano, J. C. (2003). "Thermodynamics of solar energy converters". Ch. 3 in Marti, A. and Luque, A. (Eds.), *Next Generation Photovoltaics: High efficiency through full spectrum utilization.* (Bristol and Philadelphia: Institute of Physics Publishing.)

Berntsen, T. and Fuglestvedt, J. (2008). "Global temperature responses to current emissions from transport sectors". *Proc. Natl. Acad. Sci. USA* 105, 19154.

Bethe, H. and Critchfield, C. (1938). "The formation of deuterons by proton combination". *Phys. Rev.* 54, 248.

Betz, A. (1920). "The maximum of theoretically available potential of wind by wind turbines (in German)". Zeitschrift f. d. gesamte Turbinenwesen, v. 17, Sept. 1920.

Betz, A. (1926). *Wind-Energie und ihre Ausnutzung durch Windmuhlen.* (Wind energy and its utilization through windmills.) Gottingen: Vandenhoeck & Ruprecht, reprint Oko-Buchverlag Kassel 1982.

Brand, U., Posenato, R., Came, R., Affek, K., Angiolini, L., Azmy, K. and Farabegoli, E. (2012). "The end-Permian mass extinction: a rapid volcanic CO_2 and CH_4 – climatic catastrophe". *Chemical Geology* 121, 322–3.

Brannen, P. (2017). "When Life on Earth Was Nearly Extinguished". *The New York Times* July 30 2017, p. SR2.

Bremaud, D., Rudmann, D., Bilger, G., Zogg, H., and Tiwari, A. (2005). "Towards the development of flexible CIGS solar cells on polymer films with efficiency exceeding 15%". Record of 31st IEEE Photovoltaic Specialists Conference, Lake Buena Vista, FL, p. 65.

Burst, J., Duenow, J., Albin, D., Colegrove, E., Reese, M., Agular, J., Jiang, C.-S., Patel, J., Al-Jassim, M., Kuciauskas, D., Swain, S., Ablekim, T., Lynn, K. and Metzger, W. (2016). "CdTe solar cells with open circuit voltage breaking the 1 V barrier". *Nature Energy* 1, 1.

Bytautas, L., Matsunaga, N., and Ruedenberg, K. (2010). "Accurate ab initio potential energy curve for O_2". *J. Chem. Phys.* 132, 074307.

Cardwell, D. (2017). "Off Long Island, Wind Power Tests the Waters", *The New York Times*, 22 Jan. 2017. P. BU1 of the New York Edition.

Cardwell, D. and Krauss, C. (2017). "A Big Test for Big Batteries", *The New York Times*, 14 Jan. 2017.

Carnegie, R., Gotham, D., Nderitu, D. and Preckel, P. (2013). *Utility Scale Energy Storage Systems*. (West Lafayette, IN: State Utility Forecasting Group.)

Carvin, M. and Dick, A. (2017). *The Wall Street Journal*, 6 Feb., 2017, p. A17.

Castle, K. (2007). "High-resolution vibration-rotation spectroscopy of CO_2". *J. Chem. Education* 84, 459.

Chandrasekhar, S. (1983). "On Stars, their evolution and their stability". *Nobel Lecture*, 8 Dec. 1983.

Chen, George, Miao Hao, Zhiqiang Xu, Alun Vaughan, Junzheng Cao, and Haitian Wang. (2015) "Review of high voltage direct current cables." *CSEE Journal of Power and Energy Systems* 1(2): 9–21.

Clack, C. et al (2017). "Evaluation of a proposal for reliable low-cost grid power with 100% wind, water, and solar". *Proc. Natl Acad Sciences* 114, 6722.

Clayton, D. (1983). *Principles of Stellar Evolution and Nucleosynthesis*. (Chicago: University of Chicago Press.)

Currie, J. J., Maple, J. K., Heidel, G. D., Goffri, S. and Baldo, M. A. (2008). "High-efficiency organic solar concentrators for photovoltaics" *Science* 321, 226.

Davidson, Amy. (2017). "Trump v. The Earth". *The New Yorker*, Comment Section, 10 April, 2017.

Davisson, C. and Germer, L. H. (1927). "Diffraction of electrons by a crystal of nickel". *Phys. Rev.* 30, 705.

Deb, S. K. (2010). "Recent advances and future opportunities for thin film solar cell", in "Thin–Film Solar Cells." Y. Hamakawa, ed., (Berlin: Springer.)

Debe, M. (2012). "Electrocatalyst approaches and challenges for automotive fuel cells". *Nature* 486, 43.

Deffeyes, K. S. (2001). *Hubbert's Peak*. Princeton: Princeton University Press.

Dennison, D. (1931). "The infrared spectra of polyatomic molecules part I". *Reviews of Modern Physics* 3, 280.

Dennison, D. (1940). "The infrared spectra of polyatomic molecules part II". Ibid. 12, 175.

Dietz, K., Chiocchio, S., Antipenkov, A., Federici, G., Janeschitz, G., Martin, E., Parker, R. and Tivey, R. (1995). "Engineering and design aspects related to the development of the ITER divertor". *Fusion Engineering and Design* 27, 96.

Dunn, R., Hearps, P. and Wright, M. (2012). "Molten-salt power towers: newly commercial concentrating solar storage". *Proc. IEEE* 100, 504.

Durgun, E., Ciraci, S., Zhou, W., and Yildirim, T. (2006). "Transition-metal-ethylene complexes as high capacity hydrogen-storage media". *Phys. Rev. Lett.* 97, 226102.

Elmegaard, B. and Brix, W. (2011). "Efficiency of Compressed Air Energy Storage", in: The 24th International Conference on Efficiency, Cost, Optimization, Simulation and Environmental Impact of Energy Systems. (Curran, Red Hook, NY).

Eperon, G., McGehee, M., et al (2016). "Perovskite-perovskite tandem photovoltaics with optimized bandgaps". *Science* 354, 861. or *Science DOI:* 10.1126/science.aaf 9717.

Faizal, M. and Ahmed, R. (2013). "Experimental studies on a closed cycle OTEC plant working on small temperature difference". *Renewable Energy* 51, 234.

Fletcher, E. A. and Moen, R. L., (1977). "Hydrogen and oxygen from water". *Science* 197, 1050.

Friedrich, Kurt. (2010). "Modern HVDC PLUS application of VSC in modular multilevel converter topology." In Industrial Electronics (ISIE), 2010 IEEE International Symposium on, pp. 3807–10. IEEE.

Gamow, G. (1928). "Zur quantentheorie des atomkernes (The quantum theory of atomic nuclei)" *Zeitschrift fur Physik* 51, 204.

Ganley, J., Seebauer, E. and Masei, R. (2004). "Development of a microreactor for the production of hydrogen from ammonia". *Journal of Power Sources* 137, 53.

Gasch, R. and Twele, J., (2012). *Wind Power Plants*, 2nd ed., (Springer, Heidelberg).

Gillis, J. (2017). "Carbon in Atmosphere is Rising, Even as Emissions Stabilize". *The New York Times*, 27 June 2017, p. A1.

Goody, R. and Yung Y. (1989). *Atmospheric Radiation: Theoretical Basis*, 2nd ed. (Oxford: Oxford University Press.)

Gora, E. (1959). "The rotational spectrum of ozone". *J. Molec. Spectroscopy* 3, 78.

Graetzel, M. (2001). "Photoelectrochemical cells". *Nature* 414, 338.

Greene, M. (1990). "Intrinsic concentration, effective density of states and effective mass in silicon". *J. Appl. Phys.* 67, 2944.

Grice, K., Cao, C., Love, G., Bottcher, M., Twitchett, R., Grosjean, E., Summons, R., Turgeon S., Dunning, W. and Jin, Y. (2005). "Photic Zone Euxinia During the Permian-Triassic Superanoxic Event". *Science* 307, 706.

Griffiths, D. (2005). *Introduction to Quantum Mechanics*, 2nd ed. (Saddle River, New Jersey: Pearson Education,.)

Gusev, V., Alladio, F. and Morris, F. (2003). "The basics of spherical tokamaks and progress in European research". *Plasma Phys. Control. Fusion 45*, A59.

Gustafsson, A., M. Saltzer, A. Farkas, H. Ghorbani, T. Quist, and M. Jeroense. (2014). The new 525 kV extruded HVDC cable system—World's most powerful extruded cable system. ABB, Technical. Report, 2014.

Hall, R. and Dowling, J. (1967). "Pure rotational spectrum of water vapor". *J. Chem. Phys.* 47, 2454.

Handler, C. (2016). "United Airlines is flying on biofuels. Here's why that's a really big deal." *The Washington Post*, 11 March, 2016.

Hansen, J., Sato, M., Karecha, P., Russell, G., Lea, D. and Siddall, M. (2007). "Climate change and trace gases". *Phil. Trans R. Soc. A* 365, 1925

Haxton, W. (1995). "The solar neutrino problem". *Annu. Rev. Astronomy and Astrophys.* 33, 495.

Hermann, W. (2006). "Quantifying Global Exergy Resources". *Energy* 31, 1685.

Hernandez, J. (2017) *The New York Times*, 16 January, 2017 New York Edition 1 (p. A8).

Honsberg, C., Barnett, A., and Kirkpatrick, D. (2006). "Nanostructured solar cells for high efficiency photovoltaics". Record of the 4th IEEE Worldwide Conference on Photovoltaic Energy Conversion, Vol. 2, p. 2565.

Huang, H., Uder, M., Barthelmess, R. and Dorn, J. (2008). "Application of high power thyristors in HVDC and FACTS systems." In 17th Conference of Electric Power Supply Industry (CEPSI), pp. 1–8.

Huba, J. (2006). NRL Plasma Formulary, Plasma Physics Division, Beam Physics Branch, (Washington DC: Naval Research Laboratory).

Huijgen, W., Comans, R. and Witkamp, G. (2007) . "Cost evaluation of CO_2 sequestration by aqueous mineral carbonation". *Energy Conversion and Management* 48, 1923.

Ivy, J. (2004). "Summary of Electrolytic Hydrogen Production". NREL/MP -560-36734 (Sept. 2004). National Renewable Energy Laboratory, Golden, CO, USA.

Jackson, J. (1957). "Catalysis of nuclear reactions between hydrogen isotopes by μ^- mesons". *Phys. Rev.* 106, 330.

Jacobson, M. Z., Delucchi, M., Bazouin G., Bauer, Z., Heavey, C., Fisher, E., Morris S., Piekutowski, D., Vencill, T. and Yeskoo, T. (2015a). "100% clean and renewable wind, water and sunlight (WWS) all-sector roadmaps for the 50 United States". *Energy & Environmental Science* 8, 2093.

Jacobson, M. Z., Dellucchi, M., Cameron, M. and Frew, B. (2015). "Low-cost solution to the grid reliability problem with 100% penetration of intermittent wind, water and solar for all purposes". *Proc. Natl. Acad. Sciences* 112, 15060.

Jacobson, M. Z., Delucchi, M., Cameron, M. and Frew, B. (2017). "The United States can keep the grid stable at low cost with 100% clean renewable energy in all sectors despite inaccurate claims". *Proc. Natl. Acad. Sciences* 114, E5021.

Jensen, B., Mijatovic, N. and Abrahamsen, A. (2013). "Development of superconducting wind turbine generators". *J. Renewable Sustainable Energy* 5, 023137.

Jin, J. and Chen, X. (2012). "Study on the SMES application solutions for smart grid". *Physics Procedia* 36, 902.

Joos, F., Roth, F., et al (2013). "Carbon dioxide and climate impulse response functions for the computation of greenhouse gas metrics". *Atmos. Chem. Phys.* 13, 1793.

Junghanel, M. (2007). Novel aqueous electrolyte films for hole-conduction in dye sensitized solar cells and development of an electron transport model. Dissertation, Freie University of Berlin.

Kaiho, K. and Koga, S. (2013). "Impacts of a massive release of methane and hydrogen sulfide on oxygen and ozone during the late Permian mass extinction". *Global and Planetary Change* 107, 91.

Kapur, V., Bansal, A., Le, P., Asensio, O. and Shigeoka, N. (2003). "Non-vacuum processing of CIGS solar cells on flexible polymeric substrates". *Proceedings of the Third World Conference on Photovoltaic Energy Conversion*, Vol. 1.p. 465.

Khan, J. and Bhuyan, G. (2009). "Ocean Energy: Global Technology Development Status" IEA-OES Document No.: T0104 March 2009.

Khaselev, O. and Turner, J. (1998). "A monolithic photovoltaic-photoelectrochemical device for hydrogen production via water splitting" *Science* 280, 425.

Kim, B., Dominguez-Cabellero, J., Lee, H., Friedman, D. and Menon, R. (2013). "Increased photovoltaic power output via diffractive spectrum separation". *Phys. Rev. Lett.* 110, 123901.

King, R., et al (2002). "High efficiency space and terrestrial multijunction solar cells through bandgap control in cell structures". 29th IEEE Photovoltaic Specialists Conf., New Orleans, LA, 20–24 May, 2002.

King, R., et al (2009). "40% efficient metamorphic GaInP/GaInAs/Ge multijunction solar cells". European Photovoltaic Solar Energy Conf., Hamburg, Germany, 21–25 Sept. 2009.

Kittel, C. (1986). *Introduction to Solid State Physics*, 6th ed. (Hoboken, New Jersey: John Wiley & Sons, Inc.,)

Klimont, Z., Smith, S. and Cofala, J. (2013). "The last decade of global anthropogenic sulfur dioxide: 2000-2011 emissions". *Environ. Res. Lett.* 8, 014003.

Knight, N. and Heymsfield, A. J. (1983)."Measurement and interpretation of hailstone terminal velocity". *Atmos. Sciences* 40, 1510.

Knoll, A., Barnbach, R., Canfield, D. and Grotzinger, J. (1996). "Comparative Earth history and late Permian mass extinction". *Science* 273, 452.

Kondo, M. and Matsuda, A. (2010). "Low Temperature Fabrication of Nanocrystalline-Silicon Solar Cells". Ch 8, in Y. Hamakawa, (Eed,) *Thin-film solar cells.* (Berlin: Springer.)

Kowitt, B. (2016). "How a Huge Dairy is Solving a Major Pollution Problem". Fortune, 27 January 2016.

Kwon, S., Fan, J., DaCosta, H. and Russell, A. (2011). "Factors affecting the direct mineralization of CO_2 with olivine". *J. Environ. Sci.*, 23 1233.

Lacis, A. and Hansen, J. (1974). "A parameterization of the absorption of solar radiation in the earth's atmosphere". *J. Atmospheric Sciences* 31, 118.

Lan, R. and Tao, S. (2014). "Ammonia as a suitable fuel for fuel cells". *Frontiers in Energy Research* 2, 1.

Lanchester, F. W. (1915). "A Contribution to the theory of propulsion and the screw propeller" *Trans. Inst. Naval Arch.*, Vol. LVII.

Landsberg, P. and Tonge, G. (1980). "Thermodynamic energy conversion efficiencies". *J. Appl. Phys.* 51, R1.

Lang, N. D. and Kohn, W. (1971). "Theory of metal surfaces: induced surface charge and image potential". *Phys. Rev. B3*, 1215.

Lewis, B., Cimbala, J. and Wouden, A. (2014). "27th IAHR Symposium on Hydraulic Machinery and Systems (IAHR 2014)". IOP Conf. Series: Earth and Environmental *Science* 22, 012020.

Lewis, N., and Crabtree, G. (ed.) (2005). "Basic Research Needs for Solar Energy Utilization", Report of Workshop, April 18–21, 2005. (US Department of Energy, Office of Basic Energy Sciences).

Li, X., Tian, J., Mei, M. and Li, C. (2000). "Sub-barrier fusion and selective resonant tunneling". *Phys. Rev. C* 61, 024610.

Lichtenstein, M., Derr, V. and Gallagher, J. (1966). "Millimeter-wave rotational transitions and the Stark effect of the water molecule". *J. Mol. Spectroscopy* 20, 391.

Litster, S. and McLean, G. (2004). "PEM fuel cell electrodes". *Journal of Power Sources* 130, 61.

Luque, A. and Marti, A. (1997). "Increasing the efficiency of ideal solar cells by photon induced transitions at intermediate levels". *Phys. Rev. Lett.* 78, 5014.

MacKay, D. J. (2009). *Sustainable Energy-Without the Hot Air* (Cambridge, England: UIT.)

Mandelstam, L. and Tamm, I. (1945). "The uncertainty relation between energy and time in non-relativistic quantum mechanics". *Journal of Physics* 9, 249.

Mann, M., Bradley, E. and Hughes, M. (1999). "Northern hemisphere temperatures during the past millenium". *Geophysical Research Letters* 26, 759.

Maples, B., Hand, M. and Musial, W. (2010). "Comparative assessment of direct drive high temperature superconducting generator in multi-megawatt class wind turbines". Technical Report NREL/TP-5000-49086.

Marti, A., Antolin, E., Stanley, C., Farmer, C., Lopez, N., Diaz, P., Canovas, E., Linares, P. and Luque, A. (2006). "Production of photocurrent due to intermediate-to-conduction-band transitions". *Phys. Rev. Lett.* 97, 247701.

Marti, A., Cuadre, L., and Luque, A. (2004). "Intermediate Band Solar Cells", Ch. 7, in *Next Generation Photovoltaics.* Marti. A., Luque, A. (Eds.), (Bristol and Philadelphia: Institute of Physics.)

Matsui, T., Cheung, A., Leung, K., Yoshino, K., Parkinson, W., Thorne, A., Murray, J., Ito, K. and Imajo, T. (2003). "High resolution absorption cross-section measurements of the Schumann-Runge bands of O_2 by VUV Fourier transform spectroscopy". *J. Molecular Spectroscopy* 219, 45.

Matthews, H., Gillett, N., Stott, P. and Zickfeld, K. (2009). "The proportionality of global warming to cumulative carbon emissions". *Nature* 459 829.

McInerney, F. and Wing, S. (2011). "The Paleocene-Eocene Thermal Maximum: a Perturbation of Carbon cycle, Climate and Biosphere with Implications for the Future". *Annu. Rev. Earth Planet. Sci.* 39, 489.

Mecherikunnel, A. and Richmond, J. (1980). "Spectral Distribution of Solar Radiation" NASA Technical Memorandum 82021, (Sept. 1980. Goddard Space Flight Center, Greenbelt, Maryland.)

Meinshausen, M., Meinshausen, N., Hare, W., Raper, S., Frieler, K., Knutti, R., Frame, D. and Allen, M. (2009). "Greenhouse-gas emission targets for limiting global warming to 2 C". *Nature* 458, 1158.

Merschel, F., Noe, M., Nobi, A. and Stemmie, M. (2013). "AmpaCity- Installation of advanced superconducting 10 kV system in city center replaces conventional 110 kV cables". Proc. of IEEE International Conference on Applied Superconductivity and Electromagnetic Devices ASEMD Beijing 2013 pp. 323–6.

Mitsui, T., Ito, F. and Nakamoto, Y. (1983). "Outline of the 100 kW Otec pilot plant in the Republic of Naure". IEEE Trans. Power Apparatus and Systems, PAS-102, 3167.

Miyamoto, K. (2004). *Plasma Physics and Controlled Nuclear Fusion*: (Berlin: Springer.)

Mott, N. F. (1973). "Metal-insulator Transitions", *Contemporary Physics* 14, 401.

Naranjo, B., Gimzewski, J. and Putter, S. (2005)."Observation of nuclear fusion driven by a pyro-electric crystal". *Nature* 434, 1115–17.

Navarro, M. (2001). "Mapping Sun's Potential to Power New York", *New York Times*, 16 June 2011.

Nazeeruddin, M. K., et al (2001)."Engineering of efficient panchromatic sensitizers for nanocrystalline TiO_2-based solar cells". *Journal of the American Chemical Society* 123, 1613.

Nazeeruddin, M. K., et al (2007). "A high molar extinction coefficient charge transfer sensitizer and its application in a dye-sensitized solar cell". *Journal of PhotoChemistry and PhotoBiology-A Chemistry*, 185, 331.

Negami, T., Aoyagi, T., Satoh, T. and Shimakawa, S. (2002). "Cd free CIGS solar cells fabricated by dry processes". Record of 29th IEEE Photovoltaic Specialists Conference, p. 656.

Neil, D. (2017). "Chrysler Pacifica Hybrid: The Strong Silent Type". *The Wall Street Journal* 25 March 2017, p. D11.

Nelson, D., Nehrir, M. and Wang, C. (2006). "Unit sizing and cost analysis of stand-alone hybrid-wind/PV/fuel cell power generation systems" Renewable Energy 31, 1641.

Newnham, D. and Ballard, J. (1998). "Visible absorption cross sections and integrated absorption intensities of molecular oxygen (O_2 and O_4)". *J. Geophysical Research* 103, 28801.

Olsson, P., Domain, C., and Guillemoles, J. F. (2009). "Ferromagnetic compounds for high efficiency photovoltaic conversion: the case of AlP:Cr". *Phys. Rev. Lett.* 102, 227204.

Ongena, J., Koch, R., Wolf, R. and Zohm, H. (2016). "Magnetic-confinement fusion". *Nature Physics* 12, 398.

Oppenheimer, J. (1928). "Three notes on the quantum theory of aperiodic events". *Phys. Rev.* 31, 66.

Pacheco, J., Wolf, T. and Muley N. (2013). "Incorporating supercritical steam turbines into advanced molten-salt power tower plants: feasibility and performance". Sandia Report SAND2013-1960.

Packler, M. (2008). "Latest Honda runs on hydrogen, not petroleum". *The New York Times*, 2008.

Park, D. (1992). *Introduction to the Quantum Theory*, 3d ed. (New York: McGraw-Hill).

Patchkovskii, S., Tse, J., Yurchenko, S., Zhechkov, L, Heine, T., and Seifert, G. (2005). "Graphene nanostructures as tunable storage media for molecular hydrogen". *Proceedings of the National Academy of Sciences of the United States of America* 102, 10439.

Pauling, L. and Wilson, E. B. (1935). *Introduction to Quantum Mechanics* (New York: McGraw-Hill,).

Payne, J. and Clapham M. (2012). "End-Permian mass extinction in the oceans: an ancient analog for the twenty-first century?" *Annu. Rev. Earth Planet.* Sci. 40, 89.

Peixoto, J. and Oort, A. (1992). *Physics of Climate* (New York: American Institute of Physics.)

Perret, R. (2011). "Solar Thermochemical Hydrogen Production Research (STCH)" Sandia Laboratory Report SAND2011-3622.

Peterson, E., and Hennessey, Jr, J. (1977). "On the use of power laws for estimates of wind power potential". *J. Appl. Meteorology* 17, 390.

Phillips, K. (1995). "Guide to the Sun" (Cambridge University Press: Cambridge.)

Phillips, O. M. (1977). *The Dynamics of the Upper Ocean*, 2nd ed. (Cambridge: Cambridge University Press).

Pierrehumbert, R. (2010). *Principles of Planetary Climate*. (Cambridge: Cambridge University Press.)

Pierrehumbert, R. (2011). "Infrared radiation and planetary temperature". *Physics Today*, 64, 33.

Pierrehumbert, R. (2016). "How to decarbonize? Look to Sweden". *Bull. Atomic Scientists* 72, 105, and private communication.

Pilar, F. J. (1990). *Elementary Quantum Chemistry*, 2nd ed. (Dover: McGraw-Hill).

Polinder, H., Ferreira, J., Jensen, B., Abrahamsen, A., Atallah, K. and McMahon, R. (2013). "Trends in wind turbine generator systems". *IEEE J. Emerging and Selected Topics in Power Electronics* 1, 174.

Rajagapolan, K. and Nihous, G. (2013). "Estimates of global Ocean Thermal Energy Conversion (OTEC) resources using an ocean general circulation model". *Renewable Energy* 50, 532.

Raman, A., Anoma, M., Zhu, L., Rephaeli, E. and Pan, S. (2014). "Passive radiative cooling below ambient air temperature under direct sunlight". *Nature* 515, 540.

Ramanathan, K., Noufi, R., To, B., Toung, D., Bhattarcharya, R., Contreras, M., Dhere, B, and Teeter., G. (2006). "Processing and properties of sub-micron CIGS solar cells". Proceedings of the 4th IEEE Worldwide Conference on Photovoltaic Energy Conversion, Vol. 1.

Randall, D. (2012). "Atmosphere, Clouds, and Climate". (Princeton University Press, Princeton)

Rao, A. and Rubin, E. (2002). "A Technical, Economic, and Environmental Assesment of Amine-based CO_2 Capture Technology for Power Plant Greenhouse Gas Control", *Environ. Sci. Technol.* 36, 4467.

Reed, S. "Offshore Wind Moves into Energy's Mainstream" *The New York Times*, 7 Feb. 2017, New York Edition.

Richter, B. (2010). *Beyond Smoke and Mirrors*. (Cambridge, UK: Cambridge University Press).

Roeb, M., Monnerie, N., Schmitz, M., Sattler, C., Konstandopoulos, A., Agrafiotis, C., Zasplis, V., Nalbandian, L., Steele, A. and Stobbe, P. "Thermochemical Production of hydrogen from water by metal oxides fixed on ceramic substrates", World Hydrogen Energy Conference WHEC 16, 13–15, Lyon, France.

Rolfs, E. and Rodney, W. (1988). *Cauldrons in the Cosmos: Nuclear Astrophysics* (Chicago: University of Chicago Press).

Rostalski, J., and Meissner, D. (2000). "*Solar Energy Materials and Solar Cells*" 61, 87.

Ruggiero, A. (1992). Nuclear Fusion of Protons with Boron. Conference: "Prospects for Heavy-Ion Inertial Fusion". Crete, Sept. 1, 1992. Brookhaven National Lab., Accelerator Physics Technical Note No. 48.

Sachs, Jeffrey (2016). "A Grand Bargain on Energy" *The Wall Street Journal* 2–3 Jan., 2016, p. C2.

Salter, S. H. (1974). "Wave power". *Nature* 249, 720.

Sargent and Lundy, (2003)."Assessment of parabolic trough and power tower solar technology cost and performance forecasts". Report NREL/SR-550-34440, Oct. 2003.

Schlattl, H. (2001). "Three-flavor oscillation solutions for the solar neutrino problem". *Phys. Rev.* D64, 013009.

Schmitz, G. (1956). "Theory and design of windwheels with an optimum performance". (In German). Wiss. Zeitschift der Universitat Rostock, 5, 1955/56.

Schubel, P. and Crossley, R. (2012). "Wind turbine blade design". *Energies* 5, 3425.

Schwartz, J. (2017). "Exxon Mobil lends its support to a carbon tax proposal". *The New York Times*, 20 June 2017.

Scruggs, J. and Jacob, P. (2009). "Harvesting ocean wave energy". *Science* 323 1176.

Sexton, P., Norris, R., Wilson, P., Palike, H., Westerhold, T., Rohl, U., Bolton, C. and Gibbs, S. (2011). "Eocene global warming event driven by ventilation of oceanic dissolved carbon". *Nature* 471, 349.

Sheng, Baoliang, Jonatan Danielsson, Yanny Fu, and Zehong Liu. (2010). "Converter valve design and valve testing for Xiangjiaba-Shanghai ±800kV 6400MW UHVDC power transmission project." In Power System Technology (POWERCON), 2010 International Conference on, pp. 1–5. IEEE.

Shine, K., Fuglestvedt, J., Hailemariam, K. and Stuber, N. (2005). "Alternatives to the global warming potential for comparing climate impacts of emissions of greenhouse gases". *Clim. Change* 68, 281.

Shockley, W. and Queisser, H. (1961). "Detailed-balance limit of efficiency of p-n junction solar cells". *J. Appl. Phys.* 32, 510.

Sibbitt, B., et al (2012). "The performance of a high solar fraction seasonal district heating system: Five years of operation". *Energy Procedia*, 30, 856.

Siegel, R. and Howell, J. (2002). *Thermal Radiation Heat Transfer* (New York: Taylor and Francis.)

Silva, R., et al (2013). "Global premature mortality due to anthropogenic outdoor air pollution". *Environ. Res. Lett.* 8, 134005.

Sneep, M. and Ubachs, W. (2005). "Direct measurement of the Rayleigh scattering cross section in various gases". *J. Quant. Spectroscopy and Radiative Transfer*, 92, 203.

Spindle, B. and Smith R. (2016). "Texas' Latest Gusher: Wind and Sun". *The Wall Street Journal*, 29 August, 2016, p. A1.

Stewart, R. and Higgins, T. (2017). "Tesla to build world's biggest lithium battery system". *The Wall Street Journal*, 7 July, 2017.

Stokes, G. G. (1847). "On the theory of oscillatory waves", *Trans. Cambridge Phil. Soc.* 8, 44–55.

Strasser, P., Ishida, K., Sakamoto, S., Shimomura, K., Kawamura, N., Torikai, E., Iwasaki, M., and Nagamine, K. (1996). "Muon catalyzed fusion experiments on muonic deuterium atom deceleration in thin solid deuterium films". *Phys. Lett.* B 368, 32.

Streetman, B. and Banerjee S. (2009). "Solid State Electronic Devices", 6th ed." (Saddle River, New Jersey: Prentice Hall Series in Solid State Physical Electronics).

Svensen, H. Planke, S. Polozov, A. Schmidbauer, N. Corfu, F., Podladchikov, Y. and Jamtveit, B. (2009). "Siberian gas venting and the end-Permian environmental crisis". *Earth and Planetary Science Lett.* 277, 490.

Sweet, C. (2017). "How California Utilities are Managing Excess Solar Power", *The Wall Street Journal*, 4 March, 2017 (Business Section).

Sze, S. M. (1969). *Physics of Semiconductor Devices*. (Hoboken: John Wiley Publishers.)

Tanner, B. K. (1995). *Introduction to the Physics of Electrons in Solids* (Cambridge: Cambridge University Press.)

Tarascon, J. and Armand, M. (2001). "Issues and challenges facing rechargeable lithium batteries". *Nature* 414, 359.

Thomas, H., Marian, A., Chervyakov, A., Stuckrad, S., Salmieri, D. and Rubbia, C. (2016). "Superconducting transmission lines-sustainable electric energy transfer with higher public acceptance?". *Renewable and Sustainable Energy Reviews*, 55, 50.

Thuillier, G., Herse, M., Labs, D., Foujols, T., Peetermans, W., Gillotay, D., Simon, P., and Mandel, H. (2003) "The solar spectral irradiance for 200 to 2400 nm as measured by the SOLSPEC spectrometer from the Atlas and Eureca missions". *Solar Physics* 214, 1.

Trenberth, K. E., Fasullo. J. T., and Kiehl, (2009). "Earth's global energy budget." *Bull. Amer. Meteor. Soc.* 90, 311.

Turcotte, D.L. and Schubert, G. (2002). *Geodynamics*. (Cambridge: Cambridge University Press).

Turner, J. A., (2004). "Sustainable hydrogen production". *Science*, 305, 972.

Tyndall, J. (1861). "On the absorption and radiation of heat by gases and vapours." *Phil. Mag.* Series 4, 22:146, p. 169.

Venkatramani, N. (2002). "Industrial plasma torches and applications". *Curr. Sci.* 83, 254. Table 5, p. 259.

Viswanathan, V., Kintner-Meyer, M., Balducci, P. and Jin, C. (2013). "National Assessment of Energy Storage for Grid Balancing and Arbitrage, Phase II Vol. 2: Cost and Performance Characterization". Pacific Northwest National Laboratory. Report PNNL-21368.

Wang, C. and Nehrir, M. (2008). "Power Management of a Stand-Alone Wind/Photovoltaic/Fuel Cell Energy System". *IEEE Trans. Energy Conversion*, 23, 957.

Weimer, A. W. (2005). "Solar thermochemical splitting of water." Paper presented at Global Climate and Energy Project (Stanford University) 24 February, 2006.

Werdigier, J. (2010). "Huge Wind Turbine Farm Opens Off Coast of Southeast England", *The New York Times*, 24 Sept. 2010. P. B6 of the New York edition.

Weston, Kenneth C, (1992). *Energy Conversion*. (St. Paul, MN: West Publishing Co.).

Wilkinson, P. and Mulliken, R. (1957). "Dissociation processes in oxygen above 1750 A". *Astrophysical Journal* 125, 594.

Williams, D. (2004). "Sun Fact Sheet" NASA.

Woody, Todd. (2009). "U.S. Company and China Plan Solar Project", *New York Times*, 9 Sept. 2009.

Woody, Todd. (2011). "Solar on the Water", *New York Times*, 19 April 2011.

Würfel, P. (2003). "Thermodynamics of solar energy converters". In Marti, A. and Luque, A. (Eds.) Ch. 3 in *Next Generation Photovoltaics: High efficiency through full spectrum utilization*, (Bristol and Philadelphia: Institute of Physics Publishing.)

Wurfel, P. (2009). *Physics of Solar Cells*, 2nd ed. (Wiley-VCH, Berlin).

Yoshino, K. Freeman, D. and Parkinson W. (1984). "Atlas of the Schumann-Runge absorption bands of O_2 in the wavelength region 175-205 nm". *J. Phys. Chem. Ref.* Data 13, 207.

Zachos, J. Rohl, U. Schellenberg, S. Sluijs, A. Hodell, D. Kelly, D. Thomas, E. Nicolo, M. Raffi, I. Lourens, L. McCarren, H. and Kroon, D. (2005). "Rapid Acidification of the Ocean During the Paleocene-Eocene Thermal Maximum". *Science* 308, 1611.

Zamfrescue, C. and Dincer, I. (2009). "Ammonia as a green fuel and hydrogen source for vehicular applications" *Fuel Process Technol*, 90, pp. 729–37.

Zhao, J. Wang, A. Altermatt, P. and Green, M. A. (1995). "Twenty-four percent efficient silicon solar cells with double layer antireflection coatings and reduced resistance loss". *Appl. Phys. Lett.* 66, 3636 (1995).

Zhao, J., Wang, A., Green, M. and Ferrazza, F. (1998). "19.8% efficient "honeycomb"textured multicrystalline and 24.4% efficient monocrystalline silicon solar cells". *Appl. Phys. Lett.* 73, 1991.

Zirker, J. (2002). *Journey from the Center of the Sun.* (Princeton: Princeton University Press.)

Zunft, S. (2015). "Adiabatic CAES: the ADELE-ING project". SCCER Heat & Electricity Storage Symposium, PSI, Villigen (CH), 5 May, 2015.

Exercises

Chapter 1

1. The power density from the Sun on a clear day is about 1000 Watts/m². The average electric power consumption in the US was 473 GW in 2015. How large an area would be needed to get all this power from sunlight, assuming the collection works 30 percent of the time (no Sun at night) and at 15 percent efficiency. Express the area as miles × miles.

2. If the total electric energy use in the US in 2015 was 4,144 TWh, what was the average power in watts? There were 323 million people, what was the average power per person? Why is this so high?

3. Suppose consumption of oil in the US is 92 million barrels per day. If a barrel (bbl) of oil is 0.159 m³ and costs $50, find the volume per year in m³ and in acre-feet. Find the value in dollars (one acre–foot is 1233.5 m³).

4. Using the relation $\lambda_m T$ = constant = 2.9 mm-K, find the wavelength at the maximum of the Planck distribution function for T = 288 K, approximately the temperature of the Earth. Find the energy of the photon at this wavelength, in eV.

5. The energy content in a gallon of gasoline is approximately 125 MJ. If a car is rated at 30 miles per gallon, how many Joules per mile does this correspond to? What is the cost per mile for this car, assuming $2 per gallon and also assuming 10 c/kWh.

6. The Chevy Volt is said to use 0.36 kWh/mi in full electric mode. How many Joules per mile does this correspond to? If the cost of electricity is 10 c/kWh, what is the cost per mile for this car?

7. The Agua Caliente solar farm in Arizona is rated at 290 MW, covers 1750 acres, and was completed in 2014. The plant has 5 million CdTe solar panels. What is the power per panel? If the Sun's intensity is 1000 W/m², what is the efficiency of this plant? One acre is 4050 m², about 207 feet on a side.

8. If the bandgap energy of silicon is 1.12 eV, numerically estimate the fraction of the Sun's spectrum that will be absorbed (only the photons having energy greater than 1.12 eV). For this numerical estimate, use the measured solar spectrum (above the atmosphere) Fig. 1.4. (The sharp dips in this spectrum come from absorption of atoms in the Sun's atmosphere, a side-effect not characteristic of "black body radiation" predicted by Planck's Law.)

9. Using Eq. (1.2) find the Rayleigh scattering cross section for a Buckyball C_{60} molecule at wavelengths 500 nm and 10 mm. Take its diameter as 0.71 nm and its index of refraction n as 6.

10. Using Eq. (1.3) find the rest energy of 1 gram of mass, in Joules. Find the approximate dollar value of this energy by converting the energy to kWh, valued at 14 c/kWh.

Chapter 2

1. The analysis of wind turbine performance does not indicate superior efficiency for larger values of rotor radius R. Why in practice are larger wind turbines being adopted? (Hint, the wind speed increases on average with increasing height.)
2. Find the temperature T needed to bring two alpha particles so close that they "touch" (this might allow them to "fuse" to form a larger particle of mass 8). Each particle has a charge positive 2e, where e = 1.6×10^{-19} Coulomb. Each particle has mass 4 atomic mass units, A = 4 (the alpha particle is composed of two protons and two neutrons, very strongly bound together by the nuclear force). The useful formula for the radius of a nucleus (like the alpha particle) is R = $R_0 A^{1/3}$ where R_0 = 1.44 fm = 1.44×10^{-15}m. Find the potential energy U= $k_c Q^2/r$, where k_c = 9×10^9 (in SI units) is the Coulomb constant. Then find the needed temperature T.
3. Using Eq. (2.1) find the radius R of the ^{235}U nucleus.
4. Using Eq. (2.5) explain why the windspeed v in the plane of an optimum wind turbine is 2/3 of the wind speed several rotor diameters ahead of the turbine. What is the efficiency of such a turbine?
5. Using the expression (2.6) for hydroelectric power, find the largest power in kW available for a volume flow of water 1 m^3/s if the dam height is 200 m. The density of water is 1000 kg/m^3 and g = 9.8 m/s^2.
6. Estimate the condensation energy (2.4) for a spherical drop of water of radius R = 10^{-6} m, taking the mass density of water as 1000 kg/m^3. Compare this energy to the change in potential mgR if the drop is raised a distance R against gravity, taking g = 9.8 m/s^2.
7. Explain why the Sun is nearly spherical, arguing from (2.4).

Chapter 3

1. The radiation outward from the Earth follows the Stefan–Boltzmann Law (2.26, 3.22). By what percentage will this radiated power change if the Earth's temperature rises by 2 K from an assumed value of 288 K?
2. If Denver is the mile-high city, use Eq. 3.32 to find the air pressure there compared to sea level. The parameter h for the Earth's atmosphere is about 8.48 km.
3. Following Section 3.35, what would the Earth's temperature become, in degrees K, if the Earth's atmosphere were totally lost, as occurred for the Moon?
4. In simple terms, why does the greenhouse effect make the Earth inhabitable?
5. What is the black-body radiated power from one square meter at 273 K and at 300 K if the emissivity is 1.0?
6. What is the speed of light in silicon, using its permittivity κ = 11.8? Look at Eq. 3.1, and multiply ε_0 by κ.
7. Use Eq. (3.5) to find the magnetic field B in Tesla at a distance 30 meters from a DC powerline carrying 22,000 A. Compare this field to the typical Earth's magnetic field of 1 oersted, that is 10^{-4} T.

8. Consider a long dc power line at 650 kV potential consisting of a cylindrical conductor of radius 8 cm, 30 meters above the ground. Make an estimate of the dc electric field at the wire surface and at the ground directly below the wire.

   a. Use Gauss's Law (3.2) to show that the radial electric field at radius R from the wire is $E = \lambda/(2\pi\varepsilon_0 R)$ where λ is the charge in Coulombs per meter along the wire.
   b. Make an *estimate* of the capacitance C per unit length L of the wire using formula (11.1): $C = 2\pi\varepsilon_0 L/\ln(b/a)$, for a cylindrical capacitor of inner radius a and outer radius b. Take a as the radius of the wire, 8 cm, and b as the distance to the ground, 30 m.
   c. Using the capacitance formula $Q = CV$ per unit length of the wire with $V = 650$ kV find λ, and the electric field values.
   d. Compare the deduced electric field at the wire surface to the breakdown field in air, typically estimated as 3 MV/m.
   e. Compare the deduced electric field at the ground with an estimate 10 kV/m for human sensing of electric field.

9. Following from Exercise 8, assuming the single conductor has length 1000 miles. Estimate the capacitance C of the long single cable 30 meters above ground. Assume the single cylindrical conductor is made of copper, taking the resistivity value given in Section 11.1.3, get the resistance R of the single 1000-mile cable. Assigning zero resistance to the return path, find the RC time constant for the 1000-mile cable.

10. Eq. (3.17) is the Bose factor, that represents the number of photons per electromagnetic mode, at temperature T, as described in the preceding text. Evaluate this Bose factor for the modes and temperature important in the cooling of the Earth by its black-body radiation into free space. These modes correspond to $\lambda = 10$ μm and temperature 287 K.

11. In Fig. 3.3 (top of figure) explain why the incoming solar radiation is stated as 341.3 W/m² while in Fig. 1.3 the area under the curve is 1366 W/m². (Hint: the projected area of the Earth is πR^2 while the surface area of the Earth is $4\pi R^2$.)

Chapter 4

1. Compare the mass density at the center of the Sun to the density of water.
2. In the physics of small particles, like electrons and alpha particles, quantum mechanics uses a wavefunction $\psi(x)$ such that $\psi^*\psi$ is the probability density for finding the particle. The particle of mass m can tunnel through a barrier, say of height (barrier energy-energy E of the particle). Find the value of tunnel probability T for an electron mass $(9.11 \times 10^{-31}$ kg), E= 1 eV, $\phi = 5$eV, and $t = 0.1$ nm. (Hint, you need to convert to SI units, so energy must be converted to joules.
3. Find the de Broglie wavelengths h/p of an electron of energy 1000 eV and a deuteron of energy 57.5 keV.

4. In Chapter 4, the power output from the Sun is analyzed from the point of view of proton-proton fusion, starting in Eqs. 4.23–4.27, then finally Eq. 4.34 for the power per cubic meter generated. This problem is to see how raising the temperature from 15 million K to 16 million K will change the power output. Recalculate the tunneling probability $T \approx 10^{-8}$ (Eqs. 4.23–4.27) and the power density (Eq. 4.34) for the new temperature $T = 1.6 \times 10^7$ K. (Consider the effect on the energy E, that affects the parameter r_2, and also on the mean speed that was 0.498×10^6 m/s.)

5. Use Eq. (4.15) to estimate the lowest, $n = 1$, energy of an electron trapped in a one dimensional well, assuming the length L of the one-dimensional well is the lattice spacing of Silicon, 0.54 nm, and the effective mass of the electron m as 0.19 times the free electron mass, 9.1×10^{-31} kg.

6. Explain simply why large nuclei like ^{238}U are believed to be nearly spherical in shape. (Each particle attracts its neighbors with a very short-range force.)

7. Explain simply why the binding energy per particle in ^{238}U is less than that of ^{56}Fe.

8. Compare the particle density in m^{-3} at the center of the Sun to the electron density in gold. (Hint, the electron density in gold is about $5.9 \times 10^{28}/m^3$.

9. Use Eq. (2.1) taking $R_0 = 1.44$ fm to find the radius and mass density of the ^{238}U nucleus, taking the mass as 238 times the proton mass 1.67×10^{-27} kg. Compare the mass density of this nucleus to the mass density at the center of the Sun and to the estimate in the text following Eq. (4.41).

Chapter 5

1. The energy of a single photon is expressed as $E = hf$ where h is Planck's constant 6.626×10^{-34} Joule–sec and f is the frequency of the light in Hertz. The Sun's spectrum peaks near wavelength $\lambda = 486$ nm. Assuming 486 nm, using the relation $c = f\lambda$, where $c = 3 \times 10^8$ m/s is the speed of light, what is the energy per photon? How many photons per second fall on 1 m^2 in full sunlight, 1000 W/m^2? If the observing area is approximated as the area of a single atom, approximated as $A = \pi a_0^2$, where $a_0 = 0.052$ nm is the Bohr radius of the hydrogen atom, how many photons per sec fall on a single hydrogen atom?

2. Using Eqs. 5.69–5.77 and values in Table 5.2, find a precise value of the Fermi energy in eV in pure Si at 300 K. Why is there a shift from the mid-gap location, $E = 0.56$ eV?

3. Using Eqs. 5.69–5.77 and values in Table 5.2, including $E_G = 1.12$ eV and $m_{dos} = 0.19$: find for pure Si:

   a. The densities of electrons and holes at 300 K
   b. The value of n_i
   c. Find the resistivity of pure silicon (You will need to evaluate the quantities N_C and N_V from Eqs. 5.72 and 5.74, respectively)

4. Consider an abrupt PN junction in Si assuming the acceptor concentration is 10^{21} m^{-3} and the donor concentration is 10^{23} m^{-3}. Do this approximately by assuming in

each case the majority carrier concentration equals the dopant concentration (e.g., $N_e = N_D$), and use results from Prob. 3 to find the Fermi energies on N and P sides. The difference of these two Fermi energies is eV_B.

5. Estimate, for the PN junction of Problem 4, the minority carrier concentrations on the N (p_n) and P (n_p) type sides using numbers from Problem 3 (n_i) and Problem 4 (majority conc. N_e, N_h), via the product rule, using the gap value 1.12 eV.

6. Make an estimate of the reverse current density $Jo=J_{rev}$ at 300 K for the PN junction described in Prob. 4,5. Use $J_o = e [D_n n_p/L_n + D_h p_n/L_h]$ using the minority carrier concentrations you got in Prob. 5. Assume $D_n = 40$ cm²/s and $L_n = 140$ μm (for minority electrons in the P-region); and $D_h = 2$ cm²/s and $L_h = 14$ μm (for minority holes in the N-region).

   a. Find the current density in A/m².
   b. By what factor does this current change if the temperature is raised by 10 K to 310 K? (Hint, you know the bandgap is 1.12 eV.)

7. A PN junction in silicon with bandgap 1.12 eV exhibits a reverse current density Jo = 1 pA/m² at 300 K. What value do you expect at 273 K for the same junction?

8. The Si PN junction of Problem 7 at 300 K is illuminated with 184 Watts/m² at photon energy 2.0 eV. Assume all of the light is absorbed and all of the resulting minority carriers cross the junction with no recombination.

   a. What is the short-circuit current density?
   b. What is the open circuit voltage at 300 K?
   c. What is the open circuit voltage at 273 K?

9. Explain in words why the resistivity of a pure and perfectly crystalline metal increases with temperature, while the resistivity of a pure crystalline semiconductor decreases with temperature.

10. Using the equations for the Bohr model of the hydrogen atom find the speed and orbital frequency of an electron in the n = 2 orbit. In the n = 2 orbit the radius is 4×0.0529 nm and the angular momentum is $h/2\pi$ where $h = 6.6 \times 10^{-34}$ J-s. The mass of the electron is $me = 9.1 \times 10^{-31}$ kg.

11. In a particular N-type semiconductor it is known that the Fermi level is 0.05 eV below the donor level. What is the chance (probability) that the donor level is occupied by an electron at 300K? In the same material, the acceptor level is known to lie 0.95 eV below the Fermi level. What is the chance that the acceptor level is occupied by an electron at 300K?

12. Using formulas (5.87–5.89 and 10.3–10.6), estimate the filling factor of the solar cell in Fig. 5.18, with short-circuit current stated as $I_{sc} = 100$ mA at 300 K. For this cell, working from the graph, $I_{mp} = 90.1$ mA, $V_{oc} = 0.448$ V and $V_{mp} = 0.395$ V.

   a. What is the filling factor taken from these numbers and the definition (10.3)?
   b. Using Eq. 10.6 calculate the FF from the quoted value $V_{oc} = 0.448$ V and $k_BT = 0.026$ eV at 300 K. How does it compare with the experimental data?

c. If the bandgap is taken as 1.12 eV for Si, what will be the new V_{oc} if the temperature is reduced to ice point 273 K? (Assume that I_{sc} is not changed.)

d. Predict the new filling factor at 273 K. By what percent will the power output be changed by going from 300 K to 273 K?

13. An exciton in Si forms between an electron of effective mass 0.1 and a hole with effective mass 0.6. Assuming the permittivity of Si is 11.8:

   a. Find the ground state energy of this exciton, using Bohr's model of the hydrogen atom.

   b. Find the Bohr radius of this exciton in nm.

   c. What is the energy change going from the n = 4 state to the n = 3 state?

14. Find the value of the Fermi function at energy $E_F + 3\ k_B T$. Repeat for energy $E_F - 3\ k_B T$.

15. Find the energy of an electron state in a system with $E_F = 3.0$ eV, such that the occupation of the state by an electron is 0.01, at 300 K.

16. Measurements on the metal lithium show that the electron states are occupied up to 4.2 eV above the bottom of the conduction band. If we identify 4.2 eV as the Fermi energy, what is the effective mass m* of the electrons, if we assume one electron per lithium ion is free to move, the atomic weight of lithium is 6.94, and the density of lithium is 530 kg/m³?

Chapter 6

1. Estimate the ion current (6.1) in the device shown in Fig. 6.1. Assume that the radius r around the tip center leading to 100 percent ionization of the deuterium is 600 nm. The assumption is that all D_2 molecules that fall on the front half of this surface (of area $2\pi r^2$) contribute 2e to the ion current. The rate of molecules crossing a surface in a dilute gas is nv/4 (molecules per unit area per unit time), where n is the number per unit volume of molecules whose rms speed is v.

   a. In a gas of deuterium molecules at temperature T = 270 K, show that the rms speed is 1294 m/s. The deuterium molecule has a mass of 4 amu.

   b. The gas is said to be at a pressure of 0.7 Pa at T = 270 K. Using the ideal gas law pV = RT (for one mole, corresponding to Avogadro's number of molecules), show that the number density n of molecules is 1.9×10^{20}/m³. On this basis show that the ion current would be about 44.5 nA (the observed value is 4nA).

   c. Verify that the mean free path of the D ion exceeds the dimension of the container, so that straight-line trajectories can be assumed. A formula for the mean free path is $\lambda = 1/n\sigma$, where the cross section can be taken as $\sigma = \pi\rho^2$ and take the radius ρ of the D_2 molecule as 0.037nm.

   d. Adapt the analysis in Chapter 4, Eq. 4.22 and following, to the case of two deuterons assumed to fuse to form an excited alpha particle, (mass 4, charge 2). Assume the kinetic energy of the single deuteron is 115 keV. Find the tunneling probability

T (Gamow factor) for this reaction. You have to find the energy in the center of mass frame of reference and use the reduced mass of the two particles in the calculation.

2. In Chapt. 4 the power output from the Sun is analyzed from the point of view of proton-proton fusion (Eqs. 4.22–4.27) then finally Eq. 4.34 for the power per cubic meter generated, which we will take as 313 W/m³. This problem is to scale this analysis to find the power density in a D-D-reaction Tokamak reactor assuming deuterium density 10^{20} m⁻³. Assume a factor of 10 increase in T from 15 million K to 150 million K. Scaling must include the r_2 parameter, setting the T reactivity to 1.0, note that the mass of the deuteron is twice that of the proton, and the temperature, that enters the velocity and also the cross section for the geometric collision. If the Tokamak has a volume 500 m³ how much power does it release in fusion reactions (assume Q = 3.5 MeV for the D-D reaction).

3. Following from the Tokamak plasma (Exercise 3) assuming deuterium density 10^{20} m⁻³, T = 150 million K: If the Tokamak has a volume 500 (489.5) m³, the fusion power, if assume Q = 3.5 MeV for the DD reaction, is about 60 megawatts. Following are questions about this situation.

   a. How much energy U is needed to heat the 500 m³ deuterium to 1.5×10^8 K, where we know thermal energy is 12.93 KeV per particle? Express this in kWh and find the dollar value assuming 14 c/kWh.

   b. What is the electron thermal velocity v in the plasma?

   c. What is the mean free path λ for scattering of the electrons by the ions? Assume the same parameter r_2 (111.3 fm at 12930 eV) determines the collision cross section, so $\lambda = 1/[n\pi(r_2)^2]$.

   d. Using the answers to B, C, find the collision frequency f = 1/τ for an electron, f = v/λ, note that τ the scattering time is 1/f and is used to find the mobility.

   e. Now find the mobility $\mu = e\tau/m$ and the electrical conductivity $\sigma = ne\mu$ (units m²/Vs). Explain why you can neglect the ions for this calculation.

   f. Find the resistivity $\rho = 1/\sigma$ in Ohm-m.

   g. Express this value as a multiple of the value for Cu metal at room temperature, $\rho = 1.68 \times 10^{-8}$ Ω-m.

   h. Consider a torus of volume 489 m³ (about 500) and this has major radius 6.2 m and minor radius 2 m. Think of this as a cylinder of length 2π 6.2 m and cross-sectional area π 2². Using the formula for the resistance R of a length L with cross-section A, R = ρL/A, find the resistance around this torus, containing the plasma as described.

   i. Suppose we want to heat this plasma at power 10 MW, find the needed voltage using P = V²/R.

   j. Using the Faraday Law, EMF = −dΦ/dt, where Φ is magnetic flux in Webers, consider a linear solenoid superconducting coil as is used in the MRI apparatus, that provides 10 T and has a bore inner diameter of 0.7 m. We can set it to go from 0 T to 10 T in a specified number of seconds. How many seconds will be needed to get the 10 MW of heating? (This solenoid will be perpendicular to the torus, running through the hole. The center hole in the torus has radius 6.2 m – 2 m = 4.2 m, so there is room for the solenoid to fit.

   k. How long will the reaction run at 84.7 MW before the deuterium is 50 percent burned to helium and has to be replaced?

4. Find the power density in a T-T reaction Tokamak reactor assuming triton density 10^{20} m^{-3} at 150 Million K. Scaling must include the r_1 and r_2 parameters, setting the *T* reactivity to 1.0, the mass of the triton is three times that of the proton. If the Tokamak has a volume 500 m^3 how much power does it release in fusion reactions (assume Q = 11.3 MeV for the TT reaction).

5. In a tokamak with D particles with magnetic field B = 1 T, find the Larmor radius of electrons and of deuterons (assume $k_B T$ = 12.93 keV per particle).

6. In a Tokamak with D-D fusion leading to T + p (triton mass 3 and proton mass 1) with energy release Q = 4.04 MeV, find the kinetic energy of the triton and the kinetic energy of the proton.

7. Following Prob. 6, B = 1 T, find the maximum Larmor radius for the triton and for the proton. Explain in what directions of motion for the reaction products there would be no magnetic deflection.

8. Find the energy in Joules and the cost in $ at 14 c/kwH to create a 1 T magnetic field in the volume of the torus, 500 m^3. Find the magnetic pressure on the wall in Bars (101 kPa). The magnetic energy density is $B^2/2\mu_0$.

9. Find the reduced mass for the muonic deuterium atom Dµ. Note that m_μ = 206.76 m_e, m_D = 2 × 1836.2 m_e, and m_{red} = mM/(m + M).

10. Find the Bohr radius and binding energy for the Dµ atom. (Do this by scaling from the H-atom values.)

11. Suppose two D nuclei come very close together, to create a nucleus of charge Z = 2 and mass 4 m_p. In this case what are the new Bohr radius and binding energy for the muon.

12. Since the binding energy of the Dµ atom is much larger than that of the deuterium atom De, explain what will happen if a beam of muons is directed through a gas of ordinary De atoms.

13. Following from Exercise 12, explain why in a dense De gas irradiated with muons one might expect formation of DDµµ molecules and associated DDµ ions.

14. Estimate the field ionization time τ of a hydrogen atom in a field of 1000 V/m. (This is related to the field ionization deuteron fusion method described in Section 6.1, see Fig. 6.1.) Follow Eqs. (6.6–6.9) in text Section 6.1.1, evaluating all quantities in SI units. (The answer, in the approximation used in the text, can be expressed as τ ~ $(10^{10})^{8.199}$ s. Compare this value with the published value quoted in the text just before Eq. (6.4)).

Chapter 7

1. Warren Buffett famously invested in a railroad that serves the state of Wyoming in the US, that has extensive coal deposits. Do a search and list the foreign countries to which the US exports the most coal. (Total U.S. exports of coal have been reported

as 36.8 million tons in the first quarter of 2017, up 60.3 percent over the first quarter of 2016).

2. Assuming all of the 36.8 million tons of coal, quoted in Exercise 1, is oxidized to CO_2 how many tons of carbon dioxide will this be? (The masses of C and O are 12 and 16, respectively. For a rough calculation assume the mass of coal is entirely carbon.) Compare this number to the quoted mass of carbon dioxide in the Earth's atmosphere, 5.2×10^{18} kg. (One short ton is 907.2 kg.)

3. Estimate the peak altitude for ozone distribution in the atmosphere from Fig. 7.16.

4. In Fig. 7.17, explain simply the origin of the sharp upward peaks in the measured outward infrared radiation flux at about 660 and 1050 cm^{-1}, that are marked with bold arrows. (Hint, it has to do with the facts that in the upper atmosphere, higher than the tropopause, the temperature increases with altitude, and the infrared radiation intensity increases with temperature. The point at which the atmosphere radiates to space, to be recorded by the satellite camera, is the point at which the atmosphere becomes opaque, looking down from space.)

5. Explain why the broad dips in outward radiated flux in Fig. 7.17 surrounding 660/cm and 1050/cm are direct indications of the greenhouse effect that increases the temperature of the Earth's surface.

6. Following the discussion after Fig. 7.10, explain how ozone is generated in the upper atmosphere.

7. What is the essential feature of a molecule that is important in strongly absorbing and emitting electromagnetic radiation?

8. Explain the relation between the Rayleigh power spectrum Eq. (7.2) and the atmospheric phenomenon called Alpenglow, mentioned in the text just before Section 1.3. What is the relevant power-law dependence upon wavelength?

9. Describe the specific molecular vibrational motions leading to broad dips in outward radiated flux shown in Fig. 7.17. The 660/cm feature comes from CO_2 (see Fig. 7.3) and the 1050/cm feature comes from O_3 (see Fig. 7.10).

Chapter 8

1. A windmill produces 50 KW when the wind speed is 13 mph. What do you expect for the power output of the same windmill at wind speed 20 mph? At constant wind speed, how will the power change if the temperature rises?

2. If the windspeed at 10 m is known to be 8 m/s, estimate the windspeed at height 98 m (see text following Fig. 8.9). Hint: use Eq. (8.1).

3. Give reasons while the wind turbine blade designed on a lift principle is superior to one based on a drag principle.

4. With an optimally designed wind turbine system, explain why the wind speed downwind from the blades is reduced to 1/3 of the speed upwind of the blades. What is the windspeed in the plane of the blade itself, predicted by the optimum Betz analysis?

5. What is the primary goal in designing the aerofoil shape for a wind turbine blade?

6. Explain how the modern wind turbine blade design is guided by the original derivation of Betz's law. Compare Eqs. (2.4) and (8.5).

7. In the technology of hydroelectric power, what is the essential difference between a Francis turbine and a Kaplan turbine? (Hint: compare Figs. 8.18 and 8.23.)

8. In Eq. (8.4) explain the presence of the factor 2/3.

9. If one thinks of a cylindrical horizontal column of air perpendicular to the blades of the horizontal wind turbine, why is the diameter of that cylindrical flow twice the diameter of the rotor, well downstream from the turbine?

10. Following from Exercise 9, what form of energy in the downstream air column is neglected in the Betz analysis? (Hint, the air column must provide angular momentum to the blades and hub as it passes the plane of the turbine, and the incoming air has no angular momentum.)

11. Roughly compare the efficiencies available in Francis and Kaplan water turbines to those available in wind turbines.

12. Considering that an aerofoil gets its maximum lift when the attack angle to the local air is around $10°$ (see curve marked C_L in Fig. 8.2, right panel) explain why the optimally designed turbine blade will have a twist of the aerofoil plane depending on radius.

13. For a wind turbine installation at the top of an apartment building, give two advantages of a Darrieux-style vertical-axis wind turbine over a horizontal-axis turbine.

Chapter 9

1. There are approximately 130 million housing units in the US with 91 million stated as detached single-family homes or trailers. Assume each of the 91 million detached units has installed a roof-top solar water heater of area 10 m². Assume the absorbed sunlight at 205 W/m² heats water at 100 percent efficiency that otherwise would be heated with electricity. Find the total average power that would be absorbed, that otherwise would come directly from the power grid. (Answer is 186.5 GW.)

2. In connection with Problem 2, compare the efficiency of the direct water heating to an alternative method that would have 10 percent efficient solar cells of area 10 m², with the electrical output being used solely to heat water. (Answer, it is 10 times more efficient.)

3. Why would you expect the efficiency of a parabolic trough solar thermal CSP facility to be lower than the central tower type of solar thermal facility?

4. Why is the efficiency of a concentrating solar thermal CSP facility typically higher than that of a photovoltaic facility?

5. Why is a thermal solar facility (Fig. 9.7) more adaptable for energy storage than a photovoltaic facility?

6. Using Eq. (9.3) estimate the ideal efficiency of a solar collector system such as shown in Fig. 9.4 if the collector temperature T_c and turbine input temperature can be maintained at 1615 K, the boiling point of liquid lithium. Assume solar temperature 6000 K and exhaust temperature 300 K.

7. Explain why a lens/mirror concentrating optical system directed at the Sun can achieve a higher power density on Earth than at the surface of the Sun. See text following (9.2) and Fig. 9.2.

8. Modern power plants use a steam turbine approximately following the Rankine cycle, see Fig. 9.5. What are typical values for the steam temperature and pressure at the inlet of the steam turbine?

9. What are typical values for the efficiency of the modern Rankine-cycle re-heat steam turbine used in power plants?

Chapter 10

1. A tandem solar cell is made of three materials with band gaps 1.6 eV, 1.0 eV and 0.6 eV. What is an upper limit to the open circuit voltage that could be obtained from such a cell?

2. Under 1000 W/m² illumination the three PN junctions in a particular tandem cell are known individually to provide short circuit currents 7.8 mA/cm², 12.3 mA/cm² and 9.8 mA/cm². What is the maximum short circuit density that the cell can provide in tandem operation?

3. What is the type of solar cell in the large installation at Nellis Air Force Base, shown in Fig. 10.6?

4. A semiconductor with bandgap 1.5 eV is doped to form an abrupt PN junction.

   a. On the N-side the Fermi level is at 1.4 eV and on the P-side the Fermi level is 0.1 eV. What is the built-in voltage V_B of the junction, in Volts?

   b. If the junction is operated as a solar cell at 300 K, what is the open circuit voltage in Volts if the short circuit current is 44 times the reverse current?

   c. What is the maximum open circuit voltage in Volts that can be obtained at 300 K from this junction?

5. A strongly illuminated Si solar cell exhibits an open circuit voltage 0.5 Volts at 300 K. The cell has a reverse current density of 1 nA/m² at 300 K. What short circuit current density does the cell have under these conditions?

6. Calculate the width W of the depletion region for the junction shown in Fig. 10.1, using formulas in Chapter 5. Explain why light absorbed in a wider region can contribute to the current from the cell.

7. The dye-based concentration system of Fig. 10.29 is optimized if the fluorescent emission photon energy matches the band-gap energy of the solar cells mounted at the edges of the glass plates. Make an argument that if the fluorescence spectrum is a sharp peak at the bandgap energy of the receiving solar cells, the efficiency of converting the fluorescent peak can be 100 percent.

8. From Fig. 10.21 estimate the highest possible efficiency of a two-junction tandem cell at 1000 Sun's illumination.

9. In simple terms explain why the efficiency of a solar cell is increased if the input light is concentrated. Hint: consider the effect of increased illumination, represented by larger J_{SC}, in Eq. (10.3).

10. In Fig. 10.7 write an equation for the thickness t of the ARC antireflection coating designed for $\lambda = 560$ nm making use of the index of refraction of the layer, narc.

11. What feature of the band structure in Fig. 10.45 indicates that the material is ferromagnetic?
12. If the size of a quantum dot, see Eqs. (10.13–10.14), is decreased from 6 nm to 5 nm, how will the wavelength of the emitted light change when the dot is illuminated by ultraviolet light?
13. In a "drifted germanium detector" what electric charge in Coulombs, Eq. (10.11), would flow upon absorption of a 50 keV photon? Take the energy gap for Ge at T = 0 from Table 5.2.
14. In the device shown in Fig. 10.31 what useful roles are played by the transparent glass "optical rod" that takes light from the secondary mirror and sends it to the multi-junction solar cell?
15. Spectral splitting cells send high energy photons to cells with high bandgaps and low energy photons to cells with lower bandgaps. Compare the device shown in Fig. 10.27 with the devices described in Fig. 28 (Kim et al, 2013) with regard to (a) ultimate efficiency and (b) cost. How is the spectrum separated and sent to different cell detectors in the method of Kim et al?
16. The devices in Fig. 10.32 are based on tandem cells of the type manufactured by Spectrolab (the Boeing Company) shown in Fig. 10.22a. In the design of an inexpensive system suitable for a hot sunny climate, what design features in Fig. 10.32 address the temperature sensitivity of the cell efficiency and minimize the difficulty and cost in doing at least a one-dimensional tracking of the Sun?
17. In the manufacture of the tandem cells by Spectrolab (see Fig. 10.22a) explain how the atomic epitaxy in construction of the whole tandem structure has been achieved at lower cost than would be required in the conventional molecular beam (MBE) epitaxy method. What kind of molecules are used to carry in the heavy metal atoms, and why do these molecules offer an advantage?
18. In Fig. 10.3, compare the optical absorption coefficients of Si and GaAs at about 1.5 eV photon energy. If the incident intensity is 1.00, for light onto GaAs at this energy at 300K (solid curve) what is the intensity at a depth of one micrometer?
19. Looking at Fig. 10.5, explain why the rear metal contact is at small P+ sites rather than as a broad metal contact to the rear of the cell.
20. In Fig. 10.9 explain simply why solar cells made of small microcrystals have smaller open circuit voltage than large grain sizes or single crystals.

Chapter 11

1. Why is a single junction solar cell inadequate for a water splitting device?
2. For a hydrogen powered auto, what advantage is there in using hydrogen gas with a fuel cell driving an electric motor, over using hydrogen gas directly as fuel in an internal combustion engine?
3. From text Sect. 11.7, find the estimate of solar power in GW that could be extracted utilizing the rooftops of the City of New York. What percent of the peak power usage could be provided, and what percent of the total annual electricity of the city could be provided by the rooftop cells?

4. Following from Exercise 3, take the predicted solar power from the rooftops of New York, 5.85 GW, as a peak value on a sunny day at $1000\,W/m^2$. If the total area of the City is 308.9 sq. mi, what fraction of that area would be covered by solar cells on rooftops, if their efficiency is 15 percent?

5. At the 5 GW transmission power level, compare the needed corridor width and tower height of an above-ground UHVDC transmission line to an underground XLPE (cross-linked polyethylene) cable system. See Fig. 11.11b.

6. A flywheel kinetic energy storage device such as installed at Beacon NY can store up to 90 MJ of energy, capable of 100 kW transient power output, with an off-duty loss rate as low as 2 percent per day. What are the basic design features that make the loss rate so small? (See Fig. 11.15.)

7. What is the nominal minimum voltage needed to decompose water releasing hydrogen and oxygen? (See Eq. 11.6.)

8. Following Fig. 11.30 state comparisons for predicted driving cost per mile and needed size of storage tank, for vehicles powered by gasoline, liquid hydrogen and liquid ammonia. (For liquid ammonia, there are two possible tracks: burning directly in an ICE internal combustion engine, or using a fuel cell driving an electric motor.)

9. Fig. 11.34 predicts the global temperature rise (a) after 100 years and (b) after 20 years, for a one-year pulse of greenhouse gas emissions at the 2008 level. Why does the predicted change from animal husbandry show up more strongly on the 20-year prediction than on the 100-year prediction? (Hint: it appears that ammonia, the primary emission involved, has a relatively short remanence time in the atmosphere.)

10. Following Fig. 11.34b estimate how much sulfur dioxide would be needed in a one-year dose to cancel the predicted overall temperature change of the globe after 20 years. Note from Fig. 11.34b that a negative predicted change of about 8 mK is shown as a corollary of a one year dose at the 2008 level of sulfur dioxide emissions from "industry" and "energy" sectors, at the bottom of the Figure. Approximately 6 times this amount of SO_2 could be argued, from inspecting this Fig. 11.34b, to cancel the total positive temperature change contributions shown in the Figure. According to Klimont et al (2013), sulfur dioxide is a primary precursor of anthropogenic aerosols that cool the climate, by increasing the albedo slightly, and the total global emission of sulfur dioxide in 2008 is shown (see Klimont et al, 2013, Fig. 1, left panel) as approximately 117 Tg = 117 million tonnes of sulfur dioxide. If we take the required canceling dose of sulfur dioxide to be 702 million tonnes, six times the anthropogenic side-effect dose per year in 2008, based on Fig. 11.34b, how many B1-bomber flights, at payload about 34 tons, per day, would be needed to administer this dose in one year? (Answer is 56,600, making this proposed geo-engineering method impractical on this rough analysis, that assumes the sulfur is delivered at ground level. However, according to David Keith "A Case for Climate Engineering", MIT Press 2013, p. 5, an equivalent 4 flights will be enough, if micron-sized sulfuric acid droplets are delivered at height 20 km, where the residence time is around a year.)

11. Make an estimate of the anthropogenic global atmospheric release of sulfur dioxide SO_2 per year based on global coal production and the average sulfur concentration

in coal. (Hint: If we take the average weight percent of sulfur in coal as 1 percent, and global coal production in 2008 (similar to 2016) as 7000 Mt, noting that the atomic masses of oxygen and sulfur are 16 and 32, respectively, we get 140 Mt of sulfur dioxide. This is in rough agreement with the 117 Mt reported by Klimont et al 2013.)

Chapter 12

1. The number of global premature deaths per year due to particulate matter PM-2.5 is given as 2.1×10^6 (Silva et al, 2013).[1] The chemical origins of particles of diameter less than 2.5 micrometers are principally carbon (as in soot, e.g., from coal and diesel engines), sulfates (from SO_2 from burning coal), and nitrates (from automobile exhausts); in ratios roughly 4:3:2. All of these can be traced back to carbon fossil fuels: oil, gasoline, coal. The economic value of a human life has been addressed and a median figure from US government agencies is $8 M[1]. If the global annual carbon added to the atmosphere is 9.8 Gt, what is the cost per ton, based on premature human mortality, of fossil carbon? (Answer is $1714 per ton.)
2. Make an estimate of the time to melt a cubic mile of ice, assuming only constant heating by the Sun. Assume the average insolation at ground level is 205 W/m² and the average reflection coefficient of snow is 0.8 (see text before Sect 3.3.5), the heat of fusion of ice is 334 MJ/kg (see text before Sect 11.2.4), and the density of ice is 1000 kg/m³. (Answer is 1.3×10^{10} s or 412 years.)
3. The cost of coal at the mine is reported as $31.83 per ton, and the EPA (US Environmental Protection Agency)[2] has adopted $37 as the "social cost of carbon". The text, just before Sect. 12.2 describes a tax proposed by a blue-ribbon panel for $40 per ton of carbon. Taking the US carbon emission as 5.27 Gt in 2016, what potential revenue would this proposed tax bring? If the proceeds are to be equally distributed to 323 million Americans, what would the personal "dividend" come to? (Answer is $653.)
4. Suppose an aerosol particle is composed of ammonium sulfate, $(NH_4)_2SO_4$ with known density 1770 kg/m³ and has diameter 2.5 µm. Use Eq. (3.37) to estimate the terminal velocity of the particle due to gravity in air. Then make an estimate of how long such a particle would take to fall from 30,000 feet to sea level. (Answers: velocity is 1.33 mm/s, and time is about 23 weeks. This is much shorter than the corresponding residence time for greenhouse gas.)
5. The civil-engineering renewable energy plan of Jacobson et al discussed in Chapter 12 assumes a large increase in "UTES": underground thermal energy storage. Look up and make a summary of the references in Section 11.2.3 related to UTES, including a new project in Bergdorf, near Hamburg, Germany being built by the large firm Siemens. Make an estimate of the available temperature range for the storage medium of rocks and concrete, as compared to the temperature range available in the molten salt thermal storage method described in connection with Fig. 9.6. (Hint: Basalt rocks can be used at 600°C.)

6. UTES is proposed in part to store temporarily un-usable electric power from a wind farm or solar farm, as well as a way to store direct solar heat. Make an estimate of the factor by which electricity-based heat added to UTES can be multiplied if a heat pump connecting to a deeper underground thermal reservoir, say at 55°F (that is 286 K) provides inlet steam at 800 K to a commercial turbine and electric generator. (Hint, the usual Carnot engine formulas can be employed as a guide for heat pumps and refrigerators. Typical electric motor efficiencies are around 76 percent at 1 hp and 94 percent at 200 hp. 1 hp is 746 W.) Compare this to the default case of direct I^2R heating of the basalt rock or concrete. (The answer is 1.56.)
7. Repeat Exercise 6 in case the temperature of the thermal storage is reduced to 400 K to simply produce hot water for houses, with no steam turbine. (This answer is 3.51.)
8. Summarize the reasoning in the text based on Fig. 12.13 that the industrial age deposit of around 1000 Gt of carbon into the atmosphere is insufficient to lead to a repeat of the PETM global hyperthermal event (see Fig. 12.11), much less the "hot-house Earth" and extinction event of Fig. 12.7.

Author Index

Figures, tables, and footnotes are indicated by an italic *f*, *t*, and *n* following the page number. The number after *n* indicates the footnote number if there is more than 1 footnote on the page.

Subject Index

Figures, table, and footnotes are indicated by an italic *f*, *t*, and *n* following the page number. The number after *n* indicates the footnote number if there is more than 1 footnote on the page.